Application of Microfluidics in Cell Manipulation and Biosensing

Application of Microfluidics in Cell Manipulation and Biosensing

Guest Editors

Yupan Wu
Ye Tian

MDPI

Basel • Beijing • Wuhan • Barcelona • Belgrade • Novi Sad • Cluj • Manchester

Guest Editors
Yupan Wu
School of Integrated Circuits
Southeast University
Nanjing
China

Ye Tian
College of Medicine and
Biological Information
Engineering
Northeastern University
Shenyang
China

Editorial Office
MDPI AG
Grosspeteranlage 5
4052 Basel, Switzerland

This is a reprint of the Special Issue, published open access by the journal *Biosensors* (ISSN 2079-6374), freely accessible at: https://www.mdpi.com/journal/biosensors/special_issues/91J5A5HI14.

For citation purposes, cite each article independently as indicated on the article page online and as indicated below:

Lastname, A.A.; Lastname, B.B. Article Title. *Journal Name* **Year**, *Volume Number*, Page Range.

ISBN 978-3-7258-5391-5 (Hbk)
ISBN 978-3-7258-5392-2 (PDF)
https://doi.org/10.3390/books978-3-7258-5392-2

Contents

About the Editors

Yupan Wu

Yupan Wu received a Ph.D degree from Harbin Institute of Technology, China, in 2020. From 2018 to 2019, he was a CSC Visiting Scholar in the Chemical and Biomolecular Engineering Department, University of Pennsylvania, USA. He is currently an associate professor in the School of Integrated Circuits at Southeast University, China. He was selected as the Young Science and Technology Star of Shaanxi Province. His research interests include MEMS sensors, cell manipulations using the electric field and acoustic field, AC electrokinetics, and droplet microfluidics. He has authored 2 book chapters and published more than 25 journal and conference papers. He has been granted more than 10 invention patents in China. He serves as a Guest Editor of *Biosensors* and *Electrophoresis*. He serves as an early-career Editorial Board Member for *Nanotechnology and Precision Engineering*, as well as *Micromachines*. He was the Special Sessions Co-Chair of ICMFLOC2021.

Ye Tian

Ye Tian is currently an associate professor (PI) in the College of Medicine and Biological Information Engineering, Northeastern University, Shenyang, China. He received his Ph.D. degree from the University of Hong Kong, Hong Kong, China, in 2018. His research interests include microfluidics, biomaterials, sensors and biosensors, flexible sensing, and POCT. He has published more than 60 papers in internationally renowned journals, including *Nature Communications*, *Nano-Micro Letters*, *Advanced Functional Materials*, etc. He has authored 1 book and has been granted more than 10 invention patents in China. He serves as a Guest Editor of *Biosensors* and *Micromachines*, and as an Associate Editor of *Engineering Reports*.

Preface

The field of microfluidics has profoundly transformed biological research and diagnostic methodologies by providing unprecedented control over the cellular microenvironment. This Reprint is dedicated to exploring the cutting-edge convergence of microfluidic technologies with advanced techniques in cell manipulation and biosensing. Its scope encompasses a wide array of topics, including high-throughput single-cell analysis, organ-on-a-chip models for drug screening, novel platforms for rare cell isolation, and the development of integrated, point-of-care diagnostic devices.

The primary aim of this Reprint is to highlight innovative research and review articles that demonstrate how microfluidic architectures enable precise, automated, and miniaturized analysis of cells and biomolecules. We seek to underscore the purpose of these advancements: to translate complex laboratory procedures into robust, accessible, and impactful tools that address real-world challenges in biology and medicine.

The motivation for this work stems from the urgent need for more sensitive, efficient, and cost-effective analytical platforms. Microfluidics meets this demand by offering advantages such as minimal reagent consumption, enhanced sensitivity, rapid analysis times, and the ability to mimic in vivo conditions with high fidelity.

This Reprint is addressed to a multidisciplinary audience of researchers, engineers, and clinicians working at the intersection of technology and life sciences. We hope it will serve as a valuable resource for experts in the field and inspire students and newcomers by showcasing the dynamic potential of microfluidics to revolutionize biosensing and our understanding of cellular mechanics.

We extend our sincere gratitude to all the authors and reviewers who have contributed their expertise to this Special Issue.

Yupan Wu and Ye Tian
Guest Editors

biosensors

MDPI

Article

Paper-Based Microfluidic Device for Extracellular Lactate Detection

Yan Nan, Peng Zuo * and Bangce Ye *

Lab of Biosystem and Microanalysis, State Key Laboratory of Bioreactor Engineering, East China University of Science & Technology, Shanghai 200237, China
* Correspondence: pzuo@ecust.edu.cn (P.Z.); bcye@ecust.edu.cn (B.Y.)

Abstract: Lactate is a critical regulatory factor secreted by tumors, influencing tumor development, metastasis, and clinical prognosis. Precise analysis of tumor-cell-secreted lactate is pivotal for early cancer diagnosis. This study describes a paper-based microfluidic chip to enable the detection of lactate levels secreted externally by living cells. Under optimized conditions, the lactate biosensor can complete the assay in less than 30 min. In addition, the platform can be used to distinguish lactate secretion levels in different cell lines and can be applied to the screening of antitumor drugs. Through enzymatic chemical conversion, this platform generates fluorescent signals, enabling qualitative assessment under a handheld UV lamp and quantitative analysis via grayscale intensity measurements using ImageJ (Ver. 1.50i) software. The paper-based platform presented in this study is rapid and highly sensitive and does not necessitate other costly and intricate instruments, thus making it applicable in resource-constrained areas and serving as a valuable tool for investigating cell lactate secretion and screening various anti-cancer drugs.

Keywords: paper-based microfluidic; lactate detection; colorimetric analysis; tumor cell; drug screening

1. Introduction

Lactate, a pivotal metabolic byproduct of anaerobic glycolysis, has garnered significant importance in clinical medicine, exercise physiology, and food science [1]. Traditionally perceived as a metabolic waste associated with muscle fatigue [2], recent research has unveiled lactate's critical roles in both physiological and pathological cellular states. Approximately 1500 mM of lactate is produced daily from various tissues, including skeletal muscle, heart, and brain [3]. It serves not only as an energy substrate, supplying energy [4], but also participates in energy transfer processes within and between tissues. Converted into pyruvate, lactate contributes to oxidative phosphorylation and ATP generation through the tricarboxylic acid cycle [5]. Beyond its role as an energy source, lactate functions as a signaling molecule, influencing processes such as tumor cell proliferation, immune evasion, and neuronal energy metabolism [6].

Disruptions in lactate production and consumption can lead to various diseases, highlighting the critical importance of measuring blood lactate concentrations for diagnosis and treatment [7]. In sports medicine, lactate serves as a signaling molecule that positively regulates metabolic processes during physical activity. Blood lactate concentration sensitively reflects changes in exercise intensity and duration, making it a vital metric for evaluating an athlete's training level [8]. Furthermore, during physical exertion, lactate serves as an alternate energy substrate for the brain, conserves glucose and stimulates the hypothalamus to regulate energy intake and neuronal activity, thereby playing a pivotal role in enhancing brain metabolism and overall function [9]. In food analysis, monitoring lactate levels is crucial for evaluating the quality, freshness, and preservation stability of various food products [10], including fruits, meats, alcoholic beverages [11], and specific fermented dairy items [12]. Given lactate's association with inflammatory states, cancer, and other health issues, its precise control and detection within the food sector are paramount.

Tumor cells undergo aerobic glycolysis and produce lactate even in the presence of sufficient oxygen, a phenomenon known as the "Warburg effect." [13]. Lactate accumulation within the tumor microenvironment significantly impacts tumor progression by fostering cell invasion and angiogenesis [14]. While lactate concentrations are tightly regulated at approximately 1.5–3 mM under normal physiological conditions [15], they can escalate to 30 mM within tumor microenvironments, potentially influencing cellular functions [16]. Tumor cells rapidly respond to metabolic signals in the environment and increase energy metabolism by regulating mitochondria-associated pathways and the TCA cycle, thereby promoting tumor growth and metastasis [17]. Consequently, lactate serves not only as a crucial player in energy metabolism but also as a significant regulator of tumor development [18]. Accurate and sensitive analysis of lactate secretion by tumor cells holds promise for early cancer diagnosis.

Due to the crucial regulatory role of lactate in tumor occurrence and development, accurate and reliable analysis and detection are essential. The main techniques for lactate detection include chromatographic analysis, fluorescence detection, luminescence methods, and electrochemical methods. Common chromatographic analysis methods used for lactate detection include high-performance liquid chromatography (HPLC), mass spectrometry (MS), gas chromatography (GC), and spectroscopic methods. These techniques quantitatively analyze lactate by separating and detecting the concentration of lactate present in the sample. Omar Kadi et al. proposed an LC-MS/MS method for concurrent detection and quantification of crucial metabolites in various cancer types like prostate cancer. This approach involves extracting and qualitatively and quantitatively analyzing glutamine, citrate, isocitrate, malate, succinate, fumarate, and lactate from body fluids, tissues, and human-derived cultured cell lines [19]. Fluorescence methods are analytical techniques based on the interaction between the analyte and a fluorescent dye or label, producing a fluorescent signal. A study [20] designed a glass capillary platform for fluorescence detection of lactate. The inner wall of the glass capillary is patterned with lipids and bovine serum albumin (BSA) modifications, utilizing electrostatic interactions to immobilize LDH on different regions of the lipid layers. It was found that the fluorescence intensity at the enzyme sites increases with the increasing L-lactate concentration, exhibiting a positive correlation. The established method has a detection limit of 4.9 μM for L-lactate. Electrochemical methods utilize the detection system to record and analyze changes in electrical signals, which are then converted into the concentration or activity of the target molecule. M. Briones et al. [21] designed an electrochemical biosensing platform by modifying a gold electrode with undoped diamond nanoparticles and employing LOx as a model enzyme for lactate detection. This platform exhibits a linear concentration range of 0.05–0.7 mM, a sensitivity of 4.0 μA/mM, and a detection limit of 15 μM. Although accurate lactate detection can be provided based on the techniques described above, these techniques require a significant investment of time, complex preparation processes, and expensive equipment, which limits their widespread use, especially in point-of-care (POC) testing in less developed regions. Biosensors currently outperform traditional detection methods [21], offering alternative solutions to circumvent the constraints of conventional techniques. Paper-based microfluidics is an emerging technology in the field of microfluidics, primarily utilizing porous materials like paper as the substrate for microfluidic chips, with capillary action driving the flow and diffusion of liquids within the paper. By designing specific hydrophobic/hydrophilic regions on the paper, the flow path and diffusion areas of the liquid can be controlled. Paper-based microfluidic chips can be used for various biochemical analyses, such as enzymatic reactions, immunoassays, and nucleic acid detection. By pre-immobilizing reagents in specific regions, the reaction can occur as the test liquid flows through, and the results can be read visually or using instruments. Paper-based chips can be stored for long periods of time due to their stabilizing properties. Due to its low cost, portability, and ease of operation, paper-based microfluidics has garnered significant attention in the field of lactate detection. Combining paper-based microfluidics with the catalytic reaction of lactate oxidase can enable the detection and analysis of lactate.

Based on the limitations of the above detection methods, this study combined the advantages of paper-based microfluidics to develop a paper-based microfluidic platform for lactate detection integrating the paper-based microarray previously developed by our team [22]. Based on the chip's ability to culture cells in three dimensions, the detection of exocrine lactate from living cells was realized. Although lactate lacks color characteristics or fluorescence absorption, its reaction with LDH and NAD^+ produces NADH through dehydrogenation, which can be detected through fluorescence [23]. This method allows lactate detection within 30 min, providing a new avenue for early tumor diagnosis. Furthermore, we investigated the differences in lactate secretion levels between tumor and normal cells, which is crucial for early tumor detection. Additionally, this portable microdevice enables imaging and colorimetric analysis of drug screening results, showing potential applications in effective drug screening and biomedical research.

2. Materials and Methods

Lactate was obtained from Shanghai Titan Scientific Co., Ltd. (Shanghai, China), LDH was obtained from Ying Xin Laboratory Equipment Co., Ltd. (Shanghai, China), and NAD^+ was obtained from Yi sheng Chemical Technology Co., Ltd. (Shanghai, China). The commercial lactate detection kit was provided by Elabscience Biotechnology Co., Ltd. (Wuhan, China), and paclitaxel was purchased from Dibai Biotechnology Co., Ltd. (Shanghai, China). Doxorubicin hydrochloride was obtained from Bide Pharmatech Ltd., (Shanghai, China). Collagen was purchased from Sigma Aldrich, Inc. (Saint Louis, MO, USA). Dulbecco's modified eagle medium (DMEM), fetal bovine serum (FBS), and trypsin-EDTA were purchased from Thermos Fisher Scientific Inc. (Waltham, MA, USA). CCK-8 was obtained from TransGen Biotech Co., Ltd. (Shanghai, China). PMHS was provided by Darui Chemicals Co., Ltd. (Shanghai, China). All other chemicals were of analytical grade and used as received.

The filter paper of Grade 1 (particle retention in liquid: 11 μM; thickness: 180 μM) and Grade 4 (particle retention in liquid: 20 to 25 μM; thickness: 200 μM) were purchased from Whatman (London, UK). An inkjet printer (TS3150) was purchased from Cannon Co., Ltd. (Shanghai, China). A 24-well microplate (Corning Inc., Corning, NY, USA) served as the medium container in a paper-based microfluidic platform. Chromogenic results were imaged using a smartphone camera.

Human breast adenocarcinoma cell line (MCF-7), hepatocellular carcinoma cell line (HepG2), Metastatic Breast-231 cell line (MB-231), human embryonic kidney 293T cell line (HEK293T), HeLa cell line, and normal human fetal liver cells (L-02) were obtained from the Cell Bank of Type Culture Collection of the Chinese Academy of Sciences (Shanghai, China). Cells were cultured using a standard cell culture technique for 2 or 3 days, followed by digestion with 0.05% trypsin for 2 min and centrifugation at $8000\times g$ for 2 min. The cell pellet was resuspended in the culture medium and saved for further use.

2.1. Production and Optimization of the Paper-Based Microfluidic Platform

This study utilized an inkjet printer to prepare paper-based chips for lactate detection. A hydrophobic ink was prepared by mixing polydimethylsiloxane (PMHS) and n-butanol in a 2:1 ratio and filtering through a 0.22 μm filter. The filtered ink was then loaded into a clean ink cartridge. The pattern design was created using Adobe Illustrator (Adobe Inc., San Jose, CA, USA) and subsequently printed onto the paper using an inkjet printer, with the printed area serving as the hydrophobic section. Following printing, the paper was subjected to a 65 °C oven to facilitate crosslinking of PMHS on the paper surface. The paper-based cell culture microfluidic platform was developed based on prior designs by our group, comprising two stacked layers of filter paper. A hydrophobic pattern was printed on the paper to delineate a closed hydrophilic zone conducive for cell growth. The two paper sheets were then bonded together using tape, aligning identical circular patterns. Subsequently, the hydrophilic channel was folded at a 90° angle to facilitate medium delivery. Finally, the paper device underwent overnight UV disinfection in preparation for subsequent experimental use.

To minimize the interference caused by the paper-based substrate's background color in the fluorescence signal detection and facilitate direct visual detection of low-concentration target product fluorescence signals, an optimization of the paper-based background color was performed. For this purpose, red, yellow, purple, and black paper sheets were used for optimization experiments, with each sheet undergoing hydrophobic treatment. The experimental process involved applying NADH standard substance (10 mM, 20 mM, 30 mM, 40 mM, and 50 mM) onto the paper sheets, followed by detecting the fluorescence signals emitted by the experimental groups under a portable ultraviolet lamp. Fluorescence signals were photographed, and the images were processed using ImageJ to convert them into grayscale values. By analyzing the linear relationship between fluorescence signals on different color paper sheets, the optimal background color was determined. To ensure the reliability of the research results, consistent image acquisition conditions were maintained throughout all experiments, thereby improving the accuracy of the detection method.

2.2. Optimization of the Paper-Based Lactate Detection Platform

The feasibility of a fluorescence assay method for lactate detection was validated through fluorescence intensity measurements on a microplate reader. The following experimental groups were analyzed: (1) 100 μL Tris-HCl buffer (pH = 7); (2) 100 μL LDH; (3) 100 μL NAD^+; (4) 50 μL LA + 50 μL LDH; (5) 50 μL LA + 50 μL NAD^+; (6) 50 μL LDH + 50 μL NAD^+; and (7) the positive control sample containing 50 μL LA + 25 μL LDH + 25 μL NAD^+. All solutions were prepared in Tris-HCl buffer and added to the detection zone. The samples were then incubated at 37 °C for 30 min, exposed to a UV lamp, and the fluorescence of the reaction products was captured by photography. The gray-scale values of the images were analyzed using ImageJ (Ver. 1.50i) software.

For the paper-based detection platform, we optimized the system based on four conditions, including the concentrations of LDH and NAD^+, reaction pH, and reaction time. Initially, the lactate concentration was set at 100 mM, NAD^+ concentration was set at 500 mM, and the pH of Tris-HCl buffer was set at 7. An LDH concentration gradient ranging from 1000 to 6000 U L^{-1} was established. The reaction mixture was added to the reaction zone of the paper chip and incubated at 37 °C for 30 min to facilitate the reaction. Subsequent observation and photography were conducted in a light-proof setting using a portable UV lamp for further analysis. Furthermore, optimization of NAD^+ concentration (450–500 mM), buffer pH (7–10), and reaction time (5–30 min) was performed. During optimization, all parameters except the one under investigation were kept constant to ensure result reliability. A consistent photographic environment was maintained throughout all experiments to improve assay accuracy.

2.3. Establishment of the Paper-Based Lactate Detection Platform

To establish the relationship between fluorescence intensity and lactate concentration, quantitative analysis was performed using lactate standards at concentrations of 100 mM, 200 mM, 300 mM, 400 mM, 500 mM, and 600 mM. By capturing images and analyzing the fluorescence signals with specialized software, standard curves were generated, enabling precise quantification of lactate concentration based on fluorescence intensity. Furthermore, a commercial lactate assay kit was used to be compared with the paper-based lactate detection method. Lactate samples at various concentrations (100, 200, 300, 400, 500, and 600 mM) were tested to generate a calibration curve, following the manufacturer's instructions. The kit uses NAD^+ as a hydrogen acceptor, with LDH facilitating the conversion of lactate to pyruvate, thereby transforming NAD^+ into NADH. In this process, N-methyl phenazine methyl sulfate transfers hydrogen to reduce nitro tetrazolium blue chloride to a purple colorant, which can be quantified by measuring the optical density.

2.4. Detection of Lactate Secreted by Tumor Cells Using the Paper-Based Platform

In this study, a 3D paper-based culture platform was developed for cell cultivation and the detection of cell-secreted lactate. Six different cell types (MCF-7, HepG2, MB231, HEK293T, HeLa, and L-02) were seeded onto the culture area and incubated at 37 °C with 5% CO_2. The reaction system, which included 5000 U L^{-1} of LDH and 500 mM NAD^+, was added to the culture zone after incubation. The mixture was then incubated at 37 °C, and images were collected and analyzed under light avoidance conditions. Cell viability in the culture area was evaluated using the CCK-8 assay, with color intensity representing cell viability and cytotoxicity.

2.5. Applications for the Paper-Based Lactate Detection Platform

Two anticancer drugs were prepared in DMEM following dissolution in DMSO, with DMSO concentrations ranging from 0.1% and 0.8%, a range deemed safe due to its toxicity being below 1%. The drug-containing medium was applied to the culture zone, saturating the paper-based chip, followed by the addition of MCF-7 cell suspension to each zone. The paper device was incubated at 37 °C for 24 h. After incubation, the lactate detection system was added to the incubated zones and further incubated at 37 °C for a predetermined time to facilitate reaction. Similarly, the CCK-8 re-agent was applied to the paper device to evaluate cell viability, providing a comprehensive assessment of the effects of the anticancer drugs on the cells.

Results were analyzed by one way analysis of variance (ANOVA) using GraphPad Prism 7 (GraphPad Software Inc., CA, USA). Data are presented as the mean ± standard deviation (SD) more than 3 independent experiments. Statistically significant was determined based on a *p*-value of less than 0.05 ($^{*}\ p < 0.05$, $^{**}\ p < 0.01$, $^{***}\ p < 0.001$, and $^{****}\ p < 0.0001$; n = 3). It mathematically determines the difference between two experimental data set, and the baseline is not due to random chance.

3. Results

3.1. Subsection

3.1.1. Principle of Lactate Detection on the Paper-Based Microfluidic Platform

Figure 1 illustrates the schematic diagram of the entire detection process. The design schematic of the chip is shown in Supplementary Materials (Figure S1). Building upon previous research in our laboratory, the reaction solution containing LDH and NAD^+ was added to the reaction zone on the paper-based platform, which was used for 3D cell culture. Following co-incubation and redox reactions, the plate was placed into a customized light-shielding device. The fluorescent product was then illuminated with a handheld UV lamp, and the information was captured using a smartphone camera. Finally, ImageJ software was utilized for data quantification. Since the detection on this platform occurred at the endpoint of the lactic acid detection reaction and image acquisition could be completed within 1 min, the ultraviolet irradiation from the portable UV lamp had minimal impact on cell viability.

Different concentrations of NADH standard solutions (10, 20, 30, 40, and 50 mM) were placed on the paper-based chips, and the captured fluorescence images were analyzed by using ImageJ through grayscale processing (Figure S2). The blue channel exhibited the highest sensitivity when collecting and analyzing image information (Figure S2B). Therefore, we chose to utilize the blue channel for signal collection and analysis. Figure S3 illustrates the correlation coefficients for each background color: $R^2 = 0.9871$ for red, $R^2 = 0.9134$ for yellow, $R^2 = 0.9774$ for purple, and $R^2 = 0.9943$ for black. As shown in Figure S3D, samples with different concentration gradients displayed distinct color changes only on black paper, minimizing interference from the paper color. Consequently, black was chosen as the optimal background color for the paper-based platform.

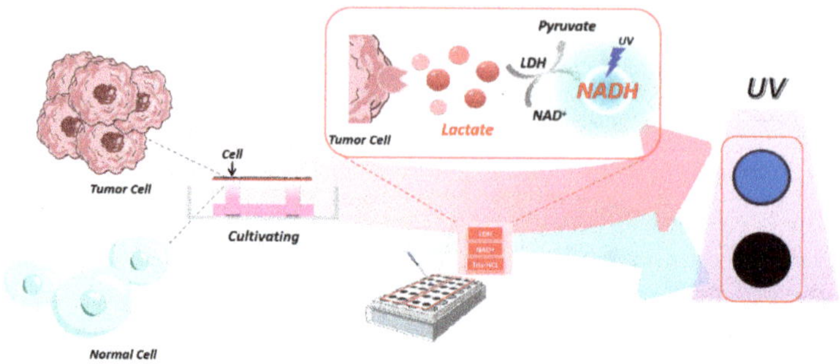

Figure 1. Principle of the paper-based microfluidic platform. This paper-based microfluidic device combines the function of both analysis and detection.

3.1.2. Validation of the Feasibility of Lactate Detection Using the Paper-Based Platform

As shown in Figure 2, lactate can only be converted to NADH and produce a visible fluorescent signal when both LDH and NAD^+ are present. Figure 2 displays seven sets of fluorescent signals evaluated using a microplate reader on the chip. The fluorescence values measured by the plate reader confirmed that the fluorescence signals generated by each group were consistent with the color intensity observed by the naked eye, indicating that the paper-based platform is suitable for lactate detection. Notably, the fluorescence intensity of the experimental group was approximately 10 times higher than that of the control group samples.

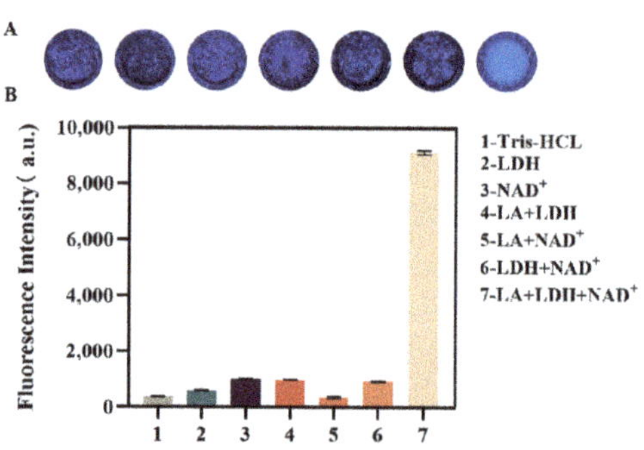

Figure 2. Validation of the feasibility of lactate detection using the paper-based platform. (**A**) The fluorescence images of the products from the reactions of different reagent combinations after 30 min (shot under a handheld UV lamp). (**B**) Fluorescence measurements of different combinations of the reagents with a microplate reader. Combination 1: the blank control; 2: LDH; 3: NAD^+; 4: LA+LDH; 5: LA+NAD^+; 6: LDH+NAD^+; 7: LA+LDH+NAD^+. All solutions were in Tris-HCl buffer.

Next, the correlation between LDH concentration and catalytic reaction was preliminarily explored. As shown in Figure 3A, when the NAD+ concentration was fixed, the fluorescence intensity gradually increased with LDH concentration ranging from 1000 U L^{-1} to 5000 U L^{-1}. However, there was a minimal change in fluorescence signal at LDH concentrations between 5000 U L^{-1} and 6000 U L^{-1}, indicating saturation at these levels. Thus, an LDH concentration of 5000 U L^{-1} was sufficient for the detection purposes. Meanwhile, the graph in Figure S4A shows the fluorescence quantification of the experimental group using a microplate reader. The results demonstrate a linear correlation between LDH concentration and fluorescence signal intensity. Similarly, Figure 3B shows the optimization of NAD^+ concentration. When the LDH concentration was fixed at 5000 U L^{-1}, adjusting the NAD^+ concentration led to changes in fluorescence in the reaction, consistent with the above results. Experimental data indicate that the saturation concentration of NAD^+

was 500 mM. Figure S4B quantitatively evaluates the fluorescence of the experimental groups using a microplate reader. The results are consistent with the linear relationship between experimental concentration and fluorescence intensity. Therefore, the optimal NAD^+ concentration for significant fluorescence response was 500 mM. Consequently, we selected 5000 U L^{-1} and 500 mM as the reaction concentrations for LDH and NAD^+, respectively. The production of NADH in the catalytic reaction led to changes in H^+ concentration, indicating that adjusting the pH with Tris-HCl affects the fluorescence results of the catalytic reaction. Figure 3C shows that when the system's pH value started to shift from pH 7 to alkaline conditions, the fluorescence intensity rapidly increased. This suggests that the detection system requires a stable pH buffer solution, so a buffer with a pH of 7 was chosen for the experiments. Furthermore, we determined the optimal reaction time for the detection method by analyzing data at different reaction times (0–50 min), as shown in Figure 3D. As can be seen from the figure, a faint fluorescence can be seen at the fifth minute, indicating that the paper-based platform can rapidly detect lactic acid. Considering the best visible effect, we set the reaction time for paper-based lactate detection to 30 min.

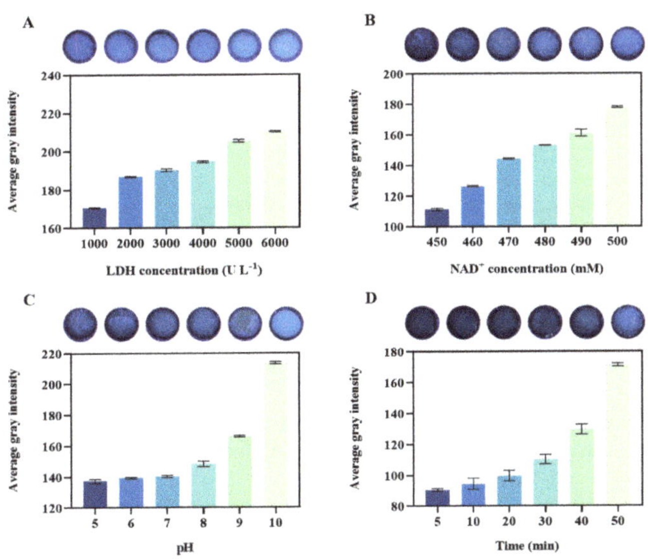

Figure 3. Optimization of the fluorescence assay for the detection of lactate. (**A**) Optimization of LDH concentration. (**B**) Optimization of NAD^+ concentration. (**C**) Optimization of buffer pH value. (**D**) Optimization of reaction time for the detection system. Each group includes fluorescence images taken under various conditions, alongside corresponding histograms showing grayscale signals. All the evaluations were carried out in triplicate (n = 3).

3.1.3. Sensitivity of the Assay

Varying lactate concentrations (100, 200, 300, 400, 500, and 600 mM) were served as reaction templates. Plotting the gray level against lactate concentration produced a linear curve with an R^2 = 0.9799 (Figure 4), indicating a strong linear relationship across each experimental group and the detection limit of the assay was 5.6699 mmol L^{-1} (S/N $\frac{1}{4}$ 3). Microplate reader analysis (Figure S5A) revealed a direct proportionality between fluorescence intensity and lactate concentration. Compared to a commercial lactate detection kit (R^2 = 0.9746; Figure S5B), our paper-based microfluidic platform exhibited a similar standard curve. This approach resulted in a standard curve with a detection limit of 5.6834 mmol L^{-1} (S/N $\frac{1}{4}$ 3), as depicted in Figure S5B. This comparison demonstrates that our paper-based microfluidic analysis achieves a comparable detection limit to that of commercial kits.

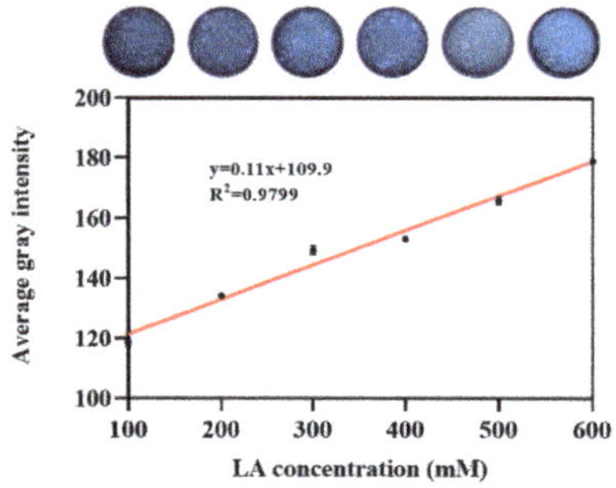

Figure 4. Sensitivity of lactate detection based on the paper-based platform. Fluorescence photographs of lactate detection at different concentrations (i.e., 100 mM, 200 mM, 300 mM, 400 mM, 500 mM, and 600 mM) on the paper-based platform are shown. The calibration curve by plotting the average grayscale values against the different lactate concentration is presented. Error bars: standard deviation from different experiments (n = 3).

3.1.4. Applications for the Paper-Based Lactate Detection Platform

We integrated a lactate detection system into our previously established 3D cell culture paper-based microfluidic platform to investigate lactate secretion variations across cell types and evaluate the platform's specificity. Feasibility of culturing MCF-7 cells on the chip was demonstrated (Figure S6A), with CCK-8 assay data showing increased cell viability with higher seeding densities. Consequently, 16,000 cells were seeded for subsequent experiments. Cell viability analysis after seeding 16,000 cells for varying durations (Figure S6B) led to a 24 h culture period being chosen to ensure sufficient cell numbers for lactate detection. Figure 5A demonstrates the assessment of lactate secretion across six cell types (normal: HEK 293T, HeLa, and L-02; tumor: MCF-7, MB231, and HepG2) using the paper-based platform, confirming its specificity. Distinct fluorescence intensities and grayscale values were observed for tumor cells, indicating varying lactate secretion levels and the platform's ability to sustain and detect continuous lactate secretion. In contrast, negligible fluorescence signals were observed for normal cells under identical culture conditions, suggesting significantly lower lactate secretion. Quantification based on the established standard curve revealed a lactate secretion rate of approximately 330 mM for tumor cells.

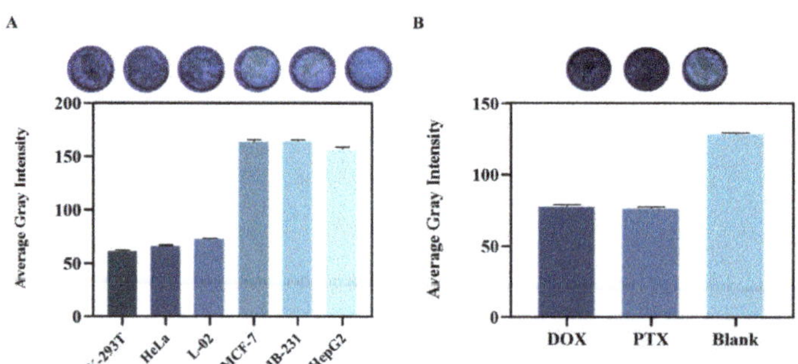

Figure 5. Applications for the paper-based lactate detection platform. (**A**) Cultivation of different types of cells on the paper-based platform and lactate detection: fluorescence images were taken after the reaction of the lactate detection system 24 h after cell culture; (**B**) Lactate secretion of MCF-7 cells under different drug treatments (doxorubicin and paclitaxel at a concentration of 10 μg mL^{-1}, as well as without treatment) were examined on the paper-based platform. Fluorescence images (arranged left to right: doxorubicin, paclitaxel, and no treatment) were used to derive grayscale values for the histograms. Error bar: standard deviation from different experiments (n = 3).

Paclitaxel and doxorubicin were employed as anticancer agents against MCF-7 cells, and the paper-based microfluidic platform facilitated quantitative analysis of lactate secretion from drug-treated cells, with color intensity directly reflecting lactate production. Figure 5B shows a notable decrease in fluorescence intensity and grayscale values in the drug-treated groups compared to in the untreated control, indicating significant reduction in lactate secretion from MCF-7 cells post-treatment ($p < 0.05$). These findings highlight the drugs' ability to markedly diminish lactate production by impeding tumor cell proliferation, demonstrating the device's sensitivity to paclitaxel and doxorubicin, corroborated by CCK8 assay data (Figure S7) showing decreased cell viability in the drug-treated groups, further underscoring the platform's utility in drug screening.

4. Conclusions

In this study, we introduced a paper-based microfluidic platform for lactate detection, which was further integrated with a wick-type paper microfluidic cell culture device. This combination enabled the detection of lactate secretion and screening of drugs targeting tumor cells. Leveraging an enzyme-assisted chemical reaction for lactate within the paper-based microfluidic platform, this approach offers biocompatibility, affordability, and operational simplicity. Others have also detected lactate secreted inside and outside of cells. Marcel Braendlein [24] used organic transistor circuits to detect lactate in tumor cell cultures, and Yuanyu Zhang [25] et al. used a multi-enzyme system in an amorphous metal—organic framework to detect lactate inside cells. Although these methods can also detect lactate, they are more expensive compared to the paper-based platform we have built and require trained specialists to operate the experiments. The method employs image acquisition for fluorescence signal collection, providing an accessible, cost-effective, efficient, and reliable technique for lactate measurement. The sensitivity and specificity of this microfluidic paper-based lactate detection platform was experimentally validated. Moreover, when combined with a smartphone-based colorimetric analysis system, the paper-based microfluidic device emerges as a powerful tool for examining lactate secretion from cells and screening various anticancer drugs.

Supplementary Materials: The following supporting information can be downloaded at: https://www.mdpi.com/article/10.3390/bios14090442/s1, Figure S1. Detail of the paper-based chip design diagram. (A) The size dimension of the 12-circle paper chip. (B) The 3D model of 12-circle paper chip. Figure S2. Optimization of Grayscale Analysis Channels. (A) Effect images of processing fluorescent products under four different grayscale channels. (B) The relation curves between grayscale values and NADH concentration under different grayscale analysis channels. Figure S3. Color optimization of paper chip. The linear relationship and correlation coefficient between the NADH concentration and the corresponding gray value of fluorescence signal in the experimental group using red paper chip (A), yellow paper chip (B), purple paper chip (C) and black paper chip (D). All the evaluations were carried out in triplicate (n = 3). * represents a significant difference among the groups with $p < 0.05$. Figure S4. Conditions and fluorescence value curves of each experimental group in the microplate reader. (A) Optimization of LDH concentration. (B) Optimization of LDH concentration. NAD+ concentration. (C) Optimization of buffer pH value. (D) Optimization of reaction time. Error bars: standard deviation from different experiments (n = 3). Figure S5. Calibration of the standard curve of fluorescence and lactate concentration using a microplate reader. Error bars: standard deviation from different experiments (n = 3). (B) Calibration of the standard curve of absorbance and lactate concentration in microplate wells using a commercial kit. Error bars: standard deviation from different experiments (n = 3). Figure S6. Feasibility verification of cell culture on paper-based platform (A) Calibration curve of different cell seeding densities and average gray intensity. (B) Calibration curve of cell culture days and average gray intensity. The average gray intensity was measured by the CCK-8 assay, and the image information was acquired by a smartphone. The results show that cell viability increases with increasing cell density. Considering the adhesion area of cells on paper, a seeding density of 16,000 cells/mL was chosen. As the culture time increased, the cell density also increased. To ensure sufficient cell density, cells were cultured for 24 h after seeding 16,000 cells for subsequent experiments. Additionally, it can be observed that our paper-based platform can be stored for at least 5 days for experimentation. All the evaluations were carried out in triplicate (n = 3). Figure S7. The cytotoxic effect of MCF-7 cells under the treatment of doxorubicin and paclitaxel: Cell viability detection after treatment with different drugs on a paper-based platform (images from left to right, doxorubicin, paclitaxel, concentration at 10 μg mL^{-1}); Histogram showing grayscale intensity, collected by Image J from the images. All evaluations were repeated three times (n = 3).

Author Contributions: Conceptualization, B.Y. and P.Z.; methodology, P.Z. and Y.N.; validation, Y.N.; investigation, Y.N.; resources, B.Y.; data curation, Y.N.; writing—original draft preparation, Y.N.; writing—review and editing, P.Z.; supervision, B.Y. and P.Z.; project administration, B.Y. and P.Z.; funding acquisition, B.Y. and P.Z. All authors have read and agreed to the published version of the manuscript.

Funding: This research was funded by National Key Research and Development Program of China grant number 2023YFC3205801, 2023YFC2812101 and Shanghai Beyond Limits Manufacturing Project.

Institutional Review Board Statement: Not applicable.

Informed Consent Statement: Not applicable.

Data Availability Statement: No new data were created.

Conflicts of Interest: The authors declare no conflict of interest.

References

1. Rattu, G.; Khansili, N.; Maurya, V.K.; Krishna, P.M. Lactate detection sensors for food, clinical and biological applications: A review. *Environ. Chem. Lett.* **2021**, *19*, 1135–1152. [CrossRef]
2. Ferguson, B.S.; Rogatzki, M.J.; Goodwin, M.L.; Kane, D.A.; Rightmire, Z.; Gladden, L.B. Lactate metabolism: Historical context, prior misinterpre-tations, and current understanding. *Eur. J. Appl. Physiol.* **2018**, *118*, 691–728. [CrossRef]
3. Adeva-Andany, M.; López-Ojén, M.; Funcasta-Calderón, R.; Ameneiros-Rodríguez, E.; Donapetry-García, C.; Vila-Altesor, M.; Rodríguez-Seijas, J. Comprehensive review on lactate metabolism in human health. *Mitochondrion* **2014**, *17*, 76–100. [CrossRef]
4. Li, X.L.; Yang, Y.Y.; Zhang, B.; Lin, X.T.; Fu, X.; An, Y.; Zou, Y.; Wang, J.-X.; Wang, Z.; Yu, T. Lactate metabolism in human health and disease. *Signal Transduct. Target. Ther.* **2022**, *7*, 305. [CrossRef]
5. Hui, S.; Ghergurovich, J.M.; Morscher, R.J.; Jang, C.; Teng, X.; Lu, W.; Esparza, L.A.; Reya, T.; Zhan, L.; Guo, J.Y.; et al. Glucose feeds the TCA cycle via circulating lactate. *Nature* **2017**, *551*, 115. [CrossRef] [PubMed]
6. Sun, S.R.; Li, H.; Chen, J.H.; Qian, Q. Lactic Acid: No Longer an Inert and End-Product of Glycolysis. *Physiology* **2017**, *32*, 453–463. [CrossRef] [PubMed]
7. Chen, P.; Zuo, H.; Xiong, H.; Kolar, M.J.; Chu, Q.; Saghatelian, A.; Siegwart, D.J.; Wan, Y. Gpr132 sensing of lactate mediates tumor-macrophage interplay to promote breast cancer metastasis. *Proc. Natl. Acad. Sci. USA* **2017**, *114*, 580–585. [CrossRef]
8. Beneke, R.; Leithäuser, R.M.; Ochentel, O. Blood Lactate Diagnostics in Exercise Testing and Training. *Int. J. Sport. Physiol.* **2011**, *6*, 8–24. [CrossRef]
9. Falkowska, A.; Gutowska, I.; Goschorska, M.; Nowacki, P.; Chlubek, D.; Baranowska-Bosiacka, I. Energy Metabolism of the Brain, Including the Cooperation between Astrocytes and Neurons, Especially in the Context of Glycogen Metabolism. *Int. J. Mol. Sci.* **2015**, *16*, 25959–25981. [CrossRef]
10. Xiao, D.; Hu, C.; Xu, X.; Lü, C.; Wang, Q.; Zhang, W.; Gao, C.; Xu, P.; Wang, X.; Ma, C. A D,L-lactate biosensor based on allosteric transcription factor LldR and amplified luminescent proximity homogeneous assay. *Biosens. Bioelectron.* **2022**, *211*, 114378. [CrossRef]
11. Dimopoulou, M.; Troianou, V.; Paramithiotis, S.; Proksenia, N.; Kotseridis, Y. Evaluation of Malolactic Starters in White and Rosé Winemaking of Moschofilero Wines. *Appl. Sci.* **2022**, *12*, 5722. [CrossRef]
12. Londero, A.; Quinta, R.; Abraham, A.G.; Sereno, R.; De Antoni, G.; Garrote, G. Inhibitory Activity of Cheese Whey Fermented with Kefir Grains. *J. Food Prot.* **2011**, *74*, 94–100. [CrossRef] [PubMed]
13. Warburg, O.; Wind, F.; Negelein, E. The metabolism of tumors in the body. *J. Gen. Physiol.* **1927**, *8*, 519–530. [CrossRef] [PubMed]
14. Ippolito, L.; Morandi, A.; Giannoni, E.; Chiarugi, P. Lactate: A Metabolic Driver in the Tumour Landscape. *Trends Biochem. Sci.* **2019**, *44*, 153–166. [CrossRef] [PubMed]
15. Walenta, S.; Wetterling, M.; Lehrke, M.; Schwickert, G.; Sundfør, K.; Rofstad, E.K.; Mueller-Klieser, W. High lactate levels predict likelihood of metastases, tumor recurrence, and restricted patient survival in human cervical cancers. *Cancer Res.* **2000**, *60*, 916–921. [PubMed]
16. Brizel, D.M.; Schroeder, T.; Scher, R.L.; Walenta, S.; Clough, R.W.; Dewhirst, M.W.; Mueller-Klieser, W. Elevated tumor lactate concentrations predict for an increased risk of metastases in head-and-neck cancer. *Int. J. Radiat. Oncol.* **2001**, *51*, 349–353. [CrossRef] [PubMed]
17. Zappasodi, R.; Serganova, I.; Cohen, I.J.; Maeda, M.; Shindo, M.; Senbabaoglu, Y.; Watson, M.J.; Leftin, A.; Maniyar, R.; Verma, S.; et al. CTLA-4 blockade drives loss of TReg stability in glycolysis-low tumours. *Nature* **2021**, *591*, 652–658. [CrossRef] [PubMed]
18. Wang, Z.H.; Embaye, K.S.; Yang, Q.; Qin, L.; Zhang, C.; Liu, L.; Zhan, X.; Zhang, F.; Wang, X.; Qin, S. Establishment and validation of a prognostic signature for lung adenocarcinoma based on metabolism-related genes. *Cancer Cell Int.* **2021**, *21*, 1–16.

19. Al Kadhi, O.; Melchini, A.; Mithen, R.; Saha, S. Development of a LC-MS/MS Method for the Simultaneous Detection of Tricarboxylic Acid Cycle Intermediates in a Range of Biological Matrices. *J. Anal. Methods Chem.* **2017**, *2017*, 1–12. [CrossRef]
20. Yang, S.S.; Lu, F.; Liu, Y.C.; Ning, Y.; Tian, S.; Zuo, P.; Ji, X.; He, Z. Quantum dots-based hydrogel microspheres for visual determination of lactate and simultaneous detection coupled with microfluidic device. *Microchem. J.* **2021**, *171*, 106801. [CrossRef]
21. Shoji, A.; Takahashi, Y.; Osato, S.; Sugawara, M. An enzyme-modified capillary as a platform for simultaneous fluorometric detection of D-glucose and L-lactate. *J. Pharm. Biomed. Anal.* **2019**, *163*, 1–8. [CrossRef] [PubMed]
22. Fu, S.X.; Zuo, P.; Ye, B.C. A Novel Wick-Like Paper-Based Microfluidic Device for 3D Cell Culture and Anti-Cancer Drugs Screening. *Biotechnol. J.* **2021**, *16*, e2000126. [CrossRef] [PubMed]
23. Qifeng, X.; Edward, S.Y. Indirect fluorescence determination of lactate and pyruvate in single erythrocytes by capillary electrophoresis. *J. Chromatogr. A* **1994**, *661*, 287–295.
24. Braendlein, M.; Pappa, A.M.; Ferro, M.; Lopresti, A.; Acquaviva, C.; Mamessier, E.; Malliaras, G.G.; Owens, R.M. Lactate Detection in Tumor Cell Cultures Using Organic Transistor Circuits. *Adv. Mater.* **2017**, *29*, 1605744. [CrossRef]
25. Zhang, Y.Y.; Xu, L.J.; Ge, J. Multienzyme System in Amorphous Metal-Organic Frameworks for Intracellular Lactate Detection. *Nano Lett.* **2022**, *22*, 5029–5036. [CrossRef]

biosensors

MDPI

Article

Concentration of Microparticles/Cells Based on an Ultra-Fast Centrifuge Virtual Tunnel Driven by a Novel Lamb Wave Resonator Array

Wei Wei [†], Zhaoxun Wang [†], Bingnan Wang, Wei Pang, Qingrui Yang * and Xuexin Duan *

State Key Laboratory of Precision Measuring Technology & Instruments, Tianjin University, Tianjin 300072, China; wei_wei@tju.edu.cn (W.W.); wangzhaoxun@tju.edu.cn (Z.W.); wangbingnan@tju.edu.cn (B.W.); weipang@tju.edu.cn (W.P.)

* Correspondence: yangqingrui@tju.edu.cn (Q.Y.); xduan@tju.edu.cn (X.D.)

† These authors contributed equally to this work and should be considered co-first authors.

Abstract: The μTAS/LOC, a highly integrated microsystem, consolidates multiple bioanalytical functions within a single chip, enhancing efficiency and precision in bioanalysis and biomedical operations. Microfluidic centrifugation, a key component of LOC devices, enables rapid capture and enrichment of tiny objects in samples, improving sensitivity and accuracy of detection and diagnosis. However, microfluidic systems face challenges due to viscosity dominance and difficulty in vortex formation. Acoustic-based centrifugation, particularly those using surface acoustic waves (SAWs), have shown promise in applications such as particle concentration, separation, and droplet mixing. However, challenges include accurate droplet placement, energy loss from off-axis positioning, and limited energy transfer from low-frequency SAW resonators, restricting centrifugal speed and sample volume. In this work, we introduce a novel ring array composed of eight Lamb wave resonators (LWRs), forming an Ultra-Fast Centrifuge Tunnel (UFCT) in a microfluidic system. The UFCT eliminates secondary vortices, concentrating energy in the main vortex and maximizing acoustic-to-streaming energy conversion. It enables ultra-fast centrifugation with a larger liquid capacity (50 μL), reduced power usage (50 mW) that is one order of magnitude smaller than existing devices, and greater linear speed (62 mm/s), surpassing the limitations of prior methods. We demonstrate successful high-fold enrichment of 2 μm and 10 μm particles and explore the UFCT's potential in tissue engineering by encapsulating cells in a hydrogel-based micro-organ with a ring structure, which is of great significance for building more complex manipulation platforms for particles and cells in a bio-compatible and contactless manner.

Keywords: concentration; ultra-fast centrifuge; Lamb wave resonator array

1. Introduction

The micro total analysis system (μTAS), also known as the Lab-on-a-Chip (LOC), is a highly integrated microsystem technology. It encapsulates functions such as sample pretreatment, execution of biochemical reactions, detection, and analysis within a single microchip, thereby efficiently completing a series of complex and precise bioanalysis and biomedical operations [1,2]. Among them, microfluidic centrifugation technology plays one of the core roles in LOC devices. Through centrifugal pumping, microvalves, micro-mixing, and separation of sample particles and fluids, it successfully constitutes indispensable sample preparation technology, as well as analysis and detection methods in the process of in vitro diagnosis and drug development [3]. Particularly in sample preprocessing, the rapid capture and enrichment of tiny objects (such as cells and particles) in the sample through microfluidic centrifugation helps to shorten the analysis cycle and greatly enhances the sensitivity and accuracy of biological detection and diagnosis [4–6]. However, due to the change in fluid characteristics under the microfluidic system, on the one hand, the viscosity effect dominates, weakening the effectiveness of the centrifugal separation mechanism. On the other hand, the low Reynolds number of the environment is not conducive to the natural formation of vortices.

Therefore, various passive and active technical strategies are developed to enhance vortex intensity to meet the needs of microfluidic centrifugation applications [7]. Passive microfluidic centrifugation technology primarily relies on the design of microchannels with specific geometric shapes, topological structures, or curvatures. This design induces vortices by utilizing the principles of inertial fluid mechanics [8,9]. Although these designs are simple and low-cost, maintaining effective fluid dynamics often requires the use of high-power pumps, which is not conducive to the miniaturization and overall integration of microfluidic control devices. In contrast, active methods depend on externally applied forces to drive the rotation within the system. These forces typically originate from electric fields [10,11], acoustic fields [12–14], and thermal fields [15]. Compared to passive methods, active methods have the advantage of real-time, precise control of vortex intensity, direction, and distribution. They are adaptable to complex, dynamic operational requirements and offer high flexibility [16,17].

The acoustic approach, as an active method in microfluidic centrifugation, offers a highly efficient and flexible solution to the challenge of vortex generation at the microfluidic scale [18]. It boasts several advantages, such as non-invasive biocompatibility, precise control through multi-parameter adjustments, strong penetration, compatibility with various microfluidic control materials, and highly localized generation of acoustic streaming [19–23]. Due to the transferability of vibrations, various waveguides [24] and acoustic black hole effects [25] can be combined to further generate acoustic fluid effects with multiple modes through wave modulation. Microfluidic centrifugation based on acoustic methods is mainly divided into two categories: indirect and direct methods. Indirect methods typically induce acoustic streaming by introducing a medium (microstructures or bubbles) that oscillates secondarily [26–28]. However, the reproducibility and robustness of streaming caused by indirect strategies are inferior to those of direct strategies. Direct strategies usually involve the direct coupling of sound waves into the fluid in the microfluidic system. In recent years, microfluidic centrifugation based on surface acoustic waves (SAWs) directly coupling into microfluidics has demonstrated the potential of centrifugation in applications such as particle concentration, separation, and micro-mixing. The pivotal objective lies in amplifying the net torque around the vertical axis to boost rotational speed and torque [29]. First, refining the electrode design and harnessing a focused SAW intensify the energy transferred into the fluid, thereby escalating the vortex velocity within the droplet [30–32]. Secondly, adjusting the distribution of the SAW, such as by introducing asymmetric or annular structures, can further enhance the streaming speed by adjusting the input power and the flow field distribution [29,33,34].

Despite their merits, these methodologies encounter several hurdles. The accurate placement of droplets poses a challenge, along with considerable energy loss due to off-axis positioning. Additionally, only a minute proportion of the input power imparted to the low-frequency SAW resonator is effectively conveyed into the fluid. Higher-frequency SAW fabrication is relatively difficult and hard to apply under high power. Therefore, it is challenging to increase the flow velocity and centrifuge speed, and the droplet capacity is typically below 10 μL [29]. For larger volumes, high streaming velocity is required. Based on higher frequency Lamb wave resonators (LWRs), rapid acoustic streaming energy conversion can be achieved, and it can be driven at a lower power [35].

In previous work, our research group has explored the acoustic streaming effects excited by a single LWR [36,37] and an array of four LWRs in a liquid environment, as well as using it for particle enrichment [35]. However, due to the presence of secondary vortices, it is challenging to form a single high-speed streaming tunnel for centrifuge applications. Therefore, in this work, we designed and manufactured a ring array composed of eight LWRs with novel structures. On one hand, this eliminates secondary vortices, allowing more energy to concentrate in the main vortex, maximizing the conversion efficiency from an acoustic wave to acoustic streaming. On the other hand, through the arrayed layout, an ultra-fast centrifuge tunnel (UFCT) is formed in the center of the microfluidic system in droplets with a larger fluid volume (50 μL). The UFCT allows the use of less power (50 mW) to achieve greater linear speed (62 mm/s), thereby achieving ultra-fast microfluidic centrifugation. Moreover, we have also explored the relationship between the fluid motion state and particle motion speed, particle size, and fluid viscosity and this impact on the UFCT, as well as the relationship with input power. We have triumphantly accomplished high-fold enrichment of 2 μm and 10 μm particles and have initially embarked on exploring the prospective utility of the UFCT in tissue engineering. Using hydrogel PEGDA as a scaffold, we encapsulated cells and formed a micro-organ with a ring structure, which contributes to the advancement of bio-compatible and non-invasive micro-assembly systems for particles and cells.

2. Materials and Methods

2.1. Composition and Working Principle of UFCT

Figure 1a is a schematic diagram of the UFCT working principle, which is composed of a PDMS chamber, the sample, and a novel LWR ring array underneath. The main purpose of the novel LWR ring array design is to focus energy effectively into the liquid by utilizing its unique ring angle and reflection-grating capabilities, thereby improving the shape of the excited vortex, generating a UFCT in the central circular area, and adapting to the subsequent particles and cell-capture assembly. The working process is as shown in Figure 1b: a low-concentration sample (about 50 μL) is applied to the sample of the chamber. When the LWR array is turned on, the UFCT instantly forms in the entire field. Particles in the entire field rotate with the vortex, and due to the combined effect of the high-speed fluid shear force of the UFCT and the acoustofluidic effect, the particles in the entire field are dragged into the central vortex for capture and enrichment in about 30 s. As most particles are dragged into the vortex, the width of the ring trajectory significantly expands until the particles fill the entire virtual tunnel and no further enrichment can be carried out. After solidification in the hydrogel, the assembly can be taken out.

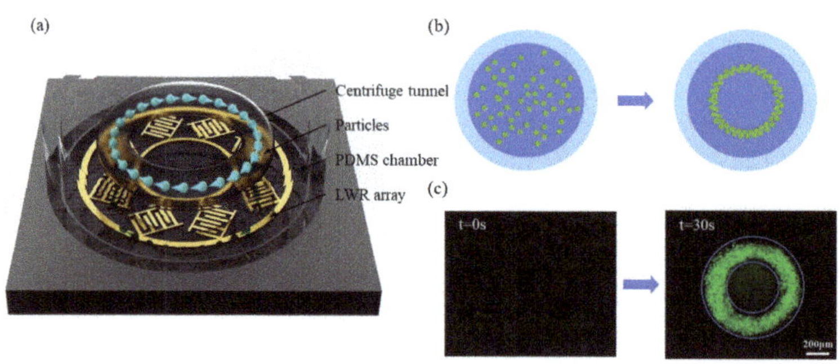

Figure 1. Working principle of UFCT for particle concentration. (**a**) Schematic diagram of UFCT for particle enrichment: the UFCT is driven by an LWR array, and the particle suspension is added to the PDMS ring as a sample reservoir. When LWR is turned on, the particles are rapidly concentrated into the central ring under the action of UFCT. (**b**,**c**) show the schematic diagram and experimental diagram of UFCT-concentrated particles after acting for 30 s and curing in hydrogel.

In previous studies, the vortices behind the LWR made the flow field distribution complex [35]. To achieve a more uniform fluid flow in the chamber and facilitate the quick enrichment of particles to the predetermined annular orbit, this paper designed a novel LWR with a new structure. To verify its feasibility, namely whether it enhances the generation and velocity of the primary vortex and reduces the formation of secondary vortices, we conducted finite element analysis (FEA) comparisons between the new structure and the traditional structure. The fluid used for simulation was water, with a density of 997 kg/m^3

and a dynamic viscosity of 1 mPa·s. For all transient studies, the relative tolerance for convergence accuracy was set to 0.001. The traditional LWR has a suspended flat-edged structure as shown in I of Figure 2a.

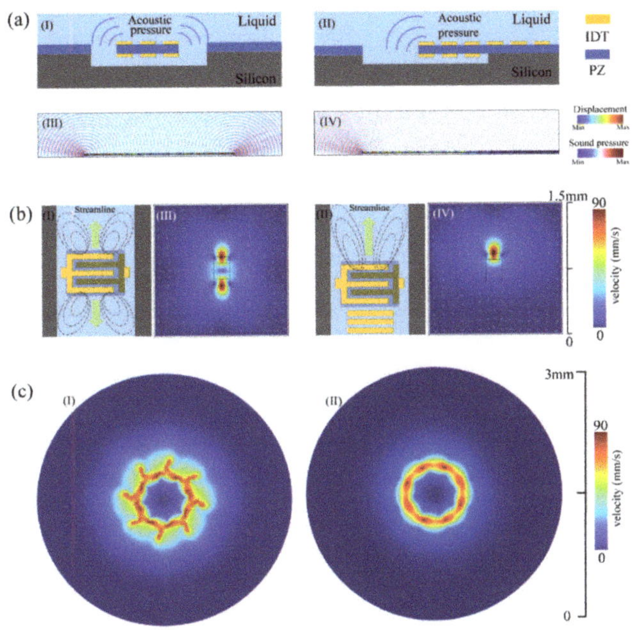

Figure 2. Principle design and simulation validation of traditional and novel LWRs: (**a**) Two-dimensional cross-sectional schematic and simulated displacement and acoustic pressure diagrams for a traditional LWR and a novel LWR. (**b**) Three-dimensional flow field propagation schematics and corresponding simulation diagrams for a traditional LWR and a novel LWR. (**c**) Schematic and simulated velocity distribution maps for traditional and novel LWR arrays.

When the radio frequency signal is connected to the electrodes, an electric field is generated perpendicular to the plane of the thin film. The particles of the piezoelectric thin film vibrate on the plane of the film surface (ignoring the thickness, as the film is sufficiently thin). Due to the strong acoustic impedance difference at the free edge, a standing wave is formed in the thin film, resulting in resonance. The resonant frequency f_r is determined by the spacing between the interdigital electrodes and the characteristics of the wave.

$$f_r = c_0\sqrt{\left(\frac{n}{2p}\right)^2 + \left(\frac{m}{2L}\right)^2} \quad (1)$$

Among them, n and m are both positive integers, p is the electrode spatial period, L is the electrode strip length, and c_0 is the wave velocity of Lamb waves in S0 mode. As shown in I and III of Figure 2a, when traditional LWRs operate in a liquid environment, the acoustic waves in the resonator propagate to the solid-liquid interface, where some of them leak into the liquid from both sides of the edge, resembling cylindrical waves. An optimized grating reflector is used to replace the air edges of the LWR so that the acoustic energy of the Lamb waves is directed from only one side of the resonator into the liquid, as indicated in II and IV of Figure 2a.

To suppress standing wave leakage, periodic structure reflectors are employed, their efficiency influenced by factors such as strip unit count and inter-strip spacing. Grating spacing adheres to $\lambda/4$ and matches IDT widths for optimal reflection, and the theory links strip numbers positively to reflected wave energy [38,39]. For miniaturization and controlled reflection, just five oppositely placed molybdenum electrodes were used on each side. Further investigation was conducted into the velocity distribution of the novel LWR and traditional LWR in liquid at resonance frequencies. The designed device was coupled into the fluid for comparison. Given that cylindrical waves are generated only on one side of the electrode bars, we extracted the first-order velocity field from the linear acoustic domain, simplified it, and subsequently introduced it into Equation (2) to obtain the volume force:

$$\langle F \rangle = -\langle \rho_0(\upsilon_1 \cdot \nabla)\upsilon_1 + \upsilon_1 \nabla \cdot \rho_0 \upsilon_1 \rangle = -\rho_0 \langle (\upsilon_1 \cdot \nabla)\upsilon_1 + \upsilon_1 \nabla \cdot \upsilon_1 \rangle \tag{2}$$

The acoustic vibration velocity field distribution υ_1 is added to the domain where the volume force acts, with the fluid density denoted as ρ_0. The upper boundary of the fluid domain is set as an "open boundary" to simulate an infinitely large liquid environment. Specific simulation steps can be found in the Supplementary Materials. The steady laminar flow calculation results for the traditional and novel Lamb wave resonators are displayed in Figure 2b. Notably, the grating's wave reflection significantly reduces leakage into the liquid, concentrating the acoustic pressure distribution on one side of the LWR.

As observed in I and III of Figure 2b, the strongest acoustic streaming effect occurs in the vicinity of the resonator device, diminishing with increasing distance from the device, consistent with the attenuation of acoustic waves in the fluid domain and the finite extent of the volume force. Under the influence of acoustic streaming, the fluid flows along the x-axis away from the device center towards both sides, followed by fluid movement along the y-axis towards the center, replenishing the displaced fluid. A circulation forms within the liquid, with a vortex present at the corners of the LWR. In contrast, as seen in II and IV of Figure 2b, the adoption of an air-grating structure on one side renders the direction of acoustic wave leakage into the fluid unidirectional, generating volume force on only that side. Concurrently, the fluid intensity and gradient on the air-grating side have increased. Due to the unidirectionality of the volume force, the fluid moves along the y-axis in the direction of acoustic wave propagation, returning to the rear of the resonator, creating vortices on both sides. This configuration results in a more concentrated energy distribution and a reduction in the formation of secondary vortices.

Considering both the continuity of the flow field and the optimization of power consumption, we concluded that an array composed of eight devices represents the most suitable configuration (the specific design concept can be found in the Figure S1). We arranged the LWR array such that the top corners of the eight novel LWRs lay on a common circumference, with inner and outer electrode rings providing electrical connections. Each device was positioned at a 45° angle relative to its neighbors, resulting in a symmetrical distribution. In an open-flow field, a single LWR generates vortices approximately 400 μm from its top corner, while the radius of the array's inner circle is about 200 μm. This configuration ensures overlapping regions of cylindrical wave generation near the inner side of each resonator, effectively rendering the volume force a continuous force in a circumferential direction.

As depicted in Figure 2c, we conducted a three-dimensional simulation of the complex composite field. At the array's center, a primary vortex formed with the highest velocity, rotating counterclockwise. Comparing the flow fields generated by traditional and novel LWR arrays, in I of Figure 2c, the traditional LWR array still exhibited velocity distributions of secondary vortices, with speeds amounting to 70% of the primary vortex. Conversely, in II of Figure 2c, the novel array collectively excited 16 vortices, all of which coalesced into the dominant primary vortex. The resulting primary vortex exhibited a more uniform velocity distribution, essentially eliminating the influence of secondary vortices.

2.2. Fabrication and Setup of UFCT

Reagents: All reagents were purchased from commercial suppliers without further purification. Monodisperse fluorescent microspheres (polystyrene microspheres, ~10 mg/mL) with a diameter of 2 μm were obtained from Aladdin Industrial Corporation (Shanghai, China). Monodisperse fluorescent microspheres (polystyrene microspheres, ~10 mg/mL) with diameters of 5 μm and 10 μm were provided by the Baseline Chromatography Technology Development Center (Tianjin, China). SU-8 2025 photoresist was purchased from Suzhou KeYi Materials Microtech Co., Ltd. (Suzhou, China). Polydimethylsiloxane (PDMS, Sylgard 184) was sourced from Dow Chemical Company (Lake Jackson, TX, USA). The synthetic hydrogel used in this paper is 30 wt% solution of polyethylene glycol diacrylate with a molecular weight of 400 (Laysan Bio. Inc., Arab, AL, USA).

Manufacturing of Lamb Wave Resonators: The Lamb wave resonators used in this study were fabricated using standard MEMS processes. The substrate was a silicon wafer cleaned with piranha solution (a mixture of concentrated sulfuric acid and 30% hydrogen peroxide). A cavity was etched into the substrate using reactive ion etching, followed by chemical vapor deposition to fill it with phosphosilicate glass (PSG), which was then polished away using chemical mechanical polishing; the PSG within the cavity served as a sacrificial layer. Next, a molybdenum film (200 nm) was deposited on the sacrificial layer as the bottom electrode via magnetron sputtering, followed by deposition of an aluminum nitride layer (1.5 μm) and another molybdenum film (200 nm, top electrode) on the sacrificial layer using the same method. Since the deposited aluminum nitride layer covered the sacrificial layer, the former was etched using potassium hydroxide wet etching and plasma etching to expose the latter for subsequent processing. Finally, a gold film was deposited by physical vapor deposition, which was then lifted off to serve as electrical interconnections. The prepared sacrificial layer was etched using diluted hydrofluoric acid to release the resonant cavity, forming a cantilever structure.

Manufacturing of Micro-chamber: (1) Mixing PDMS base and curing agent (10:1) and casting in a mold; (2) Vacuum degassing to remove bubbles; (3) Curing at 80 °C for 1.5 h; (4) Peeling, measuring height, and punching with 2 mm and 1 mm punches for a 1 mm internal diameter; (5) Cleaning in ethanol and water, drying with nitrogen, and storing in a sterile Petri dish.

Experimental Platform Setup: The experimental platform was assembled by fixing the device onto an RF test board (EVB), connecting the electrodes to signal lines and ground lines using gold wires, and attaching a circular PDMS chamber with a diameter of 1 mm to the substrate as a sample container. The setup included essential equipment such as a signal generator (Agilent N2181B) for producing RF signals and providing input power to the device, a power amplifier (Qualwave QPAR1R53337) with a gain of 33 dBm operating in the 100–500 MHz frequency band, a power isolator (EPool QCIB-350–430-S) working in the 350–430 MHz band to prevent damage to the device from excessive power, and an SMA interface connecting the device to the isolator using standard 50 Ω RF transmission lines. The instruments were connected according to the circuit diagram. Additional equipment included a bright-field microscope equipped with a CCD camera for observing particle trajectories within the microfluidic chamber and a high-speed camera capturing images at 6000 fps to calculate and track particle motion states, thereby estimating particle velocities.

Cell Staining: We stained cells to assess cell aggregation, following these specific steps: (1) Prepare a 1 mmol/L stock solution of calcein and dilute it to 50 μmol/L using PBS buffer. (2) Add 10% of the volume of the cell culture medium with the calcein solution to the cell culture medium containing cells at a density of 1×10^6 per mL. (3) Incubate the cells at 37 °C for 15–30 min. (4) Wash the cells thrice with PBS buffer to remove extracellular calcein dye, completing the preparation of the cell sample.

3. Results and Discussion

Particles subjected to a UFCT field primarily experience acoustic radiation force (ARF), acoustic streaming force (ASF), and centrifugal force (CF). The acoustofluidic dimensionless parameter κ (defined as $\kappa = \pi d_p/\lambda$) [40], representing the relative relationship between acoustic wave properties and particle size, is a key parameter for analyzing the influence of traveling surface acoustic waves on particle motion in an acoustofluidic field. When $\kappa < 1$, acoustic streaming dominates; when κ > 1, ARF prevails. Given f = 380 MHz, for d_p = 2 μm and 10 μm, κ is approximately 1.6 and 8, respectively, both greater than 1. Consequently, for UFCT, the ARF is dominant compared to the ASF. Moreover, since ARF typically scales proportionally with particle volume while resistance scales with particle radius, the ARF is more pronounced for larger particles. Thus, the 10 μm particle experiences a stronger ARF than the 2 μm particle.

Additionally, CF is also an important force that cannot be overlooked, influencing the motion of particles within vortices. The vortex velocity is related to the amplitude of the acoustic wave, which determines the magnitude of the CF. It is well-known that the CF is directly proportional to the square of the angular velocity of the centrifugal motion (CF $\propto \omega^2$). When particles rotate at high speeds within the vortex, the CF increases. When this force becomes sufficient to overcome the acoustic vortex resistance, it begins to dominate the particle's movement. The specific UFCT particle-capture process is described as follows: the LWR array couples into the fluid, generating first-order acoustic pressure effects and second-order acoustic streaming effects. The first-order acoustic pressure effect produces an ARF that pushes particles towards the center.

However, due to the presence of a single-sided reflector grating at ultra-high frequencies and the cooperation of a ring array, a larger portion of energy is rapidly converted, driving faster vortex motion. As a result, the CF acting on particles in the fluid rapidly increases and becomes dominant. Initially, particles move along streamlines towards the acoustic vortex center under the combined influence of CF and ARF. Upon reaching the virtual tunnel, the CF rapidly decreases. While the ARF pushes particles towards the center of the acoustic vortex, the enhanced CF acts in opposition, pushing particles away from the center. A stable particle ring forms where CF $\approx$ ARF, achieving an equilibrium state.As depicted in Figure 3a, for the simulation of particle enrichment by the UFCT, the chamber radius was set to 1.5 mm with a height of 1 mm. Subfigures (I–VI) illustrate the particle tracking simulation from initiation until the completion of enrichment, with time points at 0 s, 10 s, 20 s, and 30 s, respectively. Evidently, particles progressively accumulate from the edges towards the center. A dynamic visualization of this process can be found in the GIF S1. Figure 3a (V) and (VI) represent the oblique and side views of particle accumulation after enrichment. Particles are concentrated in a track approximately 300 μm above the device, a height that correlates with the attenuation distance of the acoustic waves generated by the device. Consequently, F_C and ARF jointly govern the ring-like motion of particles around a specific acoustic vortex streamline. The results obtained are similar to those of previous reports [29,41], but the supplied power of UFCT is found to be an order of magnitude lower. For specific comparisons, please refer to Table S1. Subsequently, the enrichment effect of UFCT on particles of different sizes (2 μm and 10 μm) was characterized by measuring the fluorescence intensity before and after UFCT activation, as shown in Figure 3b.

For this purpose, the fluorescence intensity was statistically analyzed using ImageView software, with the fluorescence intensity values plotted on the vertical axis and the normalized distance on the horizontal axis. From the normalized curve, it is readily apparent that, prior to UFCT activation, particles are uniformly distributed within the cavity. Upon UFCT activation, however, enriched fluorescent particles predominantly accumulate, with widths of approximately 100 μm and radii around 250 μm. Notably, the 10 μm particles experience a stronger acoustic radiation force compared to the 2 μm particles, making them more prone to entering the central region of the virtual tunnel. Consequently, they exhibit higher fluorescence enhancement factors (seven-fold higher for 2 μm particles), consistent with previous studies [42].

To further characterize the velocity and behavior of particles, the number of rotations made by particles along the annular track was tallied, allowing for the calculation of their rotational speed. The linear velocities of polystyrene beads with diameters of 2 μm, 5 μm, and 10 μm were then computed under varying power levels, and these experimental data were compared with simulation results, as depicted in Figure 3c. It is evident that particles' linear velocity increases rapidly with increasing power, and interestingly, under identical power settings, particles of different diameters exhibit nearly identical linear velocities. Moreover, these experimental findings closely align with the simulation results, thereby further substantiating the reliability of the experimental design. At a power level of 50 mW, a linear velocity of up to 62 mm/s was achieved.

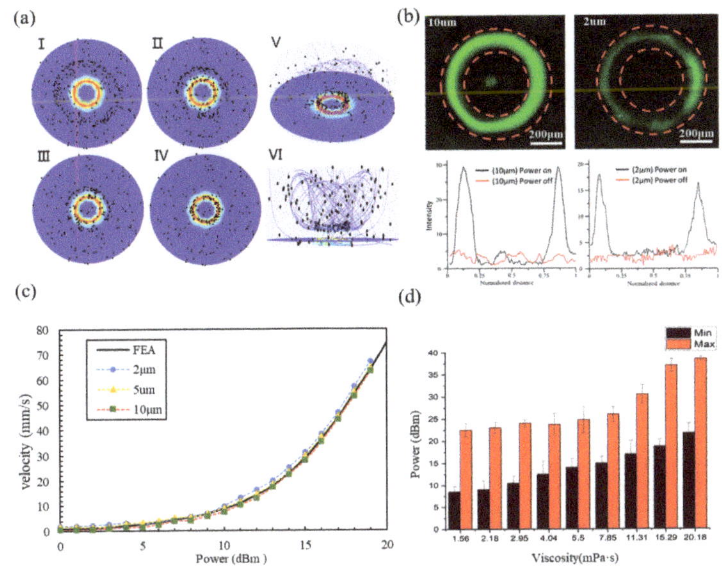

Figure 3. Particle simulation and experimental characterization based on UFCT (**a**) Simulation illustrating the particle capture process in UFCT. (**b**) Comparison of fluorescence intensity along the droplet axis before and after enrichment in UFCT, demonstrating the enrichment effect on different particles. (**c**) Graph depicting the relationship between the linear velocity of particles of various diameters and power. (**d**) Diagram showing the normal operating power range of UFCT under varied viscosities.

To further investigate the adaptability of UFCT to real samples that typically possess varying viscosities, glycerol was added to the suspension medium to adjust its viscosity within the range of 1 to 20 mPa·s, as depicted in Figure 3d. This range includes the viscosities of all body fluids, such as cerebrospinal fluid (~1 mPa·s), whole blood (4–5 mPa·s) [43], and low-molecular-weight hydrogels (more than 10 mPa·s). At lower power levels, the relatively weak fluid shear forces were insufficient to form the UFCT, whereas excessively high power could induce turbulence to destroy the UFCT. Thus, the higher the viscosity of the liquid, the greater the startup power needed to initiate particle movement along trajectories, and correspondingly, a higher power is required to displace particles from these trajectories. This also suggests that, in appropriate viscous environments, the dependence of particle or fluid motion on power decreases, implying an increased controllability of the particle dynamics. Moreover, it was demonstrated that UFCT can function effectively even in high-viscosity liquids, maintaining a wide operational power range. This finding lays a foundation for subsequent validation of cell assembly within hydrogels using UFCT technology.

Tissue engineering and organoid technologies have greatly advanced our understanding of fundamental life science issues, including tissue development, intercellular interactions, and microenvironmental regulation. They provide researchers with unique platforms for ex vivo manipulation and observation of intricate biological processes [44–46]. This section validates the cell manipulation and assembly capabilities of UFCT. We conduct experiments on microtissue assembly using HeLa cells mixed with PEGDA, following the preparation steps (I)–(IV) depicted in Figure 4a: (I) Sample injection: mix the prepared cell sample with PEGDA at a ratio of 7:3, agitating it 10 times with a pipette to ensure homogeneity. Using a pipettor, aspirate 50 μL of the PEGDA/HeLa cell mixture and inject it into the microfluidic chamber atop the device. (II) Enrichment assembly: turn on the signal generator and power amplifier, adjusting the power to 50 mW to generate UFCT and facilitate cell enrichment assembly. (III) UV curing: Once cells have aggregated and formed ring-like structures, irradiate the structure for 15 s to photopolymerize the hydrogel structure. (IV) Removal and cultivation: Gently peel the cured structured micro-tissue off the PDMS wall and transfer it to a culture dish for further cultivation.

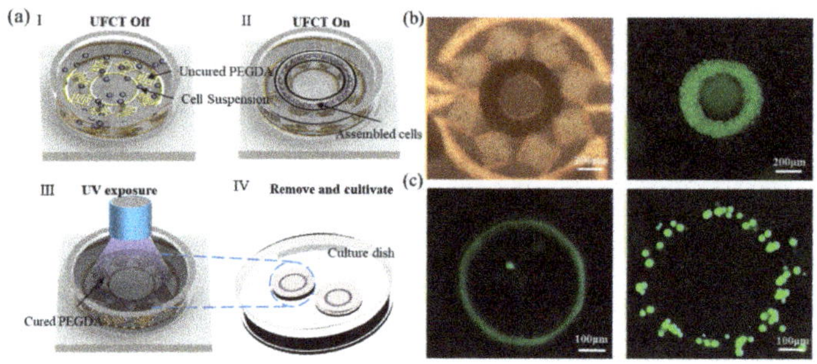

Figure 4. Cell enrichment and assembly based on UFCT. (**a**) Schematic representation of the cell enrichment assembly process facilitated by UFCT. (**b**) Comparative brightfield and fluorescence microscopy images of a 10 μm particle sample after UV curing. (**c**) The arrangement of cells before device activation and after UV exposure.

Figure 4b presents the manipulation results of a 10 μm particle after UV curing. Particles are employed to further validate the array's excellent aggregation performance in high-viscosity PEGDA solutions, with images shown under both brightfield and fluorescence microscopy. Subsequently, we proceeded with HeLa cell assembly. Prior to power activation, cells are uniformly suspended in the solution. Upon power initiation, within tens of seconds, the majority of cells are captured and rapidly rotate within the central primary vortex, as shown in Figure 4c and Video S2. The process is also documented in Video S1. Ultimately, the assembly is UV-cured; excessively long exposure times render the PEGDA hard and brittle, compromising cell viability; an irradiation time of 10–20 s is deemed appropriate. The resulting cured structure, as seen in Figure 4c, faithfully preserves the ring-like configuration established during acoustic manipulation, with a radius of approximately 250 μm. This non-contact acoustic assembly approach offers a novel perspective for three-dimensional tissue engineering.

4. Conclusions

Acoustic microcentrifugation based on acoustic streaming has attracted significant attention due to its non-contact nature, high degree of miniaturization, and integration, particularly in the biochemical domain, where it serves applications ranging from driving micro-scale mixing to pre-concentration of samples and particles [47,48]. While maintaining compactness, achieving higher efficiency in the conversion of acoustic energy to acoustic streaming and stronger local fluid gradients are persistent objectives in acoustic microcentrifugation technology. This paper addresses these goals in two ways: first, by employing higher-frequency LWRs which enable rapid conversion of acoustic energy to streaming and can be driven at lower power levels, and secondly, through the implementation of a reflective grating and an arrayed configuration, concentrating more energy in the primary vortex while eliminating secondary vortices, thus maximizing the efficiency of converting acoustic waves to streaming energy. Furthermore, this paper explores preliminary applications of acoustic microcentrifugation in sample enrichment and cell assembly, successfully demonstrating the high-speed virtual tunnel created at low power and high-fold enrichment particles. An initial investigation into the potential use of UFCT in tissue engineering is also presented. This device holds promise for controlling the width of the vortex ring by adjusting input power and frequency, or tuning the diameter of the ring through the design of different array scales and device layout angles, enabling the creation of controllable microscale assemblies for studying cell–cell interactions and cellular responses to drug molecules in pharmaceutical development [49].

Supplementary Materials: The following supporting information can be downloaded at: https://www.mdpi.com/article/10.3390/bios14060280/s1, Figure S1. Comparison of flow field simulation with different numbers of LWR Arrays. Table S1. Comparison of different mechanisms and performance of acoustofluidic in concentrating particle ring. Video S1: Cell enrichment and assembly based on UFCT. Video S2: UFCT On. GIF S1: simulation via UFCT On.

Author Contributions: W.W. and Z.W. equally contribute to this work. W.W.: Date curation (lead); Methodology (equal); Writing—original draft (lead); Writing—review & editing (equal). Z.W.: Methodology (lead); Writing—review & editing (equal). B.W.: Writing—review & editing (equal). W.P.: review & editing (lead). Q.Y.: Project administration (lead); Supervision (lead); Writing—review & editing (equal). X.D.: Conceptualization (lead). All authors have read and agreed to the published version of the manuscript.

Funding: The authors gratefully acknowledge financial support from the National Key R&D Program of China (2018YFE0118700), the National Natural Science Foundation of China (NSFC No. 62174119).

Institutional Review Board Statement: Not applicable.

Informed Consent Statement: Not applicable.

Data Availability Statement: The data that support the findings of this study are available from the corresponding author upon reasonable request.

Acknowledgments: Wei Wei and Zhaoxun Wang contributed equally to this work and should be considered co-first authors.

Conflicts of Interest: The authors declare no conflicts of interest.

References

1. Lee, S.J.; Lee, S.Y. Micro Total Analysis System (Micro-TAS) in Biotechnology. *Appl. Microbiol. Biotechnol.* **2004**, *64*, 289–299. [CrossRef] [PubMed]
2. Sackmann, E.K.; Fulton, A.L.; Beebe, D.J. The Present and Future Role of Microfluidics in Biomedical Research. *Nature* **2014**, *507*, 181–189. [CrossRef] [PubMed]
3. Gorkin, R.; Park, J.; Siegrist, J.; Amasia, M.; Lee, B.S.; Park, J.-M.; Kim, J.; Kim, H.; Madou, M.; Cho, Y.-K. Centrifugal Microfluidics for Biomedical Applications. *Lab A Chip* **2010**, *10*, 1758. [CrossRef] [PubMed]
4. Yang, Y.; Chen, Y.; Tang, H.; Zong, N.; Jiang, X. Microfluidics for Biomedical Analysis. *Small Methods* **2020**, *4*, 1900451. [CrossRef]
5. Lin, C.-T.; Kuo, S.-H.; Lin, P.-H.; Chiang, P.-H.; Lin, W.-H.; Chang, C.-H.; Tsou, P.-H.; Li, B.-R. Hand-Powered Centrifugal Microfluidic Disc with Magnetic Chitosan Bead-Based ELISA for Antibody Quantitation. *Sens. Actuators B Chem.* **2020**, *316*, 128003. [CrossRef]
6. Xiang, N.; Li, Q.; Shi, Z.; Zhou, C.; Jiang, F.; Han, Y.; Ni, Z. Low-cost Multi-core Inertial Microfluidic Centrifuge for High-throughput Cell Concentration. *Electrophoresis* **2020**, *41*, 875–882. [CrossRef] [PubMed]
7. Ahmed, H.; Ramesan, S.; Lee, L.; Rezk, A.R.; Yeo, L.Y. On-Chip Generation of Vortical Flows for Microfluidic Centrifugation. *Small* **2020**, *16*, e1903605. [CrossRef] [PubMed]
8. Tang, H.; Niu, J.; Jin, H.; Lin, S.; Cui, D. Geometric Structure Design of Passive Label-Free Microfluidic Systems for Biological Micro-Object Separation. *Microsyst. Nanoeng.* **2022**, *8*, 62. [CrossRef]
9. Bayareh, M. An Updated Review on Particle Separation in Passive Microfluidic Devices. *Chem. Eng. Process. Process Intensif.* **2020**, *153*, 107984. [CrossRef]
10. Kunti, G.; Dhar, J.; Bhattacharya, A.; Chakraborty, S. Joule Heating-Induced Particle Manipulation on a Microfluidic Chip. *Biomicrofluidics* **2019**, *13*, 014113. [CrossRef]
11. Wu, Y.; Yue, Y.; Zhang, H.; Ma, X.; Zhang, Z.; Li, K.; Meng, Y.; Wang, S.; Wang, X.; Huang, W. Three-Dimensional Rotation of Deformable Cells at a Bipolar Electrode Array Using a Rotating Electric Field. *Lab A Chip* **2024**, *24*, 933–945. [CrossRef]
12. Gu, Y.; Chen, C.; Mao, Z.; Bachman, H.; Becker, R.; Rufo, J.; Wang, Z.; Zhang, P.; Mai, J.; Yang, S.; et al. Acoustofluidic Centrifuge for Nanoparticle Enrichment and Separation. *Sci. Adv.* **2021**, *7*. [CrossRef]

13. Cui, W.; Mu, L.; Duan, X.; Pang, W.; Reed, M.A. Trapping of Sub-100 Nm Nanoparticles Using Gigahertz Acoustofluidic Tweezers for Biosensing Applications. *Nanoscale* **2019**, *11*, 14625–14634. [CrossRef] [PubMed]
14. Zhou, Y.; Wang, H.; Ma, Z.; Yang, J.K.W.; Ai, Y. Acoustic Vibration-Induced Actuation of Multiple Microrotors in Microfluidics. *Adv. Mater. Technol.* **2020**, *5*, 1–8. [CrossRef]
15. Brutin, D.; Starov, V. Recent Advances in Droplet Wetting and Evaporation. *Chem. Soc. Rev.* **2018**, *47*, 558–585. [CrossRef]
16. Li, Z.; Zhang, B.; Dang, D.; Yang, X.; Yang, W.; Liang, W. A Review of Microfluidic-Based Mixing Methods. *Sens. Actuators A Phys.* **2022**, *344*, 113757. [CrossRef]
17. Mimeau, C.; Mortazavi, I. A Review of Vortex Methods and Their Applications: From Creation to Recent Advances. *Fluids* **2021**, *6*, 68. [CrossRef]
18. Wei, W.; Wang, Y.; Wang, Z.; Duan, X. Microscale Acoustic Streaming for Biomedical and Bioanalytical Applications. *TrAC Trends Anal. Chem.* **2023**, *160*, 116958. [CrossRef]
19. Kolesnik, K.; Xu, M.; Lee, P.V.S.; Rajagopal, V.; Collins, D.J. Unconventional Acoustic Approaches for Localized and Designed Micromanipulation. *Lab A Chip* **2021**, *21*, 2837–2856. [CrossRef]
20. Lighthill, J. Acoustic Streaming. *J. Sound Vib.* **1978**, *61*, 391–418. [CrossRef]
21. Chen, Z.; Pei, Z.; Zhao, X.; Zhang, J.; Wei, J.; Hao, N. Acoustic Microreactors for Chemical Engineering. *Chem. Eng. J.* **2022**, *433*, 133258. [CrossRef]
22. Zhang, P.; Bachman, H.; Ozcelik, A.; Huang, T.J. Acoustic Microfluidics. *Annu. Rev. Anal. Chem.* **2020**, *13*, 17–43. [CrossRef] [PubMed]
23. Wu, J. Acoustic Streaming and Its Applications. *Fluids* **2018**, *3*, 108. [CrossRef]
24. Mei, J.; Friend, J. A Review: Controlling the Propagation of Surface Acoustic Waves via Waveguides for Potential Use in Acoustofluidics. *Mech. Eng. Rev.* **2020**, *7*, 19-00402. [CrossRef]
25. Liu, P.; Tian, Z.; Yang, K.; Naquin, T.D.; Hao, N.; Huang, H.; Chen, J.; Ma, Q.; Bachman, H.; Zhang, P.; et al. Acoustofluidic Black Holes for Multifunctional In-Droplet Particle Manipulation. *Sci. Adv.* **2022**, *8*, eabm2592. [CrossRef] [PubMed]
26. Zhou, Y.; Ma, Z.; Ai, Y. Submicron Particle Concentration and Patterning with Ultralow Frequency Acoustic Vibration. *Anal. Chem.* **2020**, *92*, 12795–12800. [CrossRef] [PubMed]
27. Gao, Y.; Wu, M.; Lin, Y.; Zhao, W.; Xu, J. Acoustic Bubble-Based Bidirectional Micropump. *Microfluid. Nanofluidics* **2020**, *24*, 29. [CrossRef]
28. Liu, R.H.; Yang, J.; Pindera, M.Z.; Athavale, M.; Grodzinski, P. Bubble-Induced Acoustic Micromixing. *Lab A Chip* **2002**, *2*, 151. [CrossRef]
29. Zhang, N.; Zuniga-Hertz, J.P.; Zhang, E.Y.; Gopesh, T.; Fannon, M.J.; Wang, J.; Wen, Y.; Patel, H.H.; Friend, J. Microliter Ultrafast Centrifuge Platform for Size-Based Particle and Cell Separation and Extraction Using Novel Omnidirectional Spiral Surface Acoustic Waves. *Lab A Chip* **2021**, *21*, 904–915. [CrossRef]
30. Li, Y.; Cai, S.; Shen, H.; Chen, Y.; Ge, Z.; Yang, W. Recent Advances in Acoustic Microfluidics and Its Exemplary Applications. *Biomicrofluidics* **2022**, *16*. [CrossRef]
31. Rich, J.; Cole, B.; Li, T.; Lu, B.; Fu, H.; Smith, B.N.; Xia, J.; Yang, S.; Zhong, R.; Doherty, J.L.; et al. Aerosol Jet Printing of Surface Acoustic Wave Microfluidic Devices. *Microsyst. Nanoeng.* **2024**, *10*, 2. [CrossRef] [PubMed]
32. Collins, D.J.; Neild, A.; Ai, Y. Highly Focused High-Frequency Travelling Surface Acoustic Waves (SAW) for Rapid Single-Particle Sorting. *Lab A Chip* **2016**, *16*, 471–479. [CrossRef] [PubMed]
33. Naquin, T.D.; Canning, A.J.; Gu, Y.; Chen, J.; Naquin, C.M.; Xia, J.; Lu, B.; Yang, S.; Koroza, A.; Lin, K.; et al. Acoustic Separation and Concentration of Exosomes for Nucleotide Detection: ASCENDx. *Sci. Adv.* **2024**, *10*, eadm8597. [CrossRef] [PubMed]
34. Lan, H.; Qian, J.; Liu, Y.; Lu, S.; Zhang, B.; Huang, L.; Hu, X.; Zhang, W. Swirl-like Acoustofluidic Stirring Facilitates Microscale Reactions in Sessile Droplets. *Micromachines* **2023**, *14*, 837. [CrossRef] [PubMed]
35. Zhang, H.; Tang, Z.; Wang, Z.; Pan, S.; Han, Z.; Sun, C.; Zhang, M.; Duan, X.; Pang, W. Acoustic Streaming and Microparticle Enrichment within a Microliter Droplet Using a Lamb-Wave Resonator Array. *Phys. Rev. Appl.* **2018**, *9*, 064011. [CrossRef]
36. Chen, X.; Ning, Y.; Pan, S.; Liu, B.; Chang, Y.; Pang, W.; Duan, X. Mixing during Trapping Enabled a Continuous-Flow Microfluidic Smartphone Immunoassay Using Acoustic Streaming. *ACS Sens.* **2021**, *6*, 2386–2394. [CrossRef] [PubMed]
37. Zhang, C.; Chen, X.; Wei, W.; Chen, X.; Li, Q.; Duan, X. Particle Distributions in Lamb Wave Based Acoustofluidics. *Nanotechnol. Precis. Eng.* **2024**, *7*, 023007. [CrossRef]
38. Bjurström, J.; Katardjiev, I.; Yantchev, V. Lateral-Field-Excited Thin-Film Lamb Wave Resonator. *Appl. Phys. Lett.* **2005**, *86*, 154103. [CrossRef]
39. Bjurström, J.; Yantchev, V.; Katardjiev, I. Thin Film Lamb Wave Resonant Structures—The First Approach. *Solid-State Electron.* **2006**, *50*, 322–326. [CrossRef]
40. Destgeer, G.; Cho, H.; Ha, B.H.; Jung, J.H.; Park, J.; Sung, H.J. Acoustofluidic Particle Manipulation inside a Sessile Droplet: Four Distinct Regimes of Particle Concentration. *Lab A Chip* **2016**, *16*, 660–667. [CrossRef]
41. Liu, Y.; Ji, M.; Yu, N.; Zhao, C.; Xue, G.; Fu, W.; Qiao, X.; Zhang, Y.; Chou, X.; Geng, W. Enhanced Detection in Droplet Microfluidics by Acoustic Vortex Modulation of Particle Rings and Particle Clusters via Asymmetric Propagation of Surface Acoustic Waves. *Biosensors* **2022**, *12*, 399. [CrossRef] [PubMed]
42. Han, J.L.; Hu, H.; Huang, Q.Y.; Lei, Y.L. Particle Separation by Standing Surface Acoustic Waves inside a Sessile Droplet. *Sens. Actuators A Phys.* **2021**, *326*, 112731. [CrossRef]

43. Pop, G.A.M.; Bisschops, L.L.A.; Iliev, B.; Struijk, P.C.; van der Hoeven, J.G.; Hoedemaekers, C.W.E. On-Line Blood Viscosity Monitoring in Vivo with a Central Venous Catheter, Using Electrical Impedance Technique. *Biosens. Bioelectron.* **2013**, *41*, 595–601. [CrossRef] [PubMed]
44. Atala, A.; Kasper, F.K.; Mikos, A.G. Engineering Complex Tissues. *Sci. Transl. Med.* **2012**, *4*, 160rv12. [CrossRef] [PubMed]
45. Wang, X.; Wang, Z.; Zhai, W.; Wang, F.; Ge, Z.; Yu, H.; Yang, W. Engineering Biological Tissues from the Bottom-Up: Recent Advances and Future Prospects. *Micromachines* **2021**, *13*, 75. [CrossRef]
46. Murphy, S.V.; Atala, A. 3D Bioprinting of Tissues and Organs. *Nat. Biotechnol.* **2014**, *32*, 773–785. [CrossRef]
47. Nam, J.; Lim, C.S. Micromixing Using Swirling Induced by Three-Dimensional Dual Surface Acoustic Waves (3D-DSAW). *Sens. Actuators B Chem.* **2018**, *255*, 3434–3440. [CrossRef]
48. Shilton, R.; Tan, M.K.; Yeo, L.Y.; Friend, J.R. Particle Concentration and Mixing in Microdrops Driven by Focused Surface Acoustic Waves. *J. Appl. Phys.* **2008**, *104*, 014910. [CrossRef]
49. Elliott, N.T.; Yuan, F. A Review of Three-Dimensional In Vitro Tissue Models for Drug Discovery and Transport Studies. *J. Pharm. Sci.* **2011**, *100*, 59–74. [CrossRef]

biosensors

MDPI

Communication

A Multi-Drug Concentration Gradient Mixing Chip: A Novel Platform for High-Throughput Drug Combination Screening

Jiahao Fu [1,†], Yibo Feng [1,†], Yu Sun [2], Ruiya Yi [2], Jing Tian [2,3,4], Wei Zhao [1], Dan Sun [1,3,4,*] and Ce Zhang [1,3,*]

1 State Key Laboratory of Photon-Technology in Western China Energy, Institute of Photonics and Photon-Technology, Northwest University, Xi'an 710127, China
2 Key Laboratory of Resource Biology and Biotechnology in Western China, Ministry of Education, School of Medicine, Northwest University, Xi'an 710127, China; yi-ruiya@outlook.com (R.Y.)
3 Huaxin Microfish Biotechnology Co., Ltd., Taicang 215400, China
4 Center for Automated and Innovative Drug Discovery, Northwest University, Xi'an 710127, China
* Correspondence: sund@nwu.edu.cn (D.S.); zhangce.univ@gmail.com (C.Z.)
† These authors contributed equally to this work.

Abstract: Combinatorial drug therapy has emerged as a critically important strategy in medical research and patient treatment and involves the use of multiple drugs in concert to achieve a synergistic effect. This approach can enhance therapeutic efficacy while simultaneously mitigating adverse side effects. However, the process of identifying optimal drug combinations, including their compositions and dosages, is often a complex, costly, and time-intensive endeavor. To surmount these hurdles, we propose a novel microfluidic device capable of simultaneously generating multiple drug concentration gradients across an interlinked array of culture chambers. This innovative setup allows for the real-time monitoring of live cell responses. With minimal effort, researchers can now explore the concentration-dependent effects of single-agent and combination drug therapies. Taking neural stem cells (NSCs) as a case study, we examined the impacts of various growth factors—epithelial growth factor (EGF), platelet-derived growth factor (PDGF), and fibroblast growth factor (FGF)—on the differentiation of NSCs. Our findings indicate that an overdose of any single growth factor leads to an upsurge in the proportion of differentiated NSCs. Interestingly, the regulatory effects of these growth factors can be modulated by the introduction of additional growth factors, whether singly or in combination. Notably, a reduced concentration of these additional factors resulted in a decreased number of differentiated NSCs. Our results affirm that the successful application of this microfluidic device for the generation of multi-drug concentration gradients has substantial potential to revolutionize drug combination screening. This advancement promises to streamline the process and accelerate the discovery of effective therapeutic drug combinations.

Keywords: multi-drug; mixing chip; microfluidic; drug screening

1. Introduction

Combinatorial drug treatments hold significant promise for the advancement of personalized medicine and the management of complex diseases. This approach harnesses the potential of multiple drugs to work together, creating synergies that can lead to improved treatment outcomes [1–3]. However, screening drug combinations and determining their appropriate concentrations are time-consuming and costly, and can take over 10 years and cost billions of dollars. Therefore, the development of technologies, including high-throughput screening methods [4–7], computational modeling [8–10], and microfluidic devices [11–13], is in high demand.

High-throughput cell screening typically relies on costly and complex methods such as porous plates with automated liquid transfer or manual operations, limiting their feasibility in small research centers. In contrast, microfluidic technology offers low-cost, precise control over nano-volume liquids, high integration, and efficient and cost-effective drug screening with minimal sample consumption [14–16]. Besides their requirement of few biological samples and reagents [17–19], microfluidic systems offer advantages in drug screening by creating a controlled biological microenvironment and allowing for high-resolution, real-time monitoring [20–22]. The integration of Quake's valves into the microfluidic system allows for the automated and parallel processing of a large number of samples [23–25]. In the context of combinatorial drug discovery, this might mean rapidly screening thousands of drug combinations to assess their effects on cells or biological targets [26–28].

To overcome the challenges inherent to combinatorial drug screening, we present a novel microfluidic device designed to automatically generate drug concentration gradients within an array of micro-sized chambers. This device features a symmetrical chip layout enhanced with both upper and lower microvalves at each inlet, enabling the creation of different concentration gradients for three distinct drugs. This configuration facilitates the mixing of drugs with varying doses, spanning high to low concentrations. Through the integrated control of microvalves via a bespoke Matlab program, we can orchestrate the delivery of various doses of EGF, PDGF, and FGF into the microenvironment surrounding neural stem cells (NSCs). Utilizing the expression levels of Hes-5 in individual cells as a biomarker [29,30], we uncover that introducing individual growth factors at elevated concentrations promotes the differentiation of NSCs. Conversely, when NSCs are exposed to high doses of multiple drugs, we observe a reduction in the number of differentiated cells, which suggests the promotion of stem cell self-renewal. In summary, the successful deployment of our microfluidic device for the creation of multi-drug concentration gradients represents a significant leap forward in the realm of drug combination screening. Its capacity to deliver efficient, cost-effective evaluations of drug interactions signals a major advancement in supporting the development of new pharmacological agents and paves the way for personalized treatment methodologies.

2. Materials and Methods

2.1. Fabrication of Microfluidic Chips

In this study, we prepare the control and flow layers of multi-layer polydimethylsiloxane (PDMS) chips using lithography and soft lithography techniques [31]. The PDMS chip was designed in AutoCAD and fabricated using standard UV lithography with SU-8 3025, SU-8 3075 (Microchem, Westborough, MA, USA), and AZ-50X (AZ Electronic Materials, Luxembourg) photoresists. AZ-50X is used to create liquid channels with curved profiles, which are integrated with the rectangular profile channels of SU-8 3025 and SU-8 3075 to form a fluidic layer. This layer is then aligned and bonded with the control layer, also fabricated from SU-8 3025. For the fabrication process, 50 g of PDMS (10:1 monomer to catalyst ratio) is poured over the fluidic layer mold, degassed in a vacuum oven for one hour, and then cured at 80 °C for two hours. Subsequently, a thin PDMS layer is spin-coated onto the control-layer silicon wafer at 2200 rpm and baked for 10 min. After plasma treating the fluid and control layers, the layers are aligned using the 'cross' marks on each to achieve precise multi-layer bonding. The assembled PDMS chip undergoes further plasma treatment along with a glass substrate, is bonded, and then is post-cured in an 80 °C oven for over 8 h.

2.2. Numerical Simulation

In this study, we used COMSOL Multiphysics®5.3 for numerical simulation. Based on the chip's actual dimensions, a two-dimensional schematic was generated using COMSOL software, illustrating the three channels sequentially linked to five culture chambers to delineate the simulation boundaries. Their properties are configured to confine liquid flow to within the channels without wall adhesion. The simulation utilizes a multi-physics coupling interface integrating laminar flow and mass transfer. Inlet flow parameters of 3 mm/s, 5 mm/s, and 10 mm/s are assumed, with an initial substance concentration of 1 mol/m^3 for simulation purposes.

2.3. Chip Operation

The microfluidic chips subjected to channel pressurization testing were sterilized via ultraviolet irradiation. The control channels of the chips were connected to miniature pneumatic solenoid valves (Festo, Esslingen, Germany), which were controlled using a custom MATLAB program (MathWorks, Natick, MA, USA). The optimal closing pressure of the PDMS membrane valves was determined, with the typical pressure range of the chips being between 25 and 30 psi. Prior to the chip's use, its channels were filled with PBS and degassed. Before cell culturing, the chamber was coated with fibronectin (0.25 mg/mL; Merck, Vienna, Austria), followed by a continuous rinsing of the chip with PBS or cell culture medium.

2.4. Cell Culture and Loading

NSCs are isolated and cultured from the day-16 embyro-murine forebrain of Sprague Dawley (SD) rats and Hes5-GFP/Dcx-RFP double transgenic mice, which are subjected to experimental and chip cultivation procedures according to the established protocol [32–34]. The external tube holding the cells and fresh culture medium is pressurized by air containing 5% CO_2 and delivered to the fluidic culture chamber by the programmed open–close of the PDMS membrane on valves on the chip or the on–off of the solenoid valves connected to the tube.

2.5. Image Acquisition and Data Analysis

For image acquisition, a Nikon Ti2-ECLIPSE microscope, with an automated translational stage and a digital CMOS camera (ORCA-Flash 4.0, Hamamatsu, Japan), and microscope (XDS-5, Bingyu, China) were used. Image acquisition was controlled by the Nikon microscope's software (NIS components) and a custom-written control program, respectively. Bright field and fluorescence images were captured and analyzed using a customized MATLAB program.

3. Results

3.1. Design and Characterization of Microfluidic Chips

To perform the high-throughput screening of combinatorial drugs, we propose a microfluidic chip which can generate mixtures of different drug doses with minimum effort (Figure 1). The primary material of these microfluidic chips, PDMS, is highly biocompatible, allowing for precise control of the microenvironment in long-term cell cultures and facilitating evaluations of drug screening and delivery [35–42]. The microfluidic chip is 6 cm in length, 5 cm in width, and 0.5 cm in height. It consists of multiple arrays of culture chambers (i.e., a grid of 2 × 5), which are interconnected by three microfluidic channels (Figure 1a). As shown In Figure 1a, Valves 1 and 2 function as the lower and upper inlet control valves, valve 3 (red) regulates the connection between the microfluidic channels and culture chambers, and valve 4 (green) isolates the culture chambers. To achieve precise control over the composition of combinatorial drug treatments (i.e., the direction, time and duration of drug input), all channels are independently controlled by Quake's valves [31]. For example, by opening valves 1, 3, and 4, the drug from inlet-3 will be directed through Path-2, i.e., in the down-top direction through the array of culture chambers. Switching from valve 1 to valve 2 leads to changing the flow direction to top-down (Path-1).

Numerical simulation indicates that when drug inputs are directed through array of culture chambers, they are gradually diluted (Figure 1b and Figure S1). Paths, which liquid follows through the array of chambers, determine the direction of the concentration gradient; e.g., there is a high concentration in the top chambers when following Path-1. With a defined duration and input flow rate (i.e., 4 s and 5 mm/s), approximately 22.48% of the liquid is replaced in the top chamber, i.e., 0.2248 dilution. This concentration decreases to 0.2003, 0.1716, 0.1387, and 0.1128 of the input value in the subsequent chambers (Figure S2). Since the medium exchange takes place within an area of 250 μm by 790 μm, the liquid replacements in each one of the three microfluidic channels will hardly affect each other in a culture chamber of dimensions of 250 μm by 2000 μm (Figure 1b). Therefore, a combination of three drug doses can be generated by directing the flow via either Path-1 or Path-2. The resolution of the concentration gradient is determined by the number of chambers in the array, and the complexity of the combinatorial drugs is determined by the number of culture chamber arrays and inlets used.

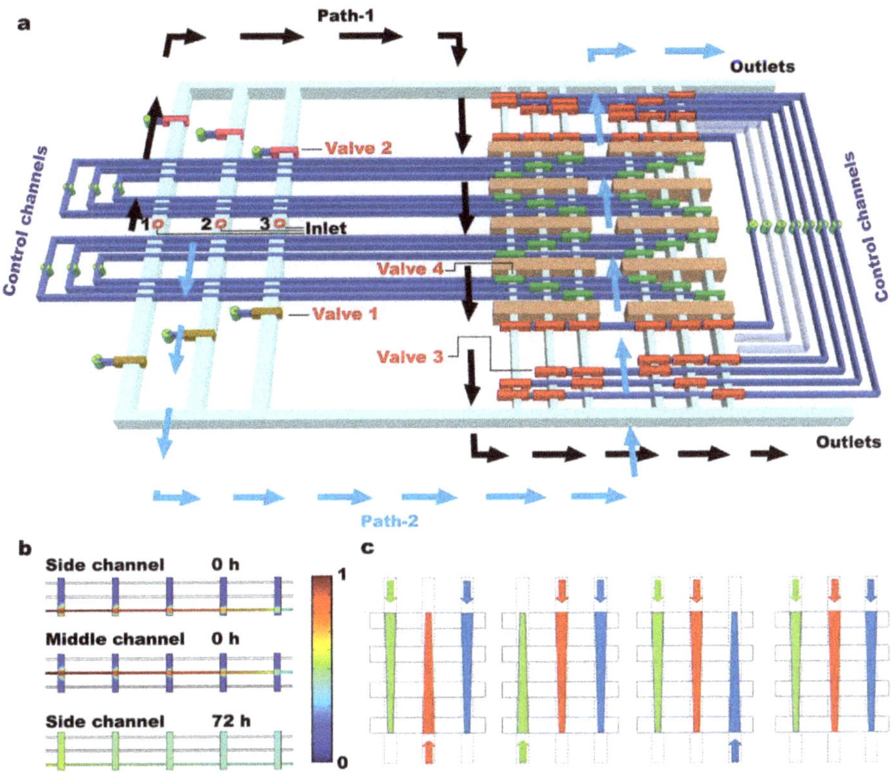

Figure 1. Design of the microfluidic device. (**a**) Three-dimensional configuration shows operational procedure of the microfluidic chip. Valves 1 (brown) and 2 (pink) function as the lower and upper inlet control valves, valve 3 (red) regulates the connection between microfluidic channels and culture chambers, and valve 4 (green) isolates the culture chambers. (**b**) Numerical simulation results indicate that when a fluid is directed through the array of culture chambers, it is gradually diluted. The direction of the concentration gradient depends on the paths, i.e., Path-1 for top-down and path-2 for the opposite direction. (**c**) Mixtures of different drug doses can be automatically generated by changing the direction of the concentration gradient, e.g., high + high + low.

To demonstrate the capacity of the proposed device, the microfluidic chip is fabricated using soft lithography and consists of two layers, i.e., the flow (red) and control layers (green) (Figures 2a and S3). Its 40 culture chambers form a grid of 8 columns × 5 rows (Figure 2b). Since each column has three independent microchannels, it requires 24 independently controllable liquid inputs to achieve the claimed screening capacity. As is shown in Figure 2c, a minimum of seven valves (through the combination calculation of C_7^3) are necessary to control the 24 liquid inlet channels (Figure S1b). Consistent with our numerical simulation, inputs of green-, red-, and blue-colored food dyes do not disturb one another, i.e., show no signs of immature mixing at the initial stage (Figure 2d and Movie S1).

By changing the direction of the concentration gradient over a period of time, the mixing of different color concentration gradients is achieved in the cavity (Figure S4 and Movie S2).

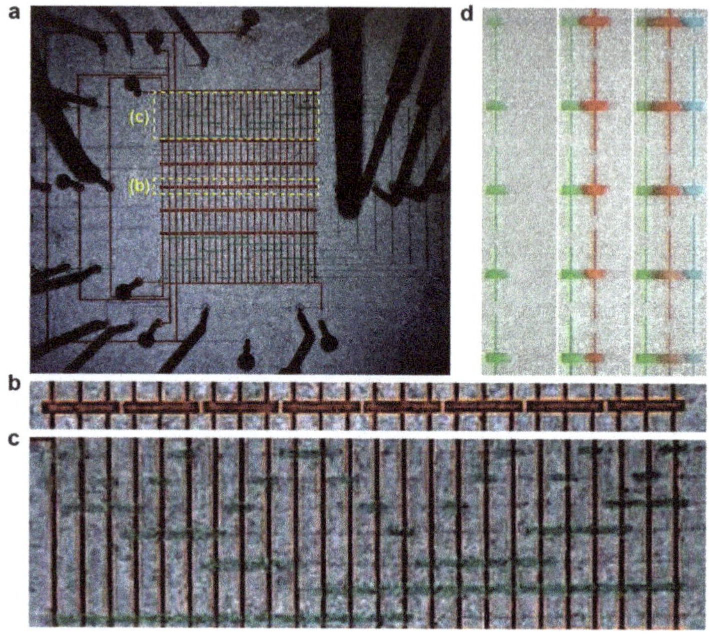

Figure 2. Operation of the microfluidic chip. (**a**) The microfluidic chip is produced, via soft lithography, with multi-layered structure, i.e., flow layers (red, such as (**b**)) and control layers (green, such as (**b**)). When pressurized by the connected tubes, the liquid in the chip can be manipulated with high accuracy. (**b**) The microfluidic chip has 8 culture chamber arrays, for which there are 5 culture chambers. These chambers are connected to 3 independently controllable microfluidic channels. (**c**) To control the liquids' passage through each microfluidic channel, a three-control-one arrangement is employed. Using this approach, 7 thin Quake's valves are needed to control 24 microfluidic channels. (**d**) Using green-, red-, and blue-colored food dyes, it is demonstrated that the concentration gradients of the three drugs can be generated independently.

3.2. Overdose of Growth Factors Leads to NSCs' Diminished Stemness

By directing cell-barring liquid through each column, NSCs are loaded into the culture chambers (Figure 3a). The number of Hes-5-positive cells is used as the indicator for NSCs' stemness [43]. As a control, NSCs are maintained in culture medium during the 48 h experiments; the NSCs maintained in different culture chambers of the same array show no observable differences in cellular behavior and differentiation, indicating that the delivered nutrients meet the needs for cell survival (Figure 3b). To induce individual growth factors, we connect inlet-1 with a growth factor, and the other two inlets with culture medium (Figure 1a). With the addition of growth factors to the culture medium (1 µg/mL PDGF; 0.2 µg/mL FGF; 500 ng/ml EGF), concentration gradients are generated when the liquid is directed to pass through the array of culture chambers, which can be estimated to be ~110 ng/mL, 100 ng/mL, 85 ng/mL, 70 ng/mL, and 55 ng/mL for EGF-1, -2, -3, -4, and -5; ~44 ng/mL, 40 ng/mL, 34 ng/mL, 28 ng/mL, and 22 ng/mL for FGF-1, -2, -3, -4, and -5; and ~220 ng/mL, 200 ng/mL, 170 ng/mL, 140 ng/mL, and 110 ng/mL for PDGF-1, -2, -3, -4, and -5, respectively. It is observed that, at the highest doses, the addition of all growth factors leads to an increased number of differentiated NSCs, i.e., the cells losing Hes-5 fluorescence (Figure 1a), while their regulatory effects towards differentiation disappear at low growth factor concentrations (Figure 3c). For example, the number of Hes-5-positive cells exceeds even the control samples at EGF-4 and EGF-5, i.e., 70 ng/mL and 55 ng/mL. That is to say that individual growth factors act beneficially at low doses and produce contrary effects at high doses, i.e., a dose-dependent biphasic response, which is often observed in pharmacological experiments [44,45].

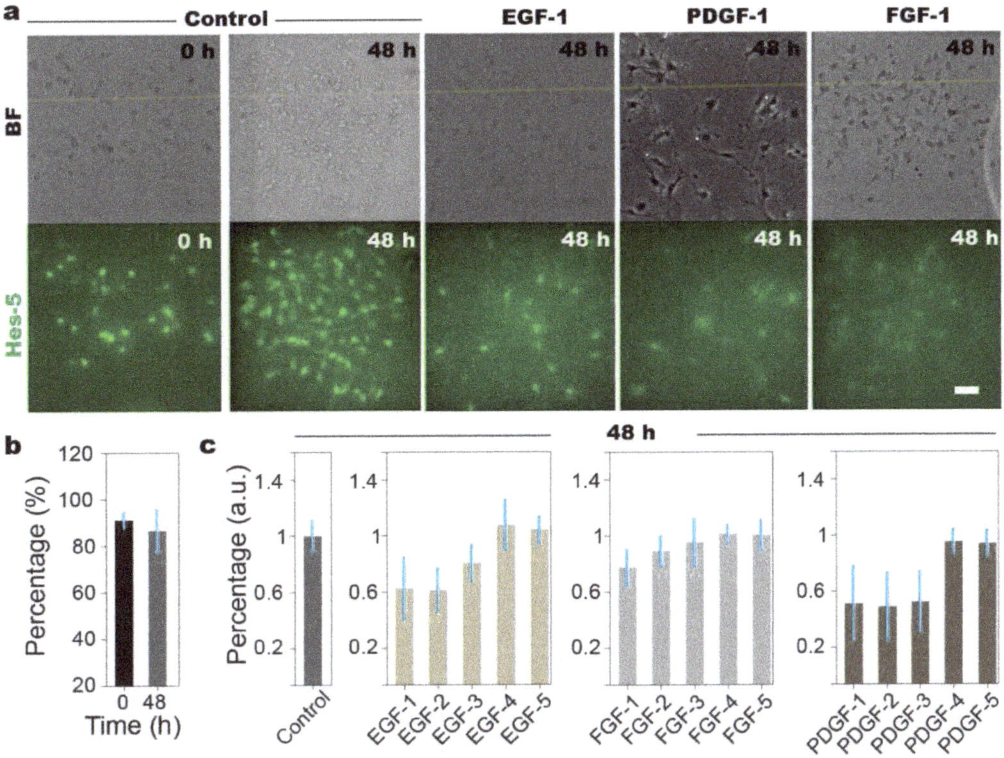

Figure 3. Live cell images reveal the regulatory effects of grow factor overdose on NSCs. (**a**) Bright field (BF) and fluorescent (Hes-5) images of NSCs when exposed to culture medium (control) and overdosed growth factors. EGF-1, -2, -3, -4, and -5 mean the high to low ends of the concentration gradient, i.e., 110 ng/mL, 100 ng/mL, 85 ng/mL, 70 ng/mL, and 55 ng/mL, respectively. FGF-1, -2, -3, -4, and -5 represent concentrations of 44 ng/mL, 40 ng/mL, 34 ng/mL, 28 ng/mL, and 22 ng/mL, respectively. PDGF-1, -2, -3, -4, and -5 represent concentrations of 220 ng/mL, 200 ng/mL, 170 ng/mL, 140 ng/mL, and 110 ng/mL, respectively. (**b**) By counting the cells losing Hes-5 fluorescence, the percentage of cells maintaining stemness can be estimated. The black and gray columns represent the results at 0 h and 48 h, respectively. Our results indicate that, with the culture medium, NSCs stemness can be well maintained. (**c**). At low growth factor concentrations, the number of differentiated NSCs is comparable to the control samples, while, at high growth factor concentrations, the percentage of differentiated NSCs reaches ~50%. Standard deviation is obtained from the average of at least 5 repeats. Scale bars denote 50 μm.

3.3. Combinatorial Treatment Reveals Logic Rules of Growth Factor Affecting NSCs Stemness

To introduce multiple growth factors into the culture chamber, two or three inlets are connected to the growth factors (Figure 1a). A combination of high and low doses are generated by directing liquid via Path-1 or Path-2 (Figure 1c), e.g., 1 + 5 + 5 represents high + low + low (Figure 4). The behavior of NSCs after being maintained on the chip under different drug conditions for 48 h reveals several rules: (1) the negative effects (i.e., differentiation) of a single growth factor overdose can be mediated by the overdosing another growth factors; (2) even though the addition of single growth factors at low concentrations helps maintain NSCs' stemness, the addition of a low-concentration growth factor generates no observable effects on either control or differentiated NSCs except for PDGF-5; and (3) NSCs tend to be irresponsive under complex environmental conditions, i.e., multiple growth factors. Notably, the differentiation of NSCs induced by overdoses of EGF and FGF can be mediated by low concentrations of PDGF, while the differentiation of NSCs induced by a PDGF overdose cannot be interfered with by low doses of EGF and FGF. These results indicate that PDGF, as a promotor of either NSCs' stemness or differentiation, has stronger effects compared to other growth factors [46]. We suspect that the signaling pathways activated by PDGF are digital, meaning that cascade reactions are triggered by trivial quantities of stimuli [47,48], while the signaling pathways of other growth factors are analog, the amplitude of which is concentration-dependent [49].

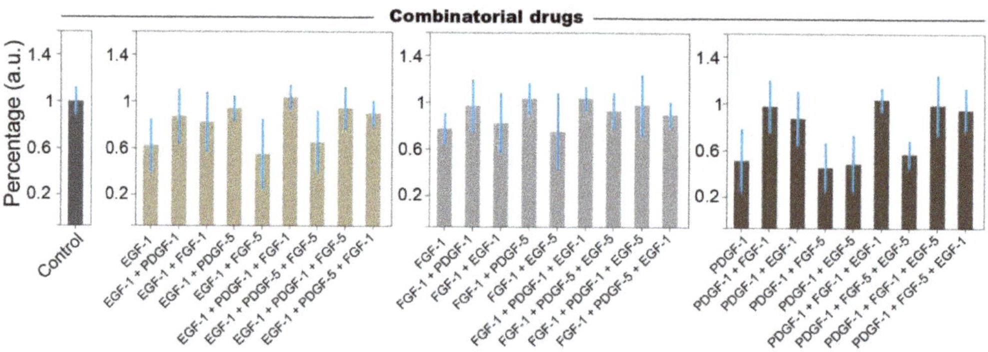

Figure 4. Regulatory effects of combinatorial drugs on NSCs' stemness. Using the number of Hes-5-positive cells as the indicator, we investigate the effects of multiple high- and low-dose growth factors on NSCs. Standard deviation is obtained from the average of at least 5 repeats.

4. Discussion

Our research presents a groundbreaking microfluidic device capable of efficiently generating multi-drug concentration gradients, revolutionizing the screening process for combinatorial drug therapies. The device's innovative design allows for the simultaneous creation of multiple drug gradients and the real-time monitoring of cellular responses, particularly NSCs. By employing this device, we have uncovered the dose-dependent biphasic responses to growth factors in the differentiation of NSCs. Our findings demonstrate that high concentrations of individual growth factors promote differentiation, whereas lower concentrations maintain NSC stemness. Moreover, the introduction of multiple growth factors reveals complex interaction patterns that modulate NSC outcomes, highlighting the importance of precise drug dosing in combinatorial therapies. The device's ability to screen thousands of drug combinations rapidly and cost-effectively holds immense potential for personalized medicine, enabling the discovery of optimal therapeutic drug combinations tailored to individual patient profiles. The integration of microfluidic technology into drug discovery streamlines the process, reduces the resources required, and accelerates the identification of effective drug synergies.

In summary, our multi-drug concentration gradient mixing chip offers a novel and powerful platform for high-throughput drug combination screening. It has the potential to significantly advance our understanding of drug interactions and their effects on cellular behavior, paving the way for the development of more effective, personalized treatment strategies for complex diseases.

Supplementary Materials: The following supporting information can be downloaded at https://www.mdpi.com/article/10.3390/bios14050212/s1, Figure S1: COMSOL simulation showing the dynamic process of drug diffusion in the array of culture chambers; Figure S2: Estimation of the concentration gradient generated in the array of cultivation chambers using numerical simulation and fluorescence intensity analysis; Figure S3: Design of the microfluidic chip using AutoCAD software; Figure S4: Comparison of results after gradient, mixing and diffusion of red, green, and blue pigments in different directions. Movie S1: Green, red, and blue food dyes are introduced into the cell culture chamber; Movie S2: Over time, the intermixing of concentration gradients of different colors (green, red, and blue) is achieved within the cavity.

Author Contributions: Investigation, J.F., Y.F. and R.Y.; Methodology, J.F., Y.S. and Y.F.; Writing—original draft. J.F., W.Z. and C.Z.; Data Curation, W.Z., D.S. and J.T.; Conceptualization, D.S.; Supervision, C.Z.; Writing—review and editing, C.Z. All authors have read and agreed to the published version of the manuscript.

Funding: This research was funded by the National Natural Science Foundation of China (51927804).

Institutional Review Board Statement: Not applicable.

Informed Consent Statement: Not applicable.

Data Availability Statement: Data underlying the results presented in this paper are not publicly available at this time but may be obtained from the authors upon reasonable request.

Conflicts of Interest: Huaxin Microfish Biotechnology Co., Ltd. had no role in the design of the study; in the collection, analyses, or interpretation of data; in the writing of the manuscript, and in the decision to publish the results.

References

1. Lawler, S.E.; Chiocca, E.A. Oncolytic Virus-Mediated Immunotherapy: A Combinatorial Approach for Cancer Treatment. *J. Clin. Oncol.* **2015**, *33*, 2812–2814. [CrossRef] [PubMed]
2. Zhang, Z.; Wang, G.; Zhong, K.; Chen, Y.; Yang, N.; Lu, Q.; Yuan, B.; Wang, Z.; Li, H.; Guo, L.; et al. A drug screening to identify novel combinatorial strategies for boosting cancer immunotherapy efficacy. *J. Transl. Med.* **2023**, *21*, 23. [CrossRef] [PubMed]
3. Vilgelm, A.E.; Johnson, D.B.; Richmond, A. Combinatorial approach to cancer immunotherapy: Strength in numbers. *J. Leukoc. Biol.* **2016**, *100*, 275–290. [CrossRef] [PubMed]
4. Normann, L.S.; Haugen, M.H.; Hongisto, V.; Aure, M.R.; Leivonen, S.K.; Kristensen, V.N.; Tahiri, A.; Engebraaten, O.; Sahlberg, K.K.; Mælandsmo, G.M. High-throughput screen in vitro identifies dasatinib as a candidate for combinatorial treatment with HER2-targeting drugs in breast cancer. *PLoS ONE* **2023**, *18*, e0280507. [CrossRef] [PubMed]
5. Flobak, Å.; Niederdorfer, B.; Nakstad, V.T.; Thommesen, L.; Klinkenberg, G.; Lægreid, A. A high-throughput drug combination screen of targeted small molecule inhibitors in cancer cell lines. *Sci. Data* **2019**, *6*, 237. [CrossRef] [PubMed]
6. Sun, X.; Vilar, S.; Tatonetti, N.P. High-throughput methods for combinatorial drug discovery. *Sci. Transl. Med.* **2013**, *5*, 205rv1. [CrossRef]
7. Held, M.A.; Langdon, C.G.; Platt, J.T.; Graham-Steed, T.; Liu, Z.; Chakraborty, A.; Bacchiocchi, A.; Koo, A.; Haskins, J.W.; Bosenberg, M.W.; et al. Genotype-Selective Combination Therapies for Melanoma Identified by High-Throughput Drug Screening. *Cancer Discov.* **2013**, *3*, 52–67. [CrossRef] [PubMed]
8. Ma, S.; Dang, D.; Wang, W.; Wang, Y.; Liu, L. Concentration optimization of combinatorial drugs using Markov chain-based models. *BMC Bioinform.* **2021**, *22*, 451. [CrossRef] [PubMed]
9. Xiao, Q.; Wang, L.; Xu, H. Application of kriging models for a drug combination experiment on lung cancer. *Stat. Med.* **2019**, *38*, 236–246. [CrossRef]
10. Lin, X.; Li, X.; Lin, X. A Review on Applications of Computational Methods in Drug Screening and Design. *Molecules* **2020**, *25*, 1375. [CrossRef]
11. Schuster, B.; Junkin, M.; Kashaf, S.S.; Romero-Calvo, I.; Kirby, K.; Matthews, J.; Weber, C.R.; Rzhetsky, A.; White, K.P.; Tay, S. Automated microfluidic platform for dynamic and combinatorial drug screening of tumor organoids. *Nat. Commun.* **2020**, *11*, 5271. [CrossRef]
12. Wan, L.; Yin, J.; Skoko, J.; Schwartz, R.; Zhang, M.; LeDuc, P.R.; Neumann, C.A. 3D Collagen Vascular Tumor-on-a-Chip Mimetics for Dynamic Combinatorial Drug Screening. *Mol. Cancer Ther.* **2021**, *20*, 1210–1219. [CrossRef] [PubMed]
13. Petreus, T.; Cadogan, E.; Hughes, G.; Smith, A.; Pilla Reddy, V.; Lau, A.; O'Connor, M.J.; Critchlow, S.; Ashford, M.; Oplustil O'Connor, L. Tumour-on-chip microfluidic platform for assessment of drug pharmacokinetics and treatment response. *Commun. Biol.* **2021**, *4*, 1001. [CrossRef]
14. Chen, L.; Ji, Y.; Li, A.; Liu, B.; Shen, K.; Su, R.; Ma, Z.; Zhang, W.; Wang, Q.; Zhu, Y.; et al. High-throughput drug screening identifies fluoxetine as a potential therapeutic agent for neuroendocrine prostate cancer. *Front Oncol.* **2023**, *13*, 1085569. [CrossRef]

15. Oudebrouckx, G.; Goossens, J.; Bormans, S.; Vandenryt, T.; Wagner, P.; Thoelen, R. Integrating Thermal Sensors in a Microplate Format: Simultaneous Real-Time Quantification of Cell Number and Metabolic Activity. *ACS Appl Mater Interfaces* **2022**, *14*, 2440–2451. [CrossRef] [PubMed]
16. Dasovich, M.; Zhuo, J.; Goodman, J.A.; Thomas, A.; McPherson, R.L.; Jayabalan, A.K.; Busa, V.F.; Cheng, S.J.; Murphy, B.A.; Redinger, K.R.; et al. High-Throughput Activity Assay for Screening Inhibitors of the SARS-CoV-2 Mac1 Macrodomain. *ACS Chem Biol.* **2022**, *17*, 17–23. [CrossRef]
17. Pham, N.; Radajewski, D.; Round, A.; Brennich, M.; Pernot, P.; Biscans, B.; Bonneté, F.; Teychené, S. Coupling High Throughput Microfluidics and Small-Angle X-ray Scattering to Study Protein Crystallization from Solution. *Anal. Chem.* **2017**, *89*, 2282–2287. [CrossRef]
18. Rodríguez-Ruiz, I.; Radajewski, D.; Charton, S.; Phamvan, N.; Brennich, M.; Pernot, P.; Bonneté, F.; Teychené, S. Innovative High-Throughput SAXS Methodologies Based on Photonic Lab-on-a-Chip Sensors: Application to Macromolecular Studies. *Sensors* **2017**, *17*, 1266. [CrossRef]
19. Ansari, M.I.H.; Hassan, S.; Qurashi, A.; Khanday, F.A. Microfluidic-integrated DNA nanobiosensors. *Biosens. Bioelectron.* **2016**, *85*, 247–260. [CrossRef]
20. Lou, C.; Yang, H.; Hou, Y.; Huang, H.; Qiu, J.; Wang, C.; Sang, Y.; Liu, H.; Han, L. Microfluidic Platforms for Real-Time in Situ Monitoring of Biomarkers for Cellular Processes. *Adv. Mater.* **2023**, *36*, e2307051. [CrossRef]
21. Carr, J.A.; Parashar, A.; Gibson, R.; Robertson, A.P.; Martin, R.J.; Pandey, S. A microfluidic platform for high-sensitivity, real-time drug screening on C. elegans and parasitic nematodes. *Lab Chip* **2011**, *11*, 2385–2396. [CrossRef]
22. Zuieva, A.; Can, S.; Boelke, F.; Reuter, S.; Schattscheider, S.; Töpfer, E.; Westphal, A.; Mrowka, R.; Wölfl, S. Real-time monitoring of immediate drug response and adaptation upon repeated treatment in a microfluidic chip system. *Arch. Toxicol.* **2022**, *96*, 1483–1487. [CrossRef]
23. Tay, S.; Hughey, J.J.; Lee, T.K.; Lipniacki, T.; Quake, S.R.; Covert, M.W. Single-cell NF-κB dynamics reveal digital activation and analogue information processing. *Nature* **2010**, *466*, 267–271. [CrossRef] [PubMed]
24. Zhang, C.; Tu, H.L.; Jia, G.; Mukhtar, T.; Taylor, V.; Rzhetsky, A.; Tay, S. Ultra-multiplexed analysis of single-cell dynamics reveals logic rules in differentiation. *Sci. Adv.* **2019**, *5*, eaav7959. [CrossRef]
25. Han, S.-I.; Han, K.-H.; Frazier, A.B.; Ferrance, J.P.; Landers, J.P. An automated micro-solid phase extraction device involving integrated \high-pressure microvalves for genetic sample preparation. *Biomed. Microdevices* **2009**, *11*, 935–942. [CrossRef]
26. Ansede, J.H.; Thakker, D.R. High-throughput screening for stability and inhibitory activity of compounds toward cytochrome P450-mediated metabolism. *J. Pharm. Sci.* **2004**, *93*, 239–255. [CrossRef] [PubMed]
27. Konkankit, C.C.; Vaughn, B.A.; MacMillan, S.N.; Boros, E.; Wilson, J.J. Combinatorial Synthesis to Identify a Potent, Necrosis-Inducing Rhenium Anticancer Agent. *Inorg. Chem.* **2019**, *58*, 3895–3909. [CrossRef]
28. Zhang, Y.; Yazid, N.B.M.; Ho, P.Y.; Hu, X.; Chen, S.; Vasoo, S.; Kanitthamniyom, P. DropCarba—An automated magnetic digital microfluidic platform for rapid phenotypic testing of carbapenemase-producing Gram-negative bacilli. *Biosens. Bioelectron.* **2023**, *225*, 115099. [CrossRef]
29. Goto, M.; Hojo, M.; Ando, M.; Kita, A.; Kitagawa, M.; Ohtsuka, T.; Kageyama, R.; Miyamoto, S. Hes1 and Hes5 are required for differentiation of pituicytes and formation of the neurohypophysis in pituitary development. *Brain Res.* **2015**, *1625*, 206–217. [CrossRef] [PubMed]
30. Manning, C.S.; Biga, V.; Boyd, J.; Kursawe, J.; Ymisson, B.; Spiller, D.G.; Sanderson, C.M.; Galla, T.; Rattray, M.; Papalopulu, N. Quantitative single-cell live imaging links HES5 dynamics with cell-state and fate in murine neurogenesis. *Nat. Commun.* **2019**, *10*, 2835. [CrossRef]
31. Unger, M.A.; Chou, H.P.; Thorsen, T.; Scherer, A.; Quake, S.R. Monolithic Microfabricated Valves and Pumps by Multilayer Soft Lithography. *Science* **2000**, *288*, 113–116. [CrossRef] [PubMed]
32. Giachino, C.; Basak, O.; Taylor, V. Isolation and manipulation of mammalian neural stem cells in vitro. *Methods Mol. Biol.* **2009**, *482*, 143–158. [PubMed]
33. Sykova, E.; Forostyak, S. Stem cells in regenerative medicine. *Laser Ther.* **2013**, *22*, 87–92. [CrossRef] [PubMed]
34. Behnan, J.; Grieg, Z.; Joel, M.; Ramsness, I.; Stangeland, B. Neuroepigenetics Gene knockdown of CENPA reduces sphere forming ability and stemness of glioblastoma initiating cells. *Neuroepigenetics* **2016**, *7*, 6–18. [CrossRef]
35. Shao, J.; Wu, L.; Wu, J.; Zheng, Y.; Zhao, H.; Jin, Q.; Zhao, J. Integrated microfluidic chip for endothelial cells culture and analysis exposed to a pulsatile and oscillatory shear stress. *Lab Chip* **2009**, *9*, 3118–3125. [CrossRef] [PubMed]
36. Mehling, M.; Tay, S. Microfluidic cell culture. *Curr. Opin. Biotechnol.* **2014**, *25*, 95–102. [CrossRef] [PubMed]
37. Kimura, H.; Nakamura, H.; Goto, T.; Uchida, W.; Uozumi, T.; Nishizawa, D.; Shinha, K.; Sakagami, J.; Doi, K. Standalone cell culture microfluidic device-based microphysiological system for automated cell observation and application in nephrotoxicity tests. *Lab Chip* **2024**, *24*, 408–421. [CrossRef] [PubMed]
38. Kwon, T.; Prentice, H.; Oliveira, J.; Madziva, N.; Warkiani, M.E.; Hamel, J.F.P.; Han, J. Microfluidic Cell Retention Device for Perfusion of Mammalian Suspension Culture. *Sci. Rep.* **2017**, *7*, 6703. [CrossRef] [PubMed]
39. Van Midwoud, P.M.; Merema, M.T.; Verpoorte, E.; Groothuis, G.M. Microfluidics enables small-scale tissue-based drug metabolism studies with scarce human tissue. *J. Lab. Autom.* **2011**, *16*, 468–476. [CrossRef]
40. Liu, X.; Zheng, W.; Jiang, X. Cell-Based Assays on Microfluidics for Drug Screening. *ACS Sens.* **2019**, *4*, 1465–1475. [CrossRef]

41. Kohl, Y.; Biehl, M.; Spring, S.; Hesler, M.; Ogourtsov, V.; Todorovic, M.; Owen, J.; Elje, E.; Kopecka, K.; Moriones, O.H.; et al. Microfluidic In Vitro Platform for (Nano)Safety and (Nano)Drug Efficiency Screening. *Small* **2021**, *17*, e2006012. [CrossRef] [PubMed]
42. Yeo, L.Y.; Chang, H.C.; Chan, P.P.; Friend, J.R. Microfluidic devices for bioapplications. *Small* **2011**, *7*, 12–48. [CrossRef] [PubMed]
43. Tian, Y.; Cao, R.; Che, B.; Sun, D.; Tang, Y.; Jiang, L.; Bai, Q.; Liu, Y.; Morozova-Roche, L.A.; Zhang, C. Proinflammatory S100A9 Regulates Differentiation and Aggregation of Neural Stem Cells. *ACS Chem. Neurosci.* **2020**, *11*, 3549–3556. [CrossRef] [PubMed]
44. Gan, D.; Cheng, W.; Ke, L.; Sun, A.R.; Jia, Q.; Chen, J.; Xu, Z.; Xu, J.; Zhang, P. Biphasic Effect of Pirfenidone on Angiogenesis. *Front. Pharmacol.* **2022**, *12*, 804327. [CrossRef] [PubMed]
45. Al-Humadi, N.H.; Ma, J.K.; Lewis, D.M.; Ma, J.Y.; Barger, M.W.; Siegel, P.D. Dose-dependent thiol and immune responses to ovalbumin challenge in Brown Norway rats. *Toxicol. Ind. Health* **2002**, *18*, 343–352. [CrossRef]
46. Jackson, E.L.; Garcia-Verdugo, J.M.; Gil-Perotin, S.; Roy, M.; Quinones-Hinojosa, A.; VandenBerg, S.; Alvarez-Buylla, A. PDGFRα-Positive B Cells Are Neural Stem Cells in the Adult SVZ that Form Glioma-like Growths in Response to Increased PDGF Signaling. *Neuron* **2006**, *51*, 187–199. [CrossRef] [PubMed]
47. Hu, J.G.; Wang, Y.X.; Wang, H.J.; Bao, M.S.; Wang, Z.H.; Ge, X.; Wang, F.C.; Zhou, J.S.; Lü, H.Z. PDGF-AA Mediates B104CM-Induced Oligodendrocyte Precursor Cell Differentiation of Embryonic Neural Stem Cells Through Erk, PI3K, and p38 Signaling. *J. Mol. Neurosci.* **2011**, *46*, 644–653. [CrossRef] [PubMed]
48. Koivisto, A.; Klinge, A.; Nedergaard, J.; Siemen, D. Regulation of the Activity of 27 pS Nonselective Cation Channels in Excised Membrane Patches from Rat Brown-Fat Cells. *Cell. Physiol. Biochem.* **1998**, *8*, 231–245. [CrossRef]
49. Mondal, D.; Pradhan, L.; Larussa, V. Signal Transduction Pathways Involved in the Lineage-Specific Differentiation of NSCs: Can the Knowledge Gained from Blood be Used in the Brain? *Cancer Investig.* **2004**, *22*, 925–943. [CrossRef]

biosensors

MDPI

Article

Multi-Functional Nano-Doped Hollow Fiber from Microfluidics for Sensors and Micromotors

Yanpeng Wang [1], Zhaoyang Wang [1], Haotian Sun [1], Tong Lyu [1], Xing Ma [2], Jinhong Guo [3,*] and Ye Tian [1,4,*]

1 College of Medicine and Biological Information Engineering, Northeastern University, Shenyang 110169, China; wangyanpeng1001@163.com (Y.W.); wangzhaoyang0908@163.com (Z.W.); s1328607240@163.com (H.S.); verapetrova@163.com (T.L.)
2 School of Materials Science and Engineering, Harbin Institute of Technology (Shenzhen), Shenzhen 518055, China; maxing@hit.edu.cn
3 School of Sensing Science and Engineering, Shanghai Jiao Tong University, Shanghai 200240, China
4 Foshan Graduate School of Innovation, Northeastern University, Foshan 528300, China
* Correspondence: guojinhong@sjtu.edu.cn (J.G.); tianye@bmie.neu.edu.cn (Y.T.)

Abstract: Nano-doped hollow fiber is currently receiving extensive attention due to its multifunctionality and booming development. However, the microfluidic fabrication of nano-doped hollow fiber in a simple, smooth, stable, continuous, well-controlled manner without system blockage remains challenging. In this study, we employ a microfluidic method to fabricate nano-doped hollow fiber, which not only makes the preparation process continuous, controllable, and efficient, but also improves the dispersion uniformity of nanoparticles. Hydrogel hollow fiber doped with carbon nanotubes is fabricated and exhibits superior electrical conductivity (15.8 S m^{-1}), strong flexibility (342.9%), and versatility as wearable sensors for monitoring human motions and collecting physiological electrical signals. Furthermore, we incorporate iron tetroxide nanoparticles into fibers to create magnetic-driven micromotors, which provide trajectory-controlled motion and the ability to move through narrow channels due to their small size. In addition, manganese dioxide nanoparticles are embedded into the fiber walls to create self-propelled micromotors. When placed in a hydrogen peroxide environment, the micromotors can reach a top speed of 615 μm s^{-1} and navigate hard-to-reach areas. Our nano-doped hollow fiber offers a broad range of applications in wearable electronics and self-propelled machines and creates promising opportunities for sensors and actuators.

Keywords: hollow fibers; nano-doped hydrogels; microfluidics; wearable sensors; micromotors

1. Introduction

Functional hydrogel fiber has attracted widespread attention due to its broad application prospects. With the advancement of technology, functional hydrogel fiber has been used for drug delivery [1,2], cell culture [3–5], tissue engineering [6,7], substance separation [8–10], wearable electronic devices [11–13], and actuators [14,15], showing excellent biocompatibility, adsorption, sensing properties, and actuation ability. However, the underlying hydrogel matrix cannot possess the above capabilities, so a series of nanoparticles have been successfully combined with polymer networks to obtain nano-doped hydrogels [16,17], including carbon nanotubes [18–20], graphene [21], metal nanoparticles [22], and transition metal carbide nanosheets [12,23]. It has been shown that the doping of nanomaterials in fiber helps to improve their mechanical properties [24]. Compared to bulk hydrogels, hollow hydrogel fibers have a larger specific surface area, which allows many reactions to proceed more fully. Nano-doped hydrogel functional fibers are receiving more and more attention because the fibers exhibit excellent weave ability, so they can be customized into arbitrary shapes with high softness.

However, conventional preparation methods for obtaining nano-doped hollow fibers have certain limitations. The conventional template method is not able to prepare ultra-long fibers continuously [25,26], and the electrostatic spinning method [27,28] is unable to ensure the nanoparticle agglomeration during the preparation process due to electrostatic force and van der Waals force [29]. The fiber impregnation method [30,31] is unable to ensure that the nanoparticles will not be dislodged during subsequent use. Microfluidics has been developed over decades as a mature technology and is widely used in many industries and basic research areas [32–36]. Since the fluids in co-flow microfluidic systems all flow in a laminar flow at low speed [37], the possibility of nanoparticle collisions is low and the adhesion between particles is weak [38], which ensures that they do not aggregate during the preparation process. However, the microfluidic techniques employed in the existing literature usually rely on an internal curing process to form hollow fiber structures [32,39]. This approach, although feasible to a certain extent, is prone to triggering channel blockage, since the curing reaction takes place inside the system, thus leading to discontinuity in the preparation process. Therefore, there is an urgent demand to use microfluidic methods to fabricate nano-doped hollow fibers in a simple, smooth, stable, continuous, well-controlled manner without system blockage.

In this study, we chose a microfluidic method to prepare nano-doped hollow fiber using coaxial glass capillary chips in a simple, smooth, stable, continuous, and well-controlled manner. Due to the curing strategy outside the microfluidic system, our microfluidic method greatly reduces system blockage. The rapid gelation reaction between sodium alginate and calcium ions ensures that the nanoparticles are firmly anchored inside the fibers, avoiding degradation of fiber properties due to displacement of nanoparticles during use. Therefore, we doped carbon nanotubes in hydrogel hollow fibers and prepared wearable fiber sensors using the hollow fibers. The hollow fibers have a high conductivity of 15.8 S m^{-1}, a high stretch rate of 342.9%, and high sensitivity at high deformation rates. Iron tetroxide nanoparticles were incorporated into hydrogels to create magnetic-driven micromotors. These micromotors not only enable trajectory-controlled motion but can also, due to their small size, pass through narrow passages, giving the fibers the ability to have adjustable physical properties with precise motion control. Additionally, to enhance the functional capabilities of these fibers, manganese dioxide (MnO_2) nanoparticles were intricately embedded within their walls, thereby transforming them into self-propelled micromotors. Upon exposure to a hydrogen peroxide environment, these micromotors demonstrated remarkable autonomous mobility, achieving a maximum velocity of 615 μm s^{-1}. This unique attribute not only facilitates their navigation through challenging and confined spaces, but also endows the fibers with heightened environmental responsiveness, autonomous motion capabilities, and an inherent mechanism for energy generation. The results show that hollow fibers containing different functional nano-dopants prepared by microfluidic technology have successfully achieved the multifunctionalities of hollow fibers. This technique not only permits the preparation of multiple functional hollow fibers on the same preparation platform, but, more importantly, it demonstrates the advancement and flexibility of microfluidics in material design and preparation. Especially in wearable electronic devices, soft robotics, and other potential applications, these hollow fibers show promising applications due to their light weight, high elasticity, and customizable chemical and physical properties, and demonstrate practicality and innovation in these fields of application.

2. Materials and Methods

2.1. Materials

Sodium alginate (NaA, Sigma Industrial Corporation, Phoenix, AZ, USA) from brown algae, polyvinyl alcohol (PVA, Mw: 13,000–23,000, hydrolyzed: 87–89%, Sigma Industrial Corporation, Phoenix, AZ, USA) and calcium chloride ($CaCl_2$ anhydrous, powder, AR, 96%, Macklin Industrial Corporation, Shanghai, China), polyethylene glycol (PEG, Mw: 6000, Aladdin Industrial Corporation, Shanghai, China), hydroxypropyl methyl cellulose (HPMC, Shandong Yousuo Chemical Technology Co., Ltd., Jinan, Shandong, China), methyl cellulose (MC, Shandong Yousuo Chemical Technology Co., Ltd., Jinan, Shandong, China), ferric oxide nanoparticle dispersion (Fe_3O_4, VK-EF01W, 20% solid content, 20 nm–30 nm, Jinan Zhiding Welding Material Co., Ltd., Jinan, Shandong, China), large diameter carboxylate multi-walled carbon nanotubes (CNTs, JCMW CC4, Shandong Jiacai Technology Co., Ltd., Jinan, Shandong, China), Lithium chloride (LiCl anhydrous, powder, AR, 98%, Aladdin Industrial Corporation, Shanghai, China), manganese dioxide (MnO_2, powder, GR, $\geq$90%, Aladdin Industrial Corporation, Shanghai, China), hydrogen peroxide (H_2O_2, Bangjian pharmaceutical chain Co., Ltd., Bangjian, China).

2.2. Methods

2.2.1. Design of the Microfluidic Chip

A micropipette puller (PUL-100, World Precision Instruments, Inc., Sarasota, FL, USA) was used to prepare the capillary-based microfluidic device. Two cylindrical capillaries (World Precision Instruments, Inc., Sarasota, FL, USA) with inner diameters of 0.58 mm and outer diameters of 1 mm were tapered by the micropipette puller to obtain an injection section and a collection section, respectively. Their tips were polished to the desired diameter using fine sandpaper. The capillary microfluidic device consists of a glass slide and a group of nested glass capillaries. Two injection needles were encapsulated at the capillary junction. Two cylindrical capillaries with tips were placed in a square glass capillary (1.4 mm $\times$ 1.1 mm, Beijing Chengteng Equipment Co., Ltd., Beijing, China), aligned coaxially, and glued together with AB glue (5 Minute Epoxy, Deli Group Co., Ltd., Ningbo, Zhejiang, China) to obtain nested capillary groups. At the end of the collection section, a petri dish was placed, filled with calcium chloride (5 wt%) solution to solidify the generated hollow fiber.

2.2.2. Fabrication of Hollow Fiber

Sodium alginate (NaA) was dissolved completely into a 2.5 wt% sodium alginate solution with heating and stirring in a constant temperature water bath at 70 °C. Subsequently, a 10 wt% polyvinyl alcohol solution was prepared. For easy observation, we added a little red pigment to the polyvinyl alcohol solution to distinguish the inner and outer phases. Next, the solutions were introduced into the microfluidic device via a syringe pump (Baoding Lange constant flow pump Co., Ltd., BaoDing, HeBei, China). Before fabrication, NaA was injected into the device to expel the air in the device and to improve the stability of the system. During fabrication, the flow rate of the inner phase should be slower than that of the outer phase. The flow rate of the outer phase was adjusted in the range of 0.5 mL h^{-1}–1.2 mL h^{-1}, and the flow rate of the inner phase was adjusted in the range of 0.1 mL h^{-1}–1.0 mL h^{-1}. Finally, a 5 wt% $CaCl_2$ solution was used as a coagulation bath to solidify the hollow fibers. The prepared fibers were extracted with tweezers and washed in ultrapure water to remove the PVA solution and obtain the hollow fibers. As a control experiment, we also fabricated hollow fibers with PEG (10 wt%), MC (2 wt%), and HPMC (1.6 wt%) as the inner phase, respectively, using the same process (Scheme 1).

Fabrication of HollowFiber-CNTs. In the preparation of CNT-doped hollow fibers, NaA (2.5 wt%), doped with 1 wt%, 2 wt%, 3 wt%, 4 wt%, and 5 wt% CNT dispersion, was used as the outer phase solution, and PVA (10 wt%) solution was used as the inner phase solution to fabricate CNT-doped hollow fiber. The other preparation procedures were the same as above. The CNT-doped hollow fibers were named HollowFiber-CNTs. The generated HollowFiber-CNTs were immersed in 1 wt% lithium chloride solution for 2 h to enhance their conductivity, and then the excess salt solution on the fiber surface was washed with ultrapure water for subsequent preparation.

Fabrication of HollowFiber-Fe_3O_4. We dispersed water-based magnetic fluid containing Fe_3O_4 nanoparticles into 2 wt% NaA solution (volume fraction 12.5%) as the outer phase [40,41] and PVA (10 wt%) as the inner phase, and the velocity ratio of the inner and outer phases was set at 0.65 mL h^{-1}:1.2 mL h^{-1} to prepare hollow fibers with a uniform distribution of magnetic nanoparticles. The other preparation procedures were the same as above. The hollow fibers with a uniform distribution of magnetic nanoparticles were named HollowFiber-Fe_3O_4.

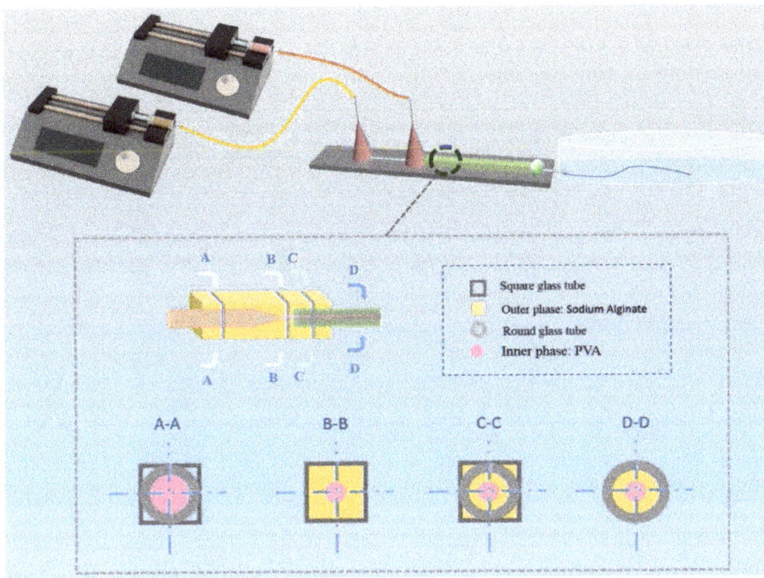

Scheme 1. Schematic diagram showing the experimental setup for the fabrication of hollow fiber. A–A, B–B, C–C, and D–D each correspond to the magnified cross-sectional views of specific combinations in different positions of the channel schematic shown above in an academic context.

Fabrication of HollowFiber-MnO_2. We utilized a vortex machine (XH-D VORTEX) and an ultrasonic machine (XM-2200ES, Komei Ultrasonic Instruments Co., Kunshan, Jiangsu, China) to homogeneously disperse the MnO_2 particles in 2 wt% NaA solution (1 wt% by volume) as the outer phase solution and PVA (10 wt%) as the inner phase. The inner phase solution was introduced into the system at a rate of 0.5 mL h^{-1} using a syringe pump, while the outer phase solution was introduced into the system at a rate of 1 mL h^{-1}. The other preparation steps were the same as described above. This fiber was named HollowFiber-MnO_2.

2.2.3. Characterization of Hollow Fiber

The cross section of the hollow fiber was observed with an inverted optical microscope (Eclipse TS2, Nikon, Tokyo, Japan). The microscale morphology of hollow fiber was further characterized by scanning electron microscopy (SEM; S4800, Hitachi, Tokyo, Japan). The crystal structure of the fibers was analyzed by X-ray diffraction (XRD; Model: ultima IV, Riigaku, Tokyo, Japan). To study the effects of flow rates and diameters of hollow fibers, the preparation process was recorded by an inverted optical microscope (Hunan Ketianjian Photoelectric Technology Co., Ltd., AcutEye-2M-169, Changsha, Hunan, China) equipped with a CCD high-speed camera (Phantom, Wayne, NJ, USA). Image analysis of the hollow fiber was performed using ImageJ 1.54c software (http://rsb.info.nih.gov/ij/, accessed on 8 March 2023).

2.2.4. Mechanical Properties of Hollow Fiber

We fixed the hollow fibers on a tensile test rig (JXLSPT-DBL, Joly Instruments Co., Ltd., Beijing, China) to test their tensile properties. The stretching speed was set to 1 mm s^{-1}. The tensile strength of the fibers was tested using a universal electronic testing machine (Model

43 MTS Criterion, Eden Prairie, MN, USA). The tensile ratio was calculated according to Equation (1):

$$Tensile = \frac{L - L_0}{L_0} \times 100\% \quad (1)$$

L_0 is the initial length of the hollow fiber, and L is the breaking length of the hollow fiber. In addition, we assembled the hollow fiber into a spiderweb-like structure to test its bending property and tied the hollow fiber into a knot to test its torsional property.

In the determination of stress–strain curves of hollow fibers with different water contents, the water content ($W\%$) of the hollow fibers was calculated according to Equation (2):

$$W\% = \frac{m_1 - m_2}{m_1} \times 100\% \quad (2)$$

m_1 is the net weight of the fiber after losing different degrees of moisture; m_2 is the dry weight of the hollow fiber. To ensure the stability of the moisture content of the hollow fibers during tensile testing, a thin protective coating of polyvinyl alcohol (PVA) was applied to the surface of the hollow fibers. The transparency of this coating ensures visibility of the fiber structure, while its barrier properties effectively slow water evaporation and ensure the moisture stability of the fiber during the test.

HollowFiber-CNTs with carbon nanotube contents of 1 wt%, 2 wt%, 3 wt%, 4 wt%, and 5 wt% were cut into 2-cm segments, respectively, and fixed on the tensile test platform to test the tensile property. The test method was the same as above.

2.2.5. Measurement of Hollow Fiber Diameter

The hollow fibers were rapidly cut into fine segments using a fine cutting tool. From these segments, a representative cross-sectional sample was obtained and imaged under an optical microscope (Eclipse TS2, Nikon, Japan). To ensure the accuracy of subsequent measurements, a graduated ruler was placed next to the sample during the imaging process to provide a scale reference. Subsequently, the scale on the ruler was measured using ImageJ 1.54c software, and the exact diameter of the fibers was determined accordingly.

2.2.6. Performance Testing of Wearable Fiber Sensors

We evaluated the performance of HollowFiber-CNTs using an electrochemical workstation (CHI760E, test resistor with an AC voltage of 0.1 V and an AC frequency of 1000 Hz). To improve the accuracy of the tests, the HollowFiber-CNTs were braided into a block form and encapsulated with titanium sheets and adhesive tape.

To evaluate the stability of HollowFiber-CNTs in different acid–base environments, we added hydrochloric acid solution with pH = 2, 4, and 5 and sodium hydroxide solution with pH = 8, respectively. After a certain period, the residual solution was removed using absorbent paper, and the relative resistance change ($\Delta R/R_0$) of the HollowFiber-CNTs was recorded continuously.

The samples were placed on a heating platform (Lebertek Instruments, Beijing, China, ZH35B). The temperature was gradually increased from 24 °C to 60 °C and then gradually decreased to 24 °C. During this period, continuous changes of $\Delta R/R_0$ were recorded to characterize the stability of the HollowFiber-CNTs over temperature changes.

We performed resistance tests on HollowFiber-CNTs with different CNT contents (1 wt%, 2 wt%, 3 wt%, 4 wt%, and 5 wt%) and calculated their conductivity (ρ) according to Equation (3):

$$\rho = \frac{RS}{L} \quad (3)$$

where R is the resistance, S is the cross-sectional area of HollowFiber-CNTs, and L is the length of HollowFiber-CNTs.

We braided the HollowFiber-CNTs to a length of 0.5 cm, connected them to the LED strip, and recorded the luminous intensity of the bulbs. Subsequently, the HollowFiber-CNTs were stretched to observe the effect of different stretching speeds on light bulb luminescence. We chose a 3.5 cm length of HollowFiber-CNTs to be fixed in the stretching test platform, set the stretching speed to 2 mm s^{-1}, recorded the changes of $\Delta R/R_0$, and obtained the sensitivity coefficient GF based on Equation (4):

$$GF = \frac{\Delta R/R_0}{\Delta L/L_0} \quad (4)$$

HollowFiber-CNTs were attached to the finger joints, wrists, elbows, knees, and throat areas, and the data changes in $\Delta R/R_0$ were monitored using an electrochemical workstation. Meanwhile, an electromyographic (EMG) signal detection device (3IT EMGC1V1) was used to record the EMG signals. The HollowFiber-CNTs were braided into 2 cm × 2 cm squares to be taped on both sides of the biceps as electrodes and then connected to EMG electromyographic sensors using wires to collect the changes of physiological electrical signals during muscle movement. Corresponding EMG signals were converted into waveforms using backyard brain V1.4. software and exported data, then plotted using MATLAB 2018a software.

2.2.7. Motion Testing of Micromotors

We investigated the motion behavior of magnetic hollow fiber micromotors using microscopy techniques. We tested the motion of magnetic hollow fiber micromotors along specific straight paths using an applied magnetic field. We sequentially preset straight paths, curved paths, complex paths, and narrow passages to observe whether the magnetic hollow fiber micromotor can move precisely and smoothly under the effect of an applied magnetic field. All tests were conducted under specific experimental conditions and parameters to ensure the reliability and reproducibility of the results.

In bubble-propelled micromotors, polydimethylsiloxane was used to make molds with complex cavities and put into the petri dish. The functionalized hollow fiber was cut into pieces and put into the cavity of a petri dish filled with H_2O_2 to test its motion. The reaction is based on the following chemical equation:

$$2H_2O_2 \overset{MnO_2}{\rightarrow} O_2 + 2H_2O \quad (5)$$

The concentration of H_2O_2 solution was varied, and the time recorded for movement over the same distance to obtain the velocity of the fiber segments at different concentrations. The data were processed using Origin 8.0 software to obtain a relationship between the concentration of the solution and the velocity of movement.

3. Results and Discussion

3.1. Fabrication and Morphology Characterization of Hollow Fibers

The hollow fibers can be fabricated according to the method shown in Scheme 1, and a clear interface (Figure 1a and Supporting Video S1) can be observed between the core solution and the shell solution. When the inner phase solution is changed to MC, HPMC, and PEG, the hollow fibers can still be prepared perfectly (Figure S1). This is because in a co-flowing microchannel, when two incompatible water phases are in contact, the growth rate of interfacial instability in the whole water jet depends on the value of interfacial tension relative to inertial force and viscous force. For this system, with an interfacial tension lower than 0.1 mN m^{-1}, the capillary force between the two water phases is usually negligible, and mass transfer does not easily occur between the two phases, so two immiscible phases can be separated by a long, straight interface [42]. After solidification, we easily rinsed the inner phase with pure water and obtained the resultant hollow fiber (Figure 1b) with a perfect hollow structure.

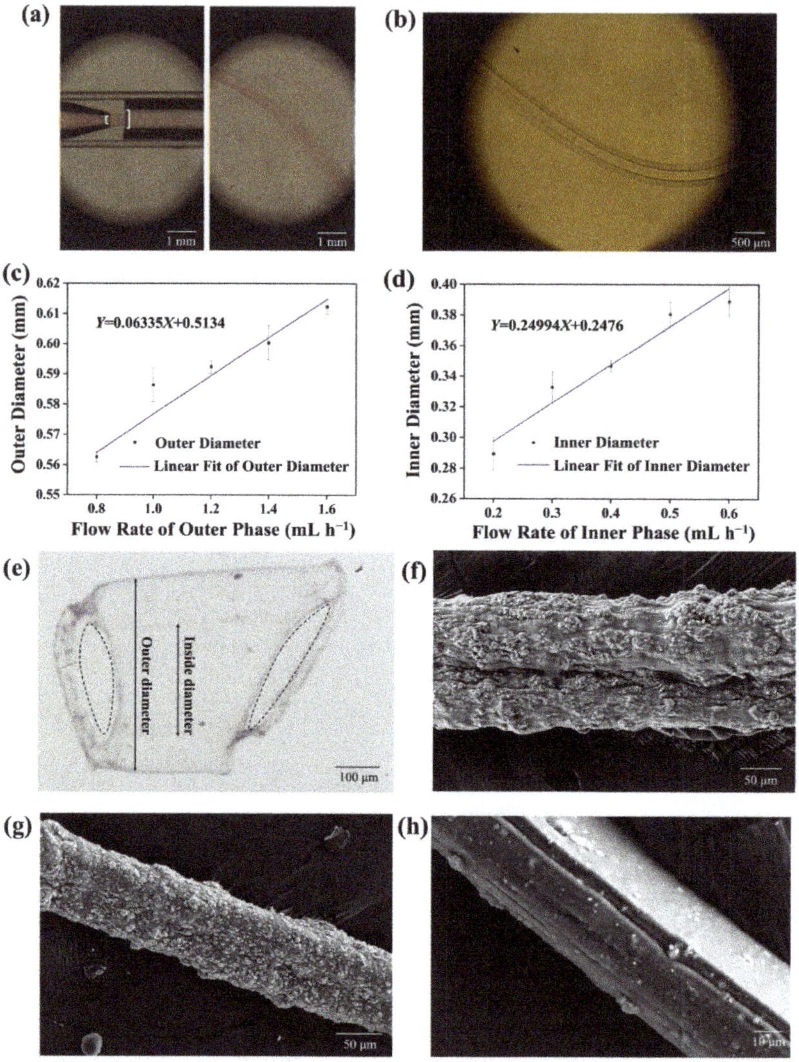

Figure 1. Fabrication and morphology properties of hollow fibers. (**a**) Partial enlarged detail of the internal structure of the microfluidic device and fresh-made hollow fiber (the inner diameter of left fine tip: 0.26 mm; the inner diameter of right thick tip: 0.58 mm). (**b**) Photograph of the resultant hollow fiber. (**c**) Relationship between the flow rate of the outer phase and the outer diameter of the hollow fiber. (**d**) Relationship between the flow rate of the inner phase and the inner diameter of the hollow fiber. (**e**) Cross-section of hollow fibers (hollow fiber inner diameter: ~0.269 mm; hollow fiber outer diameter: ~0.486 mm). (**f**) SEM image of a hollow fiber surface. (**g**) SEM image showing the Fe_3O_4 nanoparticles on the surfaces of magnetic-driven micromotors. (**h**) SEM image showing the MnO_2 nanoparticles on the surfaces of bubble-propelled micromotors. Data are presented as mean $\pm$ SD (n = 3).

By adjusting the flow rates of the inner and outer phase solutions, we found that the inner and outer diameters of the fibers were directly proportional to the inner phase flow rate and the outer phase flow rate, respectively (Figure 1c,d). The inner diameter of the fibers was inversely proportional to the outer phase flow rate (Figure S2a), and there was little effect between the outer diameter of the fibers and the inner phase flow rate (Figure S2b). This allowed us to precisely control the dual-phase flow rate and, thus, the thickness of the hollow fiber wall.

The electron microscope image of the hollow fiber cross-section shows the complete hollow structure, with the dotted line marking the hollow portion (Figure 1e). By observing HollowFiber-CNTs, HollowFiber-Fe_3O_4, and HollowFiber-MnO_2 using scanning electron microscopy, we clearly found that carbon nanotubes, iron tetroxide nanoparticles, and manganese dioxide particles were uniformly distributed on the outer phase walls. (Figure 1f–h). Intriguingly, the surface of the resultant hollow fiber demonstrated alligator cracking. Using scanning electron microscopy (SEM) to observe the surface of HollowFiber-CNTs, we found many parallel folds on the outer wall of the hollow fiber (Figure S3a,b).

This indicates that the CNTs are arranged uniformly in the hollow fiber, so the hollow fiber appears uniformly shrunken after drying. By comparing the hollow fibers' surfaces before and after doping with nanoparticles, we found that after doping with CNTs, the fiber walls showed almost no cracks, which might be the result of the interaction between carboxyl groups of CNTs and hydroxyl groups of calcium alginate through a hydrogen bond. However, compared to smooth surfaces, HollowFiber-CNTs show a folded structure on their surface. This wrinkled structure gives HollowFiber-CNTs a larger actual surface area. From the perspective of sensing applications, the increased surface area implies that more fiber surface regions are involved in the interaction when in contact with living organisms or other substances, thus increasing sensitivity to strain. This increased surface contact may lead to an increase in the number of conductive paths, further enhancing the sensitivity of the sensor. On a microscopic scale, this pleated structure provides additional contact points that are critical in human detection or other sensing applications, enhancing the stability and reliability of signal acquisition and thus optimizing the performance of the sensor.

We also performed a detailed crystal structure analysis of the three different fiber materials using X-ray diffraction (XRD) (Figure S4). Through fine resolution and comparison of the XRD patterns of the samples, we observed that the positions of the diffraction peaks were highly consistent with the characteristic diffraction peaks of CNTs, Fe_3O_4 and MnO_2, except for the diffraction peaks specific to NaA itself. This result not only verifies the successful doping of carbon nanotubes, Fe_3O_4, and MnO_2 in the calcium alginate matrix but also reconfirms that the homogeneous distribution and immobilization of these nanomaterials in calcium alginate fibers can be efficiently achieved by the microfluidic fabrication method.

3.2. Mechanical Properties of Hollow Fiber

By comparing the stretchability of hollow fibers prepared with different inner phase solutions (Figure S5a), we found that a hollow fiber prepared with PVA as the inner phase can be stretched from 5 cm to 11 cm, more than 220% of its original length (Figure S5b and Supporting Video S2), which is significantly higher than the others, showing its unique tensile strength. This may be because the residual PVA after cleaning was attached to the inner wall of the hollow fiber to improve the stretchability of the hollow fiber, as previously reported [43,44]. Therefore, PVA was used as the inner-phase solution in the subsequent preparation of hollow fiber.

We also determined the stress–strain curve of hollow fiber (Figure 2a), and the results showed that the tensile strength of the hollow fiber increased, and the stretchability decreased as the water content of the fibers decreased. The tensile strength of fully dried hollow fiber can reach up to 245 kPa. The tensile strength is highly dependent on the water content of the hollow fiber. Moreover, the resultant hollow fiber can be easily assembled into a spiderweb-like network (Figure S5c) and tied together (Figure S5d), showing its good bending and torsional properties.

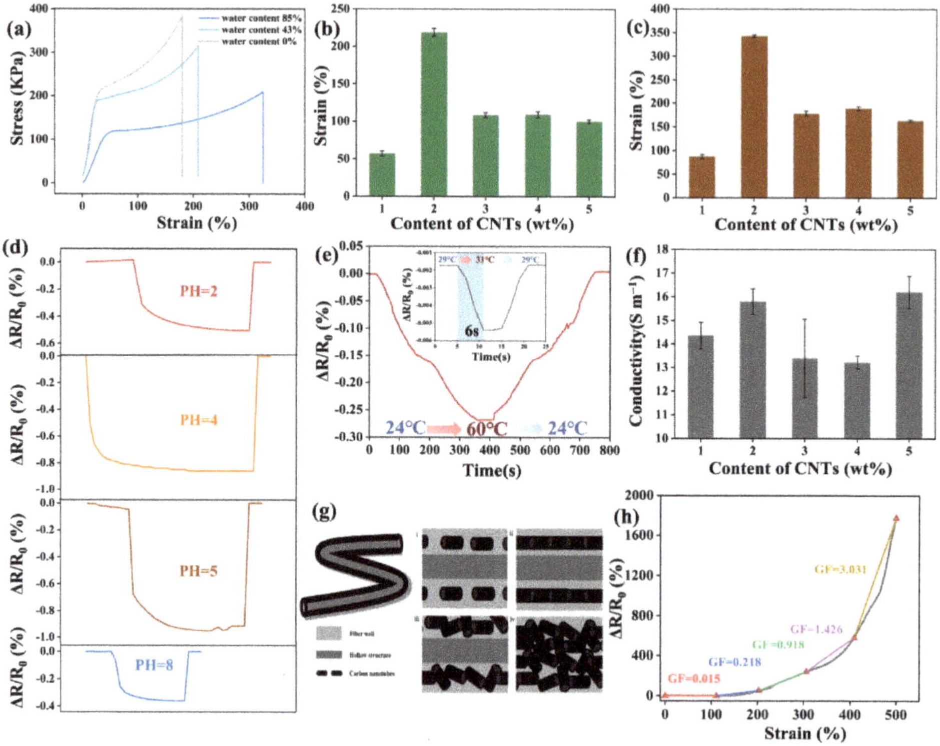

Figure 2. Property characterization and morphology of HollowFiber-CNTs. (**a**) Stress–strain curve of hollow fibers with different water contents. (**b**) Relationship between the content of CNTs and the electrical conductivity of hollow fiber. (**c**) Relationship between CNT content and hollow fiber tensile rate after soaking in LiCl. (**d**) Changes in the relative resistances of HollowFiber-CNTs at different pH values. (**e**) Changes in the relative resistance and response time of HollowFiber-CNTs at different temperatures. (**f**) Relationship between CNT content and hollow fiber tensile rate. (**g**) Alignment of CNTs with different contents inside hollow fiber. (**h**) Relationship between tensile rate and GF of HollowFiber-CNTs. Data are presented as mean ± SD (n = 3).

We then performed tensile tests on HollowFiber-CNTs with different CNT contents. When the carbon nanotube content was low, the dispersion in the fiber was good, playing the role of the enhancer. When the content reached 2%, the enhancement effect was the best. When the content continued to increase, carbon nanotubes were prone to aggregation, affecting the overall performance of the material. We found that 2% HollowFiber-CNTs exhibited the best tensile property, up to 219.05 ± 5.23% (Figure 2b). After soaking in LiCl solution, the stretchability of the HollowFiber-CNTs was significantly improved and increased to 342.9 ± 2.23% (Figure 2c).

The good mechanical characteristics of hollow fiber, together with its better wearability, provide possibilities for the subsequent development of wearable fiber sensors and self-driven water purifiers.

3.3. *Wearable Fiber Sensor*

3.3.1. Stability Test of HollowFiber-CNTs

According to the experimental results, HollowFiber-CNTs also showed good stability in acid-based solutions. The $\Delta R/R_0$ recovered to its initial level after removing the acid-based solution (Figure 2d). This characteristic can also be used to monitor the solution's pH. HollowFiber-CNTs also have good stability in a high-temperature environment. When the temperature was increased from 24 °C to 60 °C and then decreased to 24 °C the $\Delta R/R_0$ maintained a good congruent relationship with the temperature (Figure 2e). This characteristic of being able to output different $\Delta R/R_0$ signals at different temperatures can also be used for temperature monitoring. The high sensitivity, coupled with the excellent stability of HollowFiber-CNTs, is an essential property of wearable sensors when monitoring body movement.

3.3.2. Electrical Conductivity of HollowFiber-CNTs

We measured the electrical conductivity of HollowFiber-CNTs with CNT contents from 1 wt% to 5 wt%, and we found that the hollow fiber with 2% and 5% CNT contents had the highest conductivities, with conductivities of 15.8 ± 0.54 S m^{-1} and 16.2 ± 0.68 S m^{-1}, respectively (Figure 2f). Analyzing the possible reasons, the low content of CNTs means that they do not contact each other and do not form efficient conductive pathways (Figure 2g(i)). Therefore, they have low conductivity. With an increase in content, the direction of CNTs is consistent with the direction of fluid flow due to the action of shear force, and thus they can phase head to tail to form more effective conductive pathways (Figure 2g(ii)). With the increase in CNT content, the direction of the CNTs becomes uncontrollable and an effective conductive pathway cannot be formed (Figure 2g(iii)), causing decreased conductivity. But when the CNT content continues to increase, the CNTs will pile up in the hollow fiber; although the orientation is disordered, there are multiple conductive pathways, so the conductivity of the hollow fiber is enhanced (Figure 2g(iv)). At this point, if the CNT content is further increased, a high concentration of CNTs will augment both electrostatic interactions and van der Waals forces between the CNTs. This increase will make it more likely for CNTs to form agglomerates at a high volume rather than being dispersed within the matrix. The increase in spacing between the aggregated CNTs impedes electron hopping among them, thereby leading to a decrease in electrical conductivity.

We also found that the resistance changes of HollowFiber-CNTs maintain a relationship of positive correlation with the magnitude of their deformation rate. The luminescence intensity of LED light shows this phenomenon intuitively (Figure S6a): the higher the deformation rate, the darker the light, indicating greater fiber resistance. This is because before HollowFiber-CNTs are stretched, the CNTs are well connected and form a complete conductive pathway, so a lower resistance value can be maintained. When the hollow fiber segment is stretched, the CNTs doped in the hollow fiber are separated from each other, resulting in an elevated resistance value. By calculating the sensitivity coefficient at different tensile rates, we found that the GF value of this hollow fiber increased with increasing deformation (Figure 2h), illustrating its high sensitivity in detecting large joint movements.

3.3.3. Wearable Hollow Fiber Sensors for Detecting Human Motion

The resistance of HollowFiber-CNTs changes with the deformation of the fiber. The fiber deforms when it encounters an external force, which leads to a change in the conductive pathway within the fiber, resulting in a change in the resistance value. These changes in resistance can be represented by measuring the relative change in resistance ($\Delta R/R_0$), where R_0 is the initial resistance and ΔR is the change in resistance. When HollowFiber-CNTs are affixed to joints such as finger knuckles, wrists, elbows, and knees for motion detection, they deform according to the flexion angle of the joint, which leads to variations in resistance. By measuring these changes in resistance, the bending angle of the joint can be determined, thereby facilitating the detection and recognition of motion.

Due to the excellent stability of HollowFiber-CNTs, we demonstrated the potential practical application of HollowFiber-CNTs as wearable devices and prepared a comprehensive strain-sensing test on different joints in humans. It was found that HollowFiber-CNTs exhibited excellent sensitivity for the identification of human activity, including large body movements and weak signals. Figure 3a shows the detection performance of the HollowFiber-CNTs sensor for finger movement. When the finger was repeatedly flexed from 0° to 90° to

135°, the $\Delta R/R_0$ value increased correspondingly, whereas it decreased during the stabilization phase after straightening the finger. Figure 3b depicts the relationship between $\Delta R/R_0$ and bending angle as the HollowFiber-CNTs sensor monitors wrist movement. HollowFiber-CNTs sensors were also affixed to the elbow and knee joints to test their responses to large amplitude movements, showing periodic $\Delta R/R_0$ vibration curves (Figure 3c,d). The value of $\Delta R/R_0$ increased significantly with the flexion angle, so that joint movements flexing to different angles could be distinguished, showing a similar trend to finger movement.

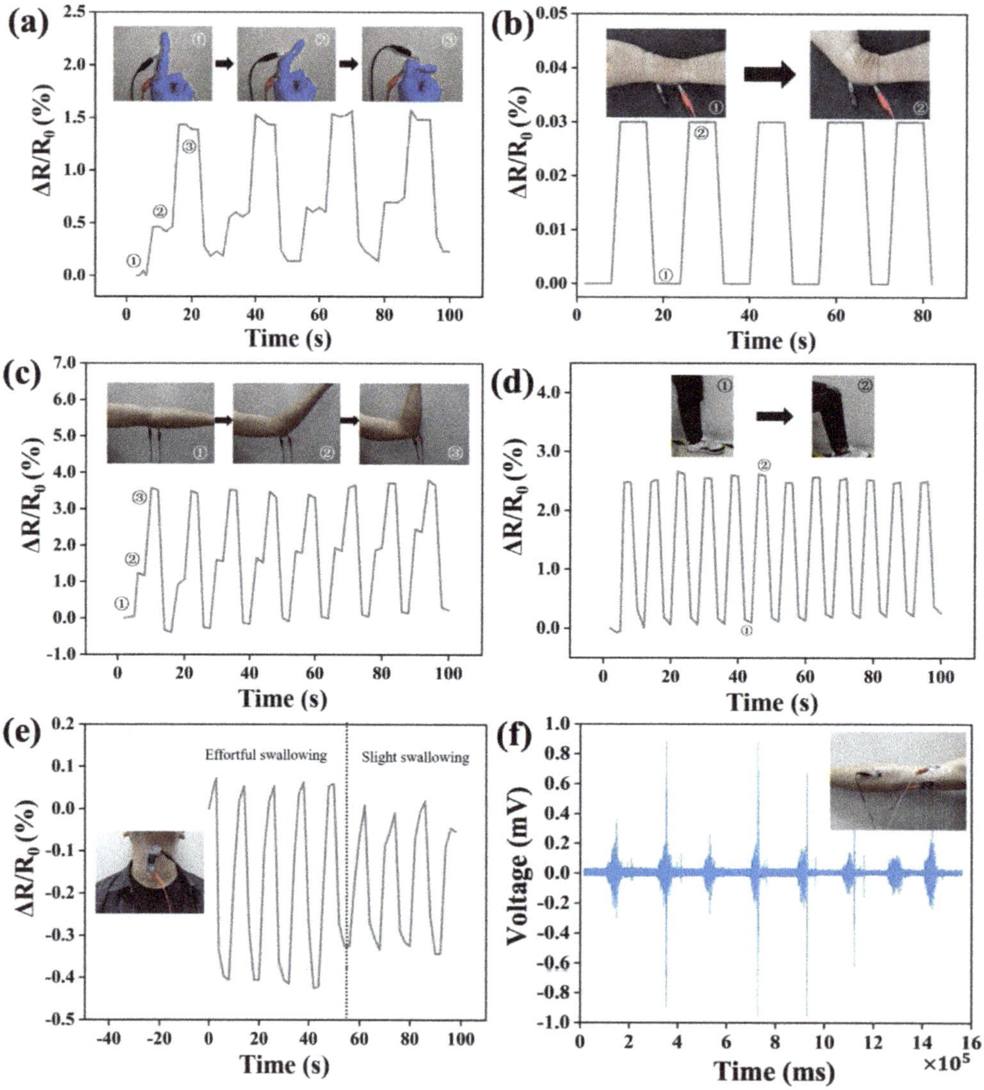

Figure 3. Applications of HollowFiber-CNTs for detecting human motion. (**a**) Relative resistance changes of HollowFiber-CNTs with finger joint bending to different angles. (**b**) Relative resistance changes of HollowFiber-CNTs with wrist bending to different angles. (**c**) Relative resistance changes of HollowFiber-CNTs with elbow bending to different angles. (**d**) Relative resistance changes of HollowFiber-CNTs with knee bending to different angles. (**e**) Relative resistance changes of HollowFiber-CNTs with varying degrees of swallowing. (**f**) HollowFiber-CNTs detection of electromyographic signals.

In addition to detecting movements, strain-sensing performance was used to monitor subtle signals, including swallowing, as well as the collection of EMG signals generated in response to muscle movements. HollowFiber-CNTs were collected on the back of an adhesive film, which was affixed to the throat site. Swallowing causes an altered laryngeal position, which in turn causes the HollowFiber-CNTs to deform. We analyzed the $\Delta R/R_0$ signal trend across various swallowing intensities (Figure 3e). The $\Delta R/R_0$ signal change amplitude was smaller when a slight swallowing movement was made, and increased when a vigorous swallowing movement was made, which was consistent with the above results when detecting large amplitude movements. While serving as an EMG signal collection electrode, HollowFiber-CNTs showed good sensing performance and could generate obvious EMG signal waveforms (Figures 3f and S6b), thus enabling the detection of muscle movement.

The above results verified the excellent sensing performance of the HollowFiber-CNTs sensor, which showed an excellent ability to identify human activities, and the profiles of these motions displayed unique morphologies and exhibited good repeatability, which guaranteed their application as wearable devices in detecting different human activities as well as physiological activities [45,46] (Table S1). To ensure that the sensors provide reliable and accurate data, they need to be used at a constant temperature to ensure environmental stability and prevent fluctuations in temperature and humidity from affecting sensor performance.

3.4. Micromotors

3.4.1. Magnetic-Driven Micromotors

Micromotors have attracted more and more attention due to their great application value [47]. The hollow fiber-based micromotor, as an emerging direction of micromotors, is playing an important role in the small-scale machine field due to its versatile hollow structure. The resultant hollow fiber with excellent performance is qualified for use in micromotor applications.

We prepared magnetic hollow fibers by introducing magnetic nanoparticles into the outer phase, which were uniformly distributed along the outer phase wall (Figure 4a). We then investigated the motion behavior of the magnetic hollow fiber micromotor, as shown in Figure 4b. To start with, we tested the motion of the magnetic hollow fiber micromotor along a specific linear path. The results revealed that the magnetic hollow fiber micromotor could travel to the intended destination along a predefined linear path under the effect of an applied magnetic field (Figure 4c). Similarly, the magnetic hollow fiber micromotor could move precisely and smoothly along a preset curved path (Figure 4d), demonstrating exceptional predictability, stability, and controllability. Moreover, we achieved precise manipulation of magnetic hollow fiber micromotors in complex channels (Figure 4e). As shown in Figure 4f, the magnetic hollow fiber micromotor was able to move vertically in a narrow channel under magnetic manipulation. Additionally, the magnetic hollow fiber micromotors could successfully achieve precise and directional motion in intricate T-shaped channels using external magnetic fields (Figure 4g and Supporting Video S3). In summary, magnetic hollow fiber micromotors offer robust maneuverability and directional mobility, thus holding great promise for various applications, such as drug delivery, targeted therapy, and small machines.

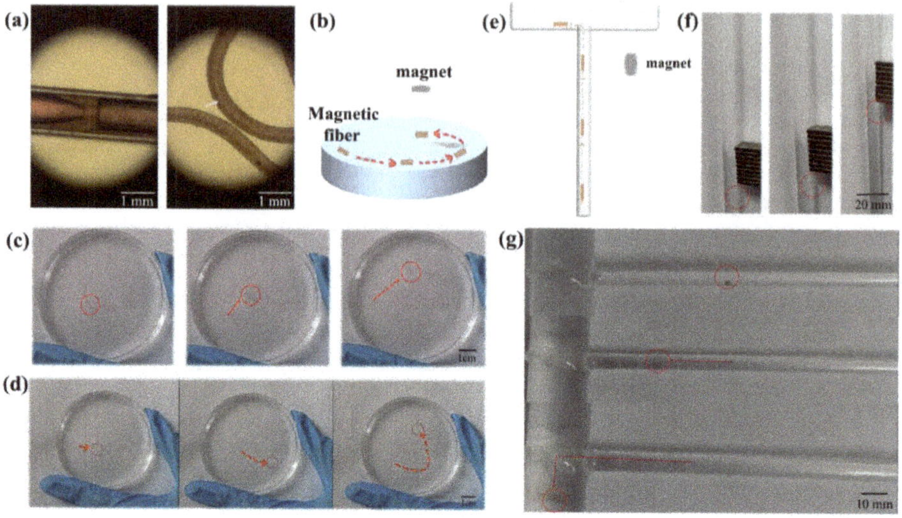

Figure 4. Directional movement of nano-doped hollow fiber. (**a**) Preparation process of magnetic hollow fibers and the resultant magnetic hollow fibers. (**b**) Schematic diagram of the magnetic control of directional movement of the directional micromotor. (**c**) Magnetic force controls the movement of a directional micromotor along a straight line. (**d**) Movement of the directional micromotor along a curve. (**e**) Schematic diagram showing the manipulation of the directional micromotor in a channel. (**f**) Manipulation of the directional micromotor in a vertical channel. (**g**) Manipulation of the directional micromotor in a T-shaped channel.

3.4.2. Bubble-Propelled Micromotors

Hydrogen peroxide (H_2O_2) has been widely used in several viable applications at low concentrations, such as a sensing platform for analyzing targets [48]. H_2O_2 is considered a powerful oxidizing agent. Calcium alginate (CaAlg) is a salt obtained by reacting alginate with calcium ions. Notably, the calcium ions in calcium alginate exhibit stability to H_2O_2 and do not participate in redox reactions. Therefore, it can be inferred that the chemical state of calcium alginate is relatively stable in H_2O_2 solution. The configuration of microtubules plays a pivotal role in enabling the generation, accumulation, and release of bubbles, which is essential for the efficacious propulsion of micromotors [49,50].

Therefore, we embedded manganese dioxide (MnO_2) particles into the walls of the hollow fibers to prepare functionalized fibers with the function of catalyzing H_2O_2 decomposition. Due to the capillary mechanism, H_2O_2 will be adsorbed by the interior of the hollow fiber and decomposed into oxygen by the catalytic effect of MnO_2. This newly generated oxygen rapidly accumulates within the limited space of the hydrogel, leading to an increase in internal pressure and creating a localized pressure gradient. Due to this pressure gradient, the hydrogel fibers are driven toward areas of lower pressure when oxygen is generated inside the fibers. The external MnO_2 also catalyzes the decomposition of H_2O_2 to produce oxygen bubbles; however, surface tension causes these bubbles to adhere tightly to the outer wall of the fiber. Since these bubbles do not form in a confined space where directional thrusts can be generated, they have less effect on the dynamics of the fibers (Figure S7a).

Therefore, the oxygen bubbles can push the hollow fiber forward, showing a good performance of the hollow fiber-based micromotor. To test the ability of bubble-driven micromotors to enter small and deep locations, we placed them inside a mold filled with hydrogen peroxide solution (Figure 5a), and the results indicated that the bubble-driven micromotor immediately spread out toward the corner and finally gathered deep in the corner. (Figure 5b and Supporting Video S4). To rule out coincidental factors, we synthesized a batch of undoped MnO_2 hollow fiber samples as a blank control (the red part in Figure 5c), which were mixed with the experimental group and put into the molds according to the experimental procedure described before. The results clearly showed that the locations of the hollow fiber segments without MnO_2 did not change, whereas the bubble-driven micromotor moved toward the periphery and finally stayed deep in the corners (Figure 5c and Supporting Video S5). By changing the concentration of the hydrogen peroxide solution, we found that the speed of the bubble-driven micromotor movement was accelerated with an increase in hydrogen peroxide concentration (Figure S7b). This is because the higher concentration of H_2O_2 can generate bubbles more quickly after it is decomposed, enabling them to push the micromotors more effectively. The movement of this bubble-driven micromotor can deliver efficient and robust performance [51,52]. The micromotor can move in 5 wt% H_2O_2 solution at a speed of 615 μm s^{-1}, which is higher than that of some other micromotors of the same type that have been reported so far (Figure S7c).

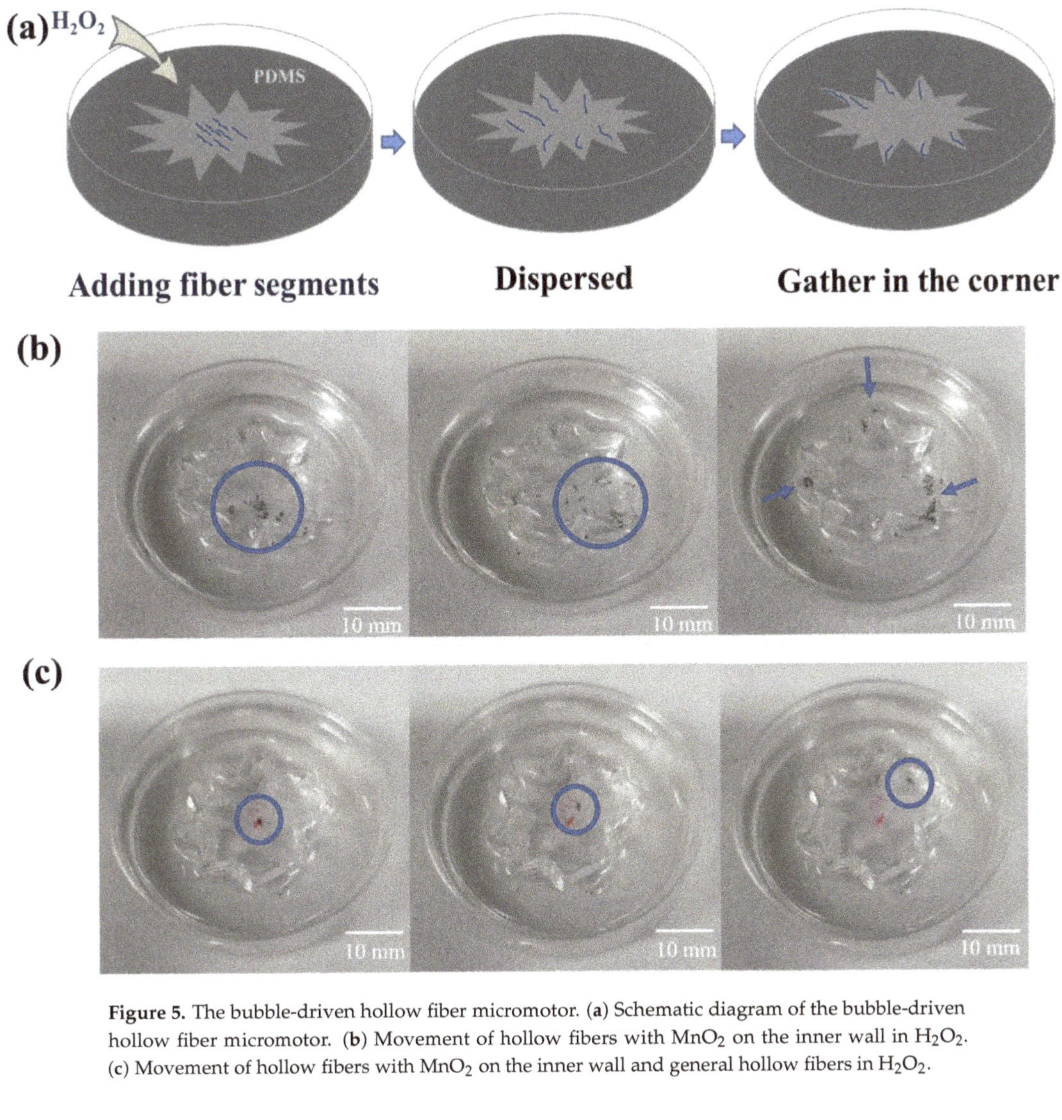

Figure 5. The bubble-driven hollow fiber micromotor. (**a**) Schematic diagram of the bubble-driven hollow fiber micromotor. (**b**) Movement of hollow fibers with MnO_2 on the inner wall in H_2O_2. (**c**) Movement of hollow fibers with MnO_2 on the inner wall and general hollow fibers in H_2O_2.

4. Conclusions

In this work, we describe a microfluidic method based on coaxial glass capillaries for fabricating nanomaterial-doped functional hollow fibers. This method features a simple, smooth, stable, continuous, and precisely controlled process that avoids system clogging. Utilizing the characteristics of microfluidic laminar flow as well as the shear forces acting in the microchannels, we achieve uniform dispersion of nanoparticles to enable hollow fiber with favorable mechanical, electrical, and magnetic operabilities. By doping CNTs in the hydrogel hollow fibers, we successfully prepare wearable sensors exhibiting superior electrical conductivity (15.8 S m^{-1}), strong flexibility (342.9%), versatility, and high sensitivity at high deformation rates. This CNT-doped hollow fiber-based wearable sensor can be used to monitor human motions and collect physiological electrical signals. As well as wrist movements, elbow and knee movements could also be detected, and minor movements, such as swallowing and finger joint movements, could also be successfully monitored. Moreover, by doping Fe_3O_4 nanoparticles into hydrogel fibers, we fabricated magnetic-driven micromotors capable of directional motion along the trajectory of a magnet. Owing to their small size, these micromotors could navigate through narrow channels. In addition, manganese dioxide nanoparticles were embedded into the hollow fiber walls to create bubble-propelled micromotors. When put into a hydrogen peroxide environment, the micromotors with MnO_2-doped fiber walls could reach high-speed self-propulsion (615 μm s^{-1}) and were capable of penetrating tiny and hard-to-reach corners. This technology provides a highly flexible and controllable preparation method, which makes it possible to precisely dope different functional materials, and not only realizes the multifunctionality of hollow fibers, but also the easy weaving and integration of these hollow fibers enables them to be used in the preparation of smart fabrics with multiple functions. This innovative technology provides a powerful tool for designing and preparing novel materials with customized functions. It sheds light on the design and development of wearable electronic devices, soft robots, micromachines, and future applications in biomedical engineering.

Supplementary Materials: The following supporting information can be downloaded at: https://www.mdpi.com/article/10.3390/bios14040186/s1, Figure S1: Fabrication and morphology of hollow fibers; Figure S2: Flow velocity versus hollow fiber diameter; Figure S3: Localized enlarged image of HollowFiber-CNT surface morphology taken by scanning electron microscope; Figure S4: XRD images of hollow fibers; Figure S5: Mechanical properties of hollow fiber; Figure S6: Electrical properties of hollow fibers; Figure S7: Bubble-propelled micromotors; Table S1: Comparison of electrical conductivity of hydrogel fibers of different networks with human detection function; Video S1: Preparation process of hollow fiber; Video S2: Stretching process of hollow fiber; Video S3: Manipulation of directional micromotor in T-shaped channel; Video S4: Movement of hollow fibers with MnO_2 on the inner wall in H_2O_2; Video S5: Movement velocity of bubble-driven hollow fiber micromotor in H_2O_2 solution with different concentration.

Author Contributions: Y.W. and Y.T. designed the project; Y.W. performed the experiments; Z.W. and T.L. analyzed the experimental data; Y.W. and Z.W. wrote the manuscript; Z.W. and H.S. drew the experimental pictures; X.M. reviewed and revised the work; Y.T. and J.G. reviewed, revised, and supervised the study. All authors have read and agreed to the published version of the manuscript.

Funding: This research was funded by the National Natural Science Foundation of China (32201179 and 62122017), Liaoning Provincial Natural Science Foundation Joint Fund (General Support Program Project) (2023-MSBA-093) and the Fundamental Research Funds for the Central Universities (N2319005) are gratefully acknowledged.

Institutional Review Board Statement: Not applicable.

Informed Consent Statement: Not applicable.

Data Availability Statement: Data are available upon reasonable request.

Conflicts of Interest: The authors declare no conflicts of interest.

References

1. Talebian, S.; Shim, I.K.; Kim, S.C.; Spinks, G.M.; Vine, K.L.; Foroughi, J. Coaxial mussel-inspired biofibers: Making of a robust and efficacious depot for cancer drug delivery. *J. Mater. Chem. B* **2020**, *8*, 5064–5079. [CrossRef] [PubMed]
2. Teixeira, M.O.; Antunes, J.C.; Felgueiras, H.P. Recent advances in fiber–hydrogel composites for wound healing and drug delivery systems. *Antibiotics* **2021**, *10*, 248. [CrossRef] [PubMed]
3. Mirani, B.; Pagan, E.; Shojaei, S.; Dabiri, S.M.H.; Savoji, H.; Mehrali, M.; Sam, M.; Alsaif, J.; Bhiladvala, R.B.; Dolatshahi-Pirouz, A. Facile method for fabrication of meter-long multifunctional hydrogel fibers with controllable biophysical and biochemical features. *ACS Appl. Mater. Interfaces* **2020**, *12*, 9080–9089. [CrossRef] [PubMed]
4. Mazari-Arrighi, E.; Okitsu, T.; Teramae, H.; Aoyagi, H.; Kiyosawa, M.; Yano, M.; Chatelain, F.; Fuchs, A.; Takeuchi, S. In vitro proliferation and long-term preservation of functional primary rat hepatocytes in cell fibers. *Sci. Rep.* **2022**, *12*, 8813. [CrossRef] [PubMed]
5. Utagawa, Y.; Ino, K.; Kumagai, T.; Hiramoto, K.; Takinoue, M.; Nashimoto, Y.; Shiku, H. Electrochemical glue for binding chitosan–alginate hydrogel fibers for cell culture. *Micromachines* **2022**, *13*, 420. [CrossRef] [PubMed]
6. Chen, J.; Zhao, Y.; Zhou, A.; Zhang, Y.; Xu, Y.; Ning, X. Alginate functionalized biomimetic 3D scaffold improves cell culture and cryopreservation for cellular therapy. *Int. J. Biol. Macromol.* **2022**, *211*, 159–169. [CrossRef] [PubMed]
7. Fatahian, S.A.; Motavalizadehkakhky, A.; Hosseiny, M.; Nouri, S.M.M.; Zhiani, R.; Sohrabpour, M.; Khanmohammadi, M. Upregulation of biochemical and biophysical properties of cell-laden microfiber, silk-hyaluronic acid composite. *Int. J. Biol. Macromol.* **2022**, *211*, 700–710. [CrossRef] [PubMed]
8. Yang, X.; Merenda, A.; Riyadh, A.-A.; Dumée, L.F.; Zhang, X.; Thang, S.H.; Pham, H.; Kong, L. Towards next generation high throughput ion exchange membranes for downstream bioprocessing: A review. *J. Membr. Sci.* **2022**, *647*, 120325. [CrossRef]
9. Li, H.-N.; Yang, J.; Xu, Z.-K. Hollow fiber membranes with Janus surfaces for continuous deemulsification and separation of oil-in-water emulsions. *J. Membr. Sci.* **2020**, *602*, 117964. [CrossRef]
10. Cao, Y.; Li, Y.-X.; Wang, M.; Xu, Z.-L.; Wei, Y.-M.; Shen, B.-J.; Zhu, K.-K. High-flux NaA zeolite pervaporation membranes dynamically synthesized on the alumina hollow fiber inner-surface in a continuous flow system. *J. Membr. Sci.* **2019**, *570*, 445–454. [CrossRef]
11. Li, M.; Chen, X.; Li, X.; Dong, J.; Zhao, X.; Zhang, Q. Wearable and robust polyimide hydrogel fiber textiles for strain sensors. *ACS Appl. Mater. Interfaces* **2021**, *13*, 43323–43332. [CrossRef] [PubMed]
12. Wang, H.; Zou, Y.; Ji, Y.; Zhong, K.; Du, X.; Du, Z.; Cheng, X.; Wang, S. Tough and extremely temperature-tolerance nanocomposite organohydrogels as ultrasensitive wearable sensors for wireless human motion monitoring. *Compos. Part A Appl. Sci. Manuf.* **2022**, *157*, 106905. [CrossRef]
13. Sun, X.; Zhu, Y.; Zhu, J.; Le, K.; Servati, P.; Jiang, F. Tough and Ultrastretchable Liquid-Free Ion Conductor Strengthened by Deep Eutectic Solvent Hydrolyzed Cellulose Microfibers. *Adv. Funct. Mater.* **2022**, *32*, 2202533. [CrossRef]
14. Chen, Q.; Huang, W.; Zhao, Q.; Zhang, L. Macromolecular Phase-Transition-Driven Motility of Hollow Hydrogel Tubes. *ACS Appl. Polym. Mater.* **2020**, *2*, 3054–3059. [CrossRef]
15. Wang, Z.; Liu, C.; Chen, B.; Luo, Y. Magnetically-driven drug and cell on demand release system using 3D printed alginate based hollow fiber scaffolds. *Int. J. Biol. Macromol.* **2021**, *168*, 38–45. [CrossRef]
16. Cha, C.; Shin, S.R.; Annabi, N.; Dokmeci, M.R.; Khademhosseini, A. Carbon-based nanomaterials: Multifunctional materials for biomedical engineering. *ACS Nano* **2013**, *7*, 2891–2897. [CrossRef]
17. Merino, S.; Martín, C.; Kostarelos, K.; Prato, M.; Vázquez, E. Nanocomposite hydrogels: 3D polymer–nanoparticle synergies for on-demand drug delivery. *ACS Nano* **2015**, *9*, 4686–4697. [CrossRef]
18. Yi, F.-L.; Meng, F.-C.; Li, Y.-Q.; Huang, P.; Hu, N.; Liao, K.; Fu, S.-Y. Highly stretchable CNT Fiber/PAAm hydrogel composite simultaneously serving as strain sensor and supercapacitor. *Compos. Part B Eng.* **2020**, *198*, 108246. [CrossRef]
19. Roy, S.; David-Pur, M.; Hanein, Y. Carbon nanotube-based ion selective sensors for wearable applications. *ACS Appl. Mater. Interfaces* **2017**, *9*, 35169–35177. [CrossRef]
20. Lu, Y.; Yue, Y.; Ding, Q.; Mei, C.; Xu, X.; Wu, Q.; Xiao, H.; Han, J. Self-recovery, fatigue-resistant, and multifunctional sensor assembled by a nanocellulose/carbon nanotube nanocomplex-mediated hydrogel. *ACS Appl. Mater. Interfaces* **2021**, *13*, 50281–50297. [CrossRef]
21. Lv, J.; Kong, C.; Yang, C.; Yin, L.; Jeerapan, I.; Pu, F.; Zhang, X.; Yang, S.; Yang, Z. Wearable, stable, highly sensitive hydrogel–graphene strain sensors. *Beilstein J. Nanotechnol.* **2019**, *10*, 475–480. [CrossRef] [PubMed]
22. Liu, W.; Chen, Q.; Huang, Y.; Wang, D.; Li, L.; Liu, Z. In situ laser synthesis of Pt nanoparticles embedded in graphene films for wearable strain sensors with ultra-high sensitivity and stability. *Carbon* **2022**, *190*, 245–254. [CrossRef]
23. Wang, Z.; Zhang, X.; Cao, T.; Wang, T.; Sun, L.; Wang, K.; Fan, X. Antiliquid-interfering, antibacteria, and adhesive wearable strain sensor based on superhydrophobic and conductive composite hydrogel. *ACS Appl. Mater. Interfaces* **2021**, *13*, 46022–46032. [CrossRef] [PubMed]
24. Yang, J.; Chen, Q.; Chen, F.; Zhang, Q.; Wang, K.; Fu, Q. Realizing the full nanofiller enhancement in melt-spun fibers of poly (vinylidene fluoride)/carbon nanotube composites. *Nanotechnology* **2011**, *22*, 355707. [CrossRef] [PubMed]
25. Wang, X.-Y.; Pei, Y.; Xie, M.; Jin, Z.-H.; Xiao, Y.-S.; Wang, Y.; Zhang, L.-N.; Li, Y.; Huang, W.-H. An artificial blood vessel implanted three-dimensional microsystem for modeling transvascular migration of tumor cells. *Lab Chip* **2015**, *15*, 1178–1187. [CrossRef] [PubMed]
26. Zhang, W.; Zhang, Y.S.; Bakht, S.M.; Aleman, J.; Shin, S.R.; Yue, K.; Sica, M.; Ribas, J.; Duchamp, M.; Ju, J. Elastomeric free-form blood vessels for interconnecting organs on chip systems. *Lab Chip* **2016**, *16*, 1579–1586. [CrossRef] [PubMed]

27. Ji, X.; Wang, P.; Su, Z.; Ma, G.; Zhang, S. Enabling multi-enzyme biocatalysis using coaxial-electrospun hollow nanofibers: Redesign of artificial cells. *J. Mater. Chem. B* **2014**, *2*, 181–190. [CrossRef] [PubMed]
28. Zhou, D.; Song, W.-L.; Fan, L.-Z. Hollow core–shell SnO_2/C fibers as highly stable anodes for lithium-ion batteries. *ACS Appl. Mater. Interfaces* **2015**, *7*, 21472–21478. [CrossRef] [PubMed]
29. Tavakoli, A.; Rahimi, K.; Saghandali, F.; Scott, J.; Lovell, E. Nanofluid preparation, stability and performance for CO_2 absorption and desorption enhancement: A review. *J. Environ. Manag.* **2022**, *313*, 114955. [CrossRef]
30. Yu, Y.; Guo, J.; Zhang, H.; Wang, X.; Yang, C.; Zhao, Y. Shear-flow-induced graphene coating microfibers from microfluidic spinning. *Innovation* **2022**, *3*, 100209. [CrossRef]
31. Hu, S.; Han, J.; Shi, Z.; Chen, K.; Xu, N.; Wang, Y.; Zheng, R.; Tao, Y.; Sun, Q.; Wang, Z.L. Biodegradable, super-strong, and conductive cellulose macrofibers for fabric-based triboelectric nanogenerator. *Nano-Micro Lett.* **2022**, *14*, 115. [CrossRef]
32. Jiang, M.-Y.; Ju, X.-J.; Deng, K.; Fan, X.-X.; He, X.-H.; Wu, F.; He, F.; Liu, Z.; Wang, W.; Xie, R. The microfluidic synthesis of composite hollow microfibers for k+-responsive controlled release based on a host–guest system. *J. Mater. Chem. B* **2016**, *4*, 3925–3935. [CrossRef] [PubMed]
33. Kim, D.; Jo, A.; Imani, K.B.C.; Kim, D.; Chung, J.-W.; Yoon, J. Microfluidic fabrication of multistimuli-responsive tubular hydrogels for cellular scaffolds. *Langmuir* **2018**, *34*, 4351–4359. [CrossRef] [PubMed]
34. Dabaghi, M.; Saraei, N.; Fusch, G.; Rochow, N.; Brash, J.L.; Fusch, C.; Selvaganapathy, P.R. Microfluidic blood oxygenators with integrated hollow chambers for enhanced air exchange from all four sides. *J. Membr. Sci.* **2020**, *596*, 117741. [CrossRef]
35. Wei, L.; Deng, N.; Wang, X.; Zhao, H.; Yan, J.; Yang, Q.; Kang, W.; Cheng, B. Flexible ordered MnS@ CNC/carbon nanofibers membrane based on microfluidic spinning technique as interlayer for stable lithium-metal battery. *J. Membr. Sci.* **2021**, *637*, 119615. [CrossRef]
36. Bacchin, P.; Snisarenko, D.; Stamatialis, D.; Aimar, P.; Causserand, C. Combining fluorescence and permeability measurements in a membrane microfluidic device to study protein sorption mechanisms. *J. Membr. Sci.* **2020**, *614*, 118485. [CrossRef]
37. Vagner, S.; Patlazhan, S.; Serra, C.; Funfschilling, D.; Kulichikhin, V. Dripping and jetting of semi-dilute polymer solutions co-flowing in co-axial capillaries. *Phys. Fluids* **2021**, *33*, 062002. [CrossRef]
38. Junzong, Z.; Haiying, Q.; Jinsheng, W. Nanoparticle dispersion and coagulation in a turbulent round jet. *Int. J. Multiph. Flow* **2013**, *54*, 22–30. [CrossRef]
39. Meng, Z.-J.; Wang, W.; Xie, R.; Ju, X.-J.; Liu, Z.; Chu, L.-Y. Microfluidic generation of hollow Ca-alginate microfibers. *Lab Chip* **2016**, *16*, 2673–2681. [CrossRef]
40. Daya, R.; Xu, C.; Nguyen, N.-Y.T.; Liu, H.H. Angiogenic hyaluronic acid hydrogels with curcumin-coated magnetic nanoparticles for tissue repair. *ACS Appl. Mater. Interfaces* **2022**, *14*, 11051–11067. [CrossRef]
41. Mahdavi, M.; Ahmad, M.B.; Haron, M.J.; Namvar, F.; Nadi, B.; Rahman, M.Z.A.; Amin, J. Synthesis, surface modification and characterisation of biocompatible magnetic iron oxide nanoparticles for biomedical applications. *Molecules* **2013**, *18*, 7533–7548. [CrossRef] [PubMed]
42. Song, Y.; Sauret, A.; Cheung Shum, H. All-aqueous multiphase microfluidics. *Biomicrofluidics* **2013**, *7*, 061301. [CrossRef]
43. Yi, Y.; Chiao, M.; Mahmoud, K.A.; Wu, L.; Wang, B. Preparation and characterization of PVA/PVP conductive hydrogels formed by freeze–thaw processes as a promising material for sensor applications. *J. Mater. Sci.* **2022**, *57*, 8029–8038. [CrossRef]
44. Wang, F.; Li, Z.; Guo, J.; Liu, L.; Fu, H.; Yao, J.; Krucińska, I.; Draczynski, Z. Highly strong, tough, and stretchable conductive hydrogels based on silk sericin-mediated multiple physical interactions for flexible sensors. *ACS Appl. Polym. Mater.* **2021**, *4*, 618–626. [CrossRef]
45. Shuai, L.; Guo, Z.H.; Zhang, P.; Wan, J.; Pu, X.; Wang, Z.L. Stretchable, self-healing, conductive hydrogel fibers for strain sensing and triboelectric energy-harvesting smart textiles. *Nano Energy* **2020**, *78*, 105389. [CrossRef]
46. Wu, H.; Wang, L.; Lou, H.; Wan, J.; Pu, X. One-step coaxial spinning of core-sheath hydrogel fibers for stretchable ionic strain sensors. *Chem. Eng. J.* **2023**, *458*, 141393. [CrossRef]
47. Fernández-Medina, M.; Ramos-Docampo, M.A.; Hovorka, O.; Salgueiriño, V.; Städler, B. Recent advances in nano-and micromotors. *Adv. Funct. Mater.* **2020**, *30*, 1908283. [CrossRef]
48. Magdanz, V.; Guix, M.; Schmidt, O.G. Tubular micromotors: From microjets to spermbots. *Robot. Biomim.* **2014**, *1*, 11. [CrossRef]
49. Yu, Y.; Guo, J.; Zou, M.; Cai, L.; Zhao, Y. Micromotors from microfluidics. *Chem.—Asian J.* **2019**, *14*, 2417–2430. [CrossRef]
50. Yuan, M.; Gong, M.; Huang, H.; Zhao, Y.; Ying, Y.; Wang, S. Bubble-propelled plasmon-reinforced Pt-ZnIn 2 S 4 micromotors for stirring-free photocatalytic water purification. *Inorg. Chem. Front.* **2022**, *9*, 5725–5734. [CrossRef]
51. Wang, H.; Zhao, G.; Pumera, M. Beyond platinum: Bubble-propelled micromotors based on Ag and MnO_2 catalysts. *J. Am. Chem. Soc.* **2014**, *136*, 2719–2722. [CrossRef] [PubMed]
52. Huang, H.; Li, J.; Yuan, M.; Yang, H.; Zhao, Y.; Ying, Y.; Wang, S. Large-Scale Self-Assembly of MOFs Colloidosomes for Bubble-Propelled Micromotors and Stirring-Free Environmental Remediation. *Angew. Chem. Int. Ed.* **2022**, *61*, e202211163. [CrossRef] [PubMed]

biosensors

MDPI

Article

A Novel Microfluidic Strategy for Efficient Exosome Separation via Thermally Oxidized Non-Uniform Deterministic Lateral Displacement (DLD) Arrays and Dielectrophoresis (DEP) Synergy

Dayin Wang [1,2,3], Shijia Yang [1,2], Ning Wang [1,2,3], Han Guo [1,2], Shilun Feng [1,2], Yuan Luo [1,2,*] and Jianlong Zhao [1,2,3,*]

[1] State Key Laboratory of Transducer Technology, Shanghai Institute of Microsystem and Information Technology, Chinese Academy of Sciences, Shanghai 200050, China; dywang@mail.sim.ac.cn (D.W.)
[2] Center of Materials Science and Optoelectronics Engineering, University of Chinese Academy of Sciences, Beijing 100049, China
[3] School of Information Science and Technology, ShanghaiTech University, Shanghai 201210, China
* Correspondence: yuanluo@mail.sim.ac.cn (Y.L.); jlzhao@mail.sim.ac.cn (J.Z.)

Abstract: Exosomes, with diameters ranging from 30 to 150 nm, are saucer-shaped extracellular vesicles (EVs) secreted by various type of human cells. They are present in virtually all bodily fluids. Owing to their abundant nucleic acid and protein content, exosomes have emerged as promising biomarkers for noninvasive molecular diagnostics. However, the need for exosome separation purification presents tremendous technical challenges due to their minuscule size. In recent years, microfluidic technology has garnered substantial interest as a promising alternative capable of excellent separation performance, reduced reagent consumption, and lower overall device and operation costs. In this context, we hereby propose a novel microfluidic strategy based on thermally oxidized deterministic lateral displacement (DLD) arrays with tapered shapes to enhance separation performance. We have achieved more than 90% purity in both polystyrene nanoparticle and exosome experiments. The use of thermal oxidation also significantly reduces fabrication complexity by avoiding the use of high-precision lithography. Furthermore, in a simulation model, we attempt to integrate the use of dielectrophoresis (DEP) to overcome the size-based nature of DLD and distinguish particles that are close in size but differ in biochemical compositions (e.g., lipoproteins, exomeres, retroviruses). We believe the proposed strategy heralds a versatile and innovative platform poised to enhance exosome analysis across a spectrum of biochemical applications.

Keywords: exosomes; deterministic lateral displacement (DLD); dielectrophoresis (DEP); microfluidic separation; thermal oxidation

1. Introduction

Exosomes, integral to intercellular communication, have attracted substantial interest across myriad biomedical research fields. These nano-sized extracellular vesicles are pivotal in mediating a myriad of physiological and pathological processes, including cancer progression, immune responses, and tissue regeneration [1–4]. These vesicles encapsulate a diverse array of biomolecules, including proteins, lipids, and nucleic acids, reflecting the molecular signature of the parent cell and thereby offering a snapshot of its health and activity [5–8]. However, the co-presence of several other types of EVs in biofluids often complicates the precise analysis of exosomes, underscoring the need for their separation from abundant sources like blood, urine, and saliva. Traditional exosome separation techniques, including ultracentrifugation, density gradient centrifugation, and methods relying on magnetic beads, are time-consuming and labor-intensive, while achieving exosome separation of low purity [9–11].

In recent years, the adaptation of microfluidics to separate and capture exosomes has gradually grown in prominence. A series of technological solutions have been developed for manipulating the fluid flow. Nanoporous membranes integrated into microfluidic chips intercept particles larger than the pore size while smaller components pass through [12]. A nanowire-based strategy achieves exosome trapping through spatial hindrance [13]. The viscoelastic flow method achieves particle sorting based on the elastic lift forces exerted on particles in a viscoelastic medium [14–16]. Deterministic lateral displacement (DLD), a label-free method, enables the sorting and enrichment of nanoparticles based on size [17–21]. Dielectrophoresis (DEP) separates particles based on their intrinsic dielectric properties rather than size [22,23]. Among these methods, DLD provides a continuous process that circumvents the drawbacks of nanoporous membranes, nanowires, and viscoelastic methods by avoiding blockages and the introduction of impurities. The high separation resolution of DLD enables it to stand out among various microfluidic separation techniques [24]. DEP proves more effective for micrometer-scale particles but shows limited efficacy with nanometer-scale particles.

In this work, we propose a novel approach via thermally oxidized DLD arrays with tapered shapes to improve the separation performance of nano-sized particles. First, by applying thermal oxidation to the column array fabricated through micrometer-scale lithography and etching, nanometer-level inter-column spacing is achieved, which significantly reduces fabrication costs by avoiding the use of high-precision lithography. Second, while previous studies have also applied thermal oxidation in DLD technology [25], we critically realized that thermal oxidation could transform vertical columns into tapered shapes, causing the separation threshold of the DLD array to deviate from its intended design. We attempt to elucidate this phenomenon with further experimental investigation. The tapered structure (Figure 1a) after thermal oxidation employs the wider upper layer to exert the DLD effect on larger impurity particles, while both the upper and lower layers allow exosomes to pass through unimpeded. We used mixtures of polystyrene nanoparticles and exosomes to verify the separation performance of the device, resulting in >90% purity from collected samples. In addition, we aim for a design capable of separating extracellular vesicles (EVs) of similar sizes (e.g., lipoproteins, retroviruses, and exoparticles) that combines dielectrophoresis (DEP) and DLD. By designing a droplet-shaped structure to replace the traditional cylindrical structure of DLD (Figure 1b), we tested the feasibility of such a design through a simulation study. The overall strategy in our work demonstrated significant promise as a multifunctional and label-free methodology for exosome separation, offering enhanced specificity in particle discrimination through the integration of size-based and electrical-property-based separation mechanisms.

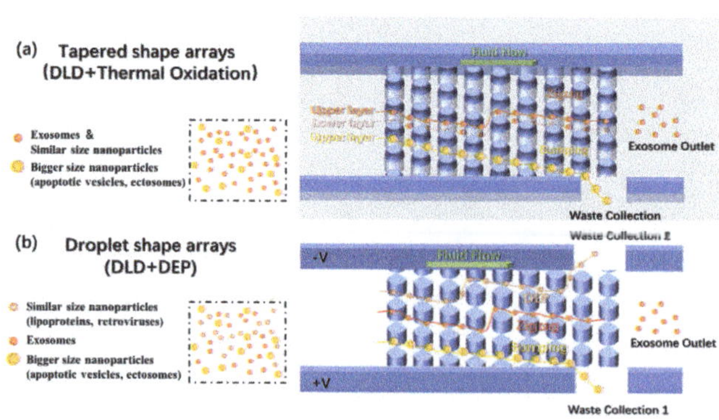

Figure 1. The overall design of the chips. (**a**) Schematic of the tapered shape DLD arrays fabricated after thermal oxidation showing how exosomes and larger nanoparticles flow through. (**b**) Schematic of the droplet-shaped arrays combining DEP with DLD arrays, showing how exosomes, bigger nanoparticles, and similarly sized nanoparticles flow through.

2. Materials and Methods

2.1. Theory

2.1.1. DLD

DLD represents a size-based particle separation methodology and exhibits significant potential for the continuous, label-free separation of particles that are smaller than the microfluidic array's feature size. A DLD device incorporates a particle separation region, which consists of an array composed of microposts or columns crafted from a rigid material [26], as depicted in Figure 2a. The width of the gaps, perpendicular to the direction of fluid flow, is denoted as G, and the center-to-center spacing between the posts is represented by λ. The lateral displacement between adjacent rows of posts is referred to as δ. The crucial parameter in the design of a DLD device is the shift fraction (ε), which is defined as

$$\varepsilon = \delta/\lambda, \quad (1)$$

The operational principle of the DLD device hinges on a critical size threshold (D_C). Particles exceeding this threshold size are laterally displaced and follow bumping mode, a trajectory along the 'blue line' of the array, whereas particles below this threshold size flow in zigzag mode, the red line, with no displacement occurring. However, before establishing the calculation model for determining D_C, we first need to conduct a fluid dynamics analysis of the fluid within the device. Below, we will introduce the Navier–Stokes equations and the Reynolds Number.

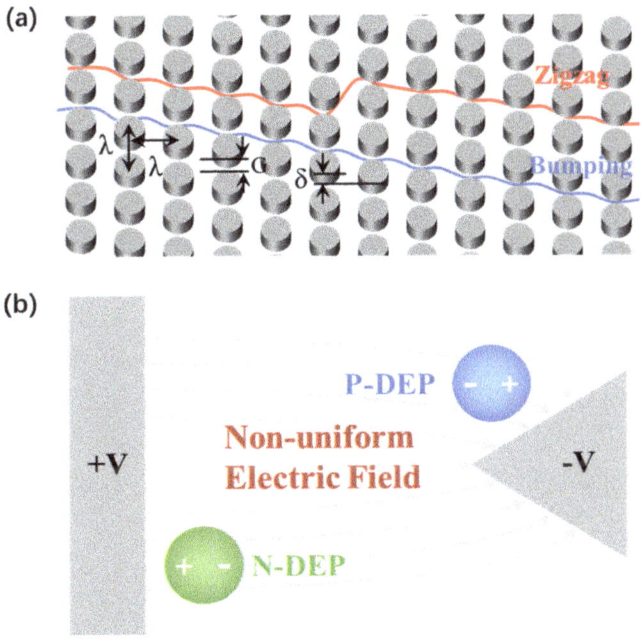

Figure 2. (**a**) Schematic of the DLD arrays that illustrates the array parameters of the pillar gap size, G, pillar pitch, λ, and row-to-row shift, δ, and the two different modes separate particles flow in. (**b**) Schematic of dielectrophoresis, showing how n-DEP particles and p-DEP particles act in a non-uniform electric field.

A large number of phenomena for fluid can be described by the Navier–Stokes equation [26]:

$$\rho\frac{\partial\vec{v}}{\partial t} + \rho(\vec{v}\cdot\nabla)\vec{v} = -\nabla p + \mu\nabla^2\vec{v} \quad (2)$$

where ρ is the density of the fluid, $\vec{v}$ is the velocity vector, p is the pressure, and μ is viscosity. The Reynolds Number (Re) serves as a measure that contrasts inertial forces against viscous forces, thereby quantifying the respective significance of these forces in influencing fluid flow. It is defined as [26]:

$$Re = \frac{\rho v L}{\mu} \tag{3}$$

where v is the speed of the fluid (0.25 mm/s in our experiment) and L (approximately 1 mm in our design) is a characteristic length. The ρ and μ for 1xPBS used in our experiment are close to those for water (10^3 kg/m^3 and 10^{-3}Pa·s). The calculated Re in our experiment is 0.25 < 1, which means that viscous damping rapidly dissipates kinetic energy (both translational and rotational) from a fluid element, rendering inertial forces negligible and thus permitting their exclusion from the Navier–Stokes equation [24]. The modified Stokes' Equation (2) is

$$\rho \frac{\partial \vec{v}}{\partial t} = -\nabla p + \mu \nabla^2 \vec{v} \tag{4}$$

which reveals a laminar flow. In this scenario, the optimal model for the critical threshold (D_C) of DLD device is defined as [26]

$$D_C = 1.4 G \varepsilon^{0.48} \tag{5}$$

2.1.2. DEP

Dielectrophoresis (DEP) describes the displacement of a particle within a non-uniform electric field [27], as depicted in Figure 2b. When the particle is in an electric field, an induced charge appears on the particle's surface, engendering a polarization aligned with the electric field. In uniform electric fields, the polarized charges experience a homogeneous force, leading to no particle displacement, unless the particle inherently possesses a net charge. However, in a non-uniform electric field, the forces diverge, thereby leading to the particle's displacement.

The direction of the DEP force depends on the comparative electrical polarizability of the particle relative to its suspending medium. When the particle's polarizability surpasses that of the medium, the DEP force aligns with the electric field gradient, which is called positive dielectrophoresis, p-DEP. On the contrary, if the electrical polarizability of the particle is lower than that of the medium, the DEP force and the displacement vector oppose the field gradient, denoting negative dielectrophoresis, n-DEP. The composition, morphology, and phenotype of the particles in conjunction with electric field frequency significantly influence its polarizability. The DEP force on a spherical particle is given by [28]

$$F_{DEP} = 2\pi\varepsilon_m r^3 Re[f_{CM}] \cdot \nabla |E_{rms}|^2 \tag{6}$$

where r is the radius of particles; ε_m is the permittivity of the medium; f_{CM} is the Clausius–Mossoti factor; and E_{rms} is the root mean square value of an electric field. $Re[f_{CM}]$ means a real part of the f_{CM}, which can be represented as follows [28]:

$$f_{CM} = \frac{\varepsilon_p^* - \varepsilon_m^*}{\varepsilon_p^* + 2\varepsilon_m^*} \tag{7}$$

where ε^* is the complex permittivity ($\varepsilon^* = \varepsilon - j\sigma/\omega$); σ is the conductivity; and ω is the electric field frequency. Subscripts p and m mean the particles and the medium, respectively. $Re[f_{CM}] > 0$ means that particles show p-DEP response (the blue particle), while $Re[f_{CM}] < 0$ means n-DEP response (the green particle).

For the analysis of f_{CM} on cells, the spherical single-shell model is commonly utilized [29,30]. This model treats the particles as homogenous spherical bodies while segmenting them into two regions: the shell and the interior. It utilizes aggregated dielectric parameters from these regions along with the suspending medium to predict the DEP response as a function of frequency [31]. Exosomes, possessing a single-layer membrane structure similar to cells, have more uniformly distributed contents [1–6]; thus, the single-shell model will better reflect their electrical properties [29,32]. Below, we will derive the function describing the relationship between the f_{CM} and the applied electric field frequency through a spherical single-shell model.

According to the model, the effective complex permittivity of exosomes ε_{ex}^* can be described as:

$$\varepsilon_{ex}^* = \frac{c_{mem}^* r \varepsilon_c^*}{c_{mem}^* r + \varepsilon_c^*} \tag{8}$$

where ε_c^* is the complex permittivity of the contents of exosomes and c_{mem}^* is the complex capacitance of exosome membrane, which can be defined as:

$$c_{mem}^* = c_{mem} + g_{mem}/j\omega \tag{9}$$

where c_{mem} and g_{mem} are the capacitance and conductance (often negligible) of the exosome membrane per unit area. Thus, the analysis of f_{CM} is defined as:

$$f_{CM}(\omega) = -\frac{\omega^2(\tau_m\tau_c^* - \tau_c\tau_m^*) + j\omega(\tau_m^* - \tau_c^* - \tau_m) - 1}{\omega^2(2\tau_m\tau_c^* + \tau_c\tau_m^*) - j\omega(\tau_m^* + \tau_c^* + 2\tau_m) - 2} \tag{10}$$

where the time constants are defined as follows: $\tau_c = \frac{\varepsilon_c}{\sigma_c}$, $\tau_c^* = \frac{c_c^* r}{\sigma_c}$, $\tau_m = \frac{\varepsilon_m}{\sigma_m}$ and $\tau_m^* = \frac{c_m^* r}{\sigma_m}$.

2.2. DLD Array Fabrication Process

The DLD array fabrication process is shown in Figure 3b. The silicon-on-insulator wafer was used to isolate the sidewall and the arrays inside after etching up the whole top silicon layer. The parameters are shown in Table S1 (in Supplementary Material). To enhance the adhesion of the photoresist to the silicon wafer, we first applied a spin coating of HMDS (Hexamethyldisilazane) adhesion promoter, using coating parameters of 2000 rpm for 20 s. The spin-coating parameters for the S1805 photoresist were set at 3000 rpm for 30 s. The photoresist is pre-baked at 110 °C for 90 s. The photolithography system (SUSS Mask Aligner MA6, SUSS MicroTec, Garching, Germany) is set to Vacuum mode, with an exposure time of 3 s. Development is conducted in ZX-238 Positive Photoresist Developer for approximately 40 s, completing the photolithography process for the pattern. Utilizing the STS etching machine, the device undergoes a deep reactive ion etching (DRIE) process to a depth of 2 μm. The etching parameters are detailed in Table S2 (in Supplementary Material). The thermal oxidation is conducted according to the standard 1.5 μm thermal oxidation protocol.

2.3. Soft Lithography and PDMS Preparation

Given the relatively shallow etching depth of the chips used in this study, to prevent collapse and subsequent blockage of the inlet and outlet areas with low aspect ratios during bonding with the PDMS layer, we employed soft lithography to fabricate molds. This process allowed for the creation of PDMS with upward-facing grooves corresponding to the regions of the inlets and outlets, ensuring their structural integrity and preventing occlusion.

The mold fabrication was carried out on standard silicon substrates. We utilized SU8-3025 photoresist for the creation of patterns. First, we dispensed 1 mL of photoresist for each inch (25 mm) of substrate diameter. Then, we spun at 500 rpm for 5–10 s with acceleration of 100 rpm/s. After that, we spun at 3000 rpm for 30 s with acceleration of 300 rpm/second. We soft baked for 10 min at 95 °C. We exposed at 200 mJ/cm^2. We post-baked for 5 min at 95 °C. Finally, we developed for 8 min in PGMEA solution. To ensure repeated use of the mold, a layer of HMDS (Hexamethyldisilazane) was sputtered onto the mold to prevent the impact on the patterns during PDMS demolding.

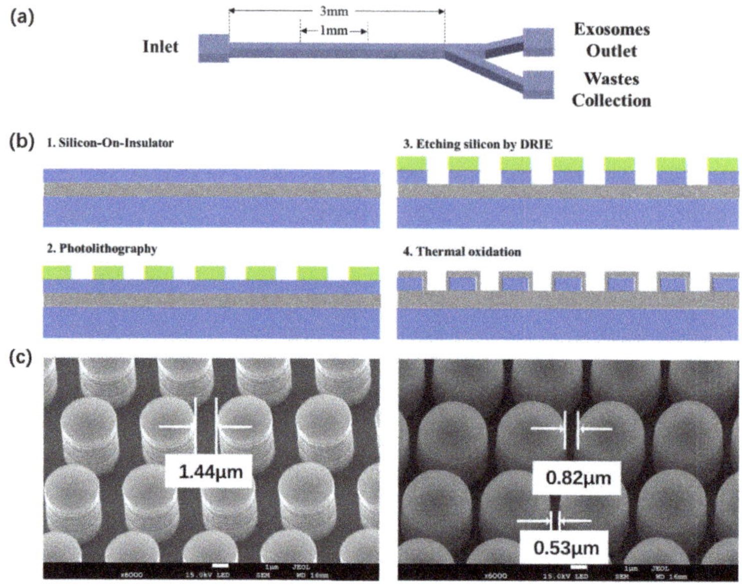

Figure 3. (**a**) Schematic of the DLD exosome separation chip. (**b**) The fabrication process flow of the non-uniform DLD arrays inviting thermal oxidation. (**c**) The SEM images of the perpendicular arrays fabricated through micron-scale lithography etching (**left**) and the tapered arrays shaped by thermal oxidation (**right**).

PDMS polymer was placed in a vacuum pump to remove internal bubbles. Subsequently, the degassed PDMS was poured onto the mold and cured on a hot plate at 110 °C for approximately 2 h before being peeled off. The chips were cleaned using acetone and isopropanol. The PDMS slabs and the chips were then treated with oxygen plasma and bonded with each other. The assembled device was placed into an oven at 60 °C for 3 h to enhance bonding. The soft lithography and PDMS preparation process flow are shown in Figure S1 (in Supplementary Material).

2.4. Particle Sample Preparation

In our work, we utilized standard polystyrene fluorescent particles of 600 nm and 100 nm (Thermo Fisher Scientific, Waltham, MA, USA). The 600 nm particles were labeled with Fluoro-Max Blue dye (ex/em: 365/447, 412/473 nm), while the 100 nm particles were marked with Fluoro-Max Red dye (ex/em: 542/612 nm). Solutions of 600 nm and 100 nm standard polystyrene fluorescent particles, with identical solid contents, were combined at a volume ratio of 200:1, equating to a quantity ratio of approximately 1:1.

2.5. DLD Experimental Instruments

Sample mixture fluid was injected into the microfluidic chips using a syringe pump (LSP02-2A, LongerPump, Baoding, China) at various constant flow rates. The fluorescent images were obtained using a biological microscope (BX53(LED), Evident, Tokyo, Japan). We employed the DAPI (ex/em: 365/473 nm) and TRITC (ex/em: 543/585 nm) excitation groups.

2.6. EV Sample Preparation

EVs were isolated from the cell culture medium of HaCaT cells. These cells were cultured in high-glucose Dulbecco's Modified Eagle's Medium (DMEM) enriched with 10% exosome-depleted fetal bovine serum (Wisent, Toronto, ON, Canada) within a 5% CO_2, 37 °C incubator (Thermo Scientific, USA). After cultivating for 48 h to achieve approximately 80% cell confluence, the medium was replaced with a serum-free variant, and the supernatant, rich in EVs, was collected after an additional 24–48 h. The isolation of EVs was performed using differential centrifugation at 4 °C. The process began with the initial centrifugation of the collected supernatant (300 mL) at 500× *g* for 5 min, followed by 2000× *g* for 30 min to eliminate cells, and then at 10,000× *g* for 60 min. Subsequently, the supernatant was transferred into ultracentrifuge tubes and spun at 120,000× *g* for 70 min at 4 °C. Post-centrifugation, the supernatant was carefully decanted, and the pellet was reconstituted in sterile 1X PBS. This resuspension was then collected to procure exosomes.

2.7. Nanoparticle Tracking Analysis (NTA)

To assess the recovery rate and purity of the isolated 100 nm polystyrene nanoparticles and exosomes, separately, Nanoparticle Tracking Analysis (NTA; NanoSight NS300, Malvern Instruments, Malvern, UK) was employed to measure the size distributions of both the initial and the processed samples. All samples were diluted in 1xPBS to concentrations ranging from 2×10^8 to 1×10^9 particles/mL to ensure optimal accuracy. The acquisition of size distribution data was performed using the NTA 3.2 Analytical Software Suite (27 November 2015, NanoSight NTA software update v3.2 | Malvern Panalytical). These measurements were executed at a controlled temperature of 20 °C.

2.8. Scanning Electron Microscope (SEM)

After etching, the four-inch wafers were sectioned into individual chips for observation with a Supra 55 (Zeiss, Jena, Germany) scanning electron microscope (SEM). The silicon on the chip exhibited excellent conductivity, allowing for observations to be conducted at room temperature (298 K) with a 30-degree tilt in a vacuum. After undergoing thermal oxidation, the surface of the chips was coated with an insulating layer of silicon dioxide. For enhanced visualization of the column sidewalls, it is advisable to sputter-coat a thin layer of gold to improve conductivity prior to SEM observation. The chips after thermal oxidation in this work did not undergo a gold sputtering process.

3. Results

3.1. DLD Chip Design and Fabrication

Figure 3a first shows the schematic of the DLD exosome separation chip. The chip consists of one inlet, a straight DLD array separation region, and two outlets (one for exosome collection, the other for waste). The whole length of the microfluidic channel is 3 mm. The length of the DLD separation area is set at 1 mm, which is determined based on the requirements for the Re number, as outlined in Equation (2). With a fluid flow rate of 0.25 mm/sec within our DLD device, maintaining the fluid as laminar flow necessitates that the length of the sorting area does not exceed 4 mm. A length of 1 mm provides sufficient margin to vary the flow rate. The width of the device is set at 90 μm. This specification arises from our designed shift fraction of 0.1, and, in order to ensure the array sufficiently separates all particles completely, the width must be less than 1000 μm length multiplied by the shift fraction of 0.1, equating to less than 100 μm. The depth of the microchannel is controlled at 2 μm through a DRIE etching process, which was identified as the optimal array depth for thermal oxidation effects. In our experimental design, we first employed standard fluorescent polystyrene nanoparticles with diameters of 100 nm and 600 nm as proxies for exosomes and larger nanoparticles present in actual bodily fluids, respectively. We therefore set the DLD separation threshold at 300 nm. According to Equation (2), we need a column-to-column spacing of ~800 nm to achieve the threshold of 300 nm with a displacement fraction of 0.1 in order to realize DLD separation between the two selected sizes.

Typically, the nano-scale linewidth necessitates the use of high-precision fabrication equipment, such as stepper lithography systems, thereby significantly increasing device production costs. To address this issue, we incorporated thermal oxidation into the array construction process. The left image of Figure 3c displays the micrometer-scale array produced through micrometer-level lithography and etching. After thermal oxidation, as shown in the right image of Figure 3c, the spacing between arrays was reduced to the nanometer scale, meeting the requirements of DLD devices for the separation of nanometer-scale particles. It is worth noting that during the thermal oxidation of the device, only the upper ends and the sides of the pillars are exposed to oxygen species. This leads to the silicon at the lower part of the pillars undergoing lateral thermal oxidation growth exclusively, whereas the silicon at the upper ends experiences both lateral and vertical thermal oxidation growth during thermal oxidation (Figure S2 in Supplementary Material). Consequently, the lateral thermal oxidation growth at the upper ends is less extensive than at the lower ends, resulting in the formation of a tapered structure. According to Equation (2) from the DLD theory section, such differential spacing between the upper and lower layers leads to distinct critical diameters (D_C) for each; as such, the arrays as a whole no longer maintain a precise separation threshold. To minimize the impact of this tapered structure on exosome separation, we tried to refine the thermal oxidation process by precisely controlling the spacing between the upper and lower layers. We conducted the same thermal oxidation process on columns with different etching depths and found that in columns of smaller depth, the non-uniformity of the upper and lower layer was less pronounced. A possible explanation is that when the columns are of smaller depth, the volume of silicon at the upper layer is smaller, leading to a reduced amount of silicon being oxidized both laterally and vertically. Moreover, oxygen can distribute more evenly among columns of smaller depth, which is beneficial for the uniformity of the overall thermal oxidation. At an etching depth of 2 μm, we achieved a relatively optimal thermal oxidation effect. The upper layer, as depicted in Figure 3c, initially spanning 1.44 μm, undergoes reduction to 0.82 μm, leading a 300 nm threshold, thereby enabling the separation of most impurity particles without exerting the DLD effect on exosomes. In contrast, the reduction of the lower layer from 1.44 μm to 0.53 μm establishes a separation threshold of 200 nm, ensuring that exosomes (below 200 nm) remain unaffected by the DLD phenomenon. It is worthwhile to stress that the utilization of thermal oxidation in device fabrication unequivocally results in the formation of such tapered structures. Previous work that also employs thermal oxidation to create nano-sized DLD pillars [25] did not recognize that such a variation in gap sizes in the vertical direction plays an important role in explaining the actual DLD behaviors of the particles. Here, we pay specific attention to this phenomenon when designing the pillar sizes and oxidation parameters. Furthermore, such a process markedly reduces the costs associated with chip fabrication by circumventing the need for high-precision lithography techniques.

3.2. DLD Separation Performance

Figure 4a,b depict the microfluidic experimental setup and the actual image of the device. A syringe pump injects sample solutions into the chip at a constant flow rate. The chip is enclosed by a PDMS layer, whose transparency allows for the observation of particle trajectories via fluorescence microscopy. During device operation, the larger nanoparticles were heavily influenced by the DLD effect, resulting in their deviation along the streamline and subsequent collection from the waste outlet. On the contrary, the smaller-sized exosomes flow directly towards the exosome outlet.

For the experiment, a mixture of 600 nm (labeled with Fluoro-Max blue dye) and 100 nm (labeled with Fluoro-Max red dye) standard fluorescent polystyrene nanoparticles was injected into the chip at a flow rate of 5×10^{-2} μL/min. Within the solution, we introduced a small quantity of Tween-20 surfactant to mitigate the adhesion of the fluorescent polystyrene nanoparticles to the PDMS layer, thereby preventing any interference with the microscopic observations. As shown in Figure 4c, during a typical device run, the streamlines of 600 nm polystyrene nanoparticles form a triangular pattern, indicating the displacement of these larger nanoparticles due to DLD, whereas the 100 nm nanoparticles pass through unaltered. We also present a dynamic video showcasing the movement of particles in Video S1.

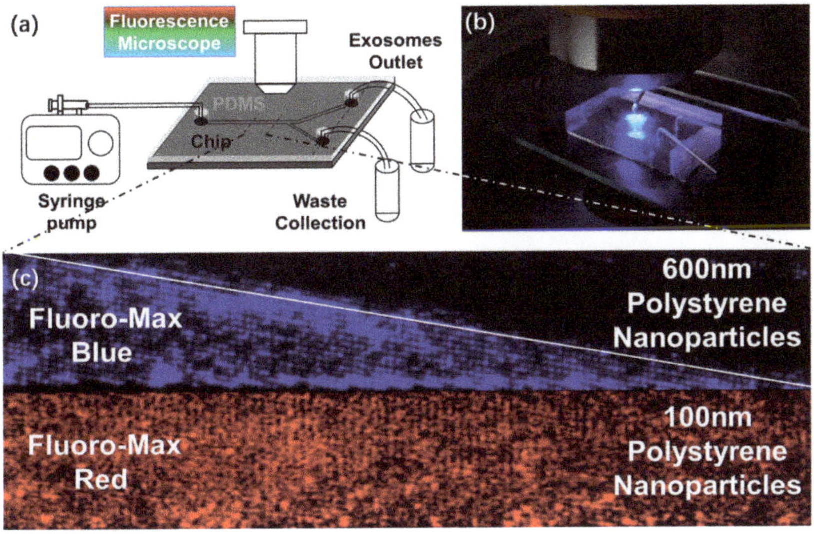

Figure 4. (**a**) Schematic of the microfluidic DLD experiment setup. (**b**) Actual image of the chip used for the DLD experiment. (**c**) Image obtained through fluorescence microscopy reveals distinct behaviors: 600 nm polystyrene nanoparticles, labeled with Fluoro-Max blue dye, exhibit DLD phenomena, whereas 100 nm polystyrene nanoparticles, marked with Fluoro-Max red dye, demonstrate no displacement.

Subsequent collection and analysis of the sample solutions from both the inlet and the exosome outlet through Nanoparticle Tracking Analysis (NTA), as demonstrated in Figure 5a, confirm that the 100 nm nanoparticles have been effectively separated from the 600 nm ones, resulting in a purity from 52.31% to 92.31% from the outlet collection. Next, we further examined the separation performance with the mixture of HaCaT cell exosomes and 600 nm nanoparticles to evaluate the chip's efficacy in a real bodily fluid context. The same experimental parameters as those used in the fluorescent polystyrene nanoparticle experiment were maintained. Our device demonstrated a significant reduction of 600 nm polystyrene particles and larger extracellular nanoparticles in the exosome sample post-separation. The purity of exosomes collected at the outlet was measured at 91.47% via NTA, a significant increase compared to 57.84% in the original sample, as shown in Figure 5b, indicating the chip's robust capability of separating exosomes from complex biological mixtures.

3.3. Simulation of the Synergistic Effect of DLD and DEP

Famously, DLD faces challenges in the separation of extracellular nanoparticles of similar sizes [14], while the DEP effect is less potent at the nano scale compared to the separation of micron-scaled particles [22]. In pursuit of enhancing the precision of separating nanoparticles of similar sizes yet distinct biochemical compositions, such as lipoproteins, exomeres, and retroviruses, we propose here a strategy to integrate these two microscopic effects. As illustrated in Figure 6a, the key point of our design is an array of nano-scale, droplet-shaped pillars. This configuration not only preserves the characteristic layout of DLD arrays but also facilitates the generation of non-uniform electric fields, a crucial factor for implementing DEP.

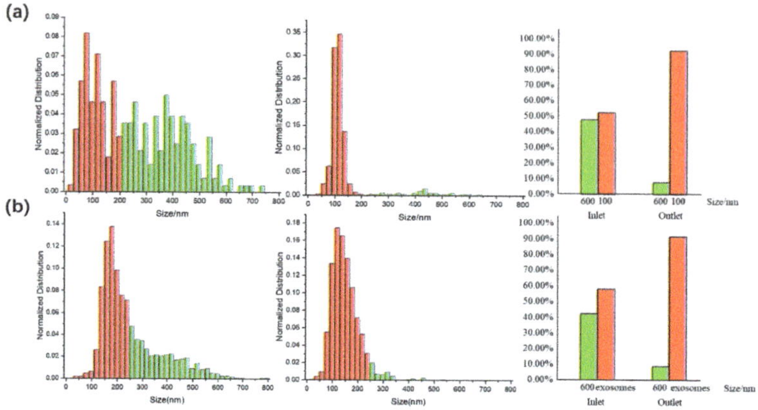

Figure 5. NTA results and the purity behavior of the two experiments. (**a**) Normalized distribution of the mixture sample of 100 and 600 nm particles before and after separation in NTA test, (**b**) Normalized distribution of the mixture sample of HaCaT cell exosomes and 600 nm particles before and after separation in NTA test.

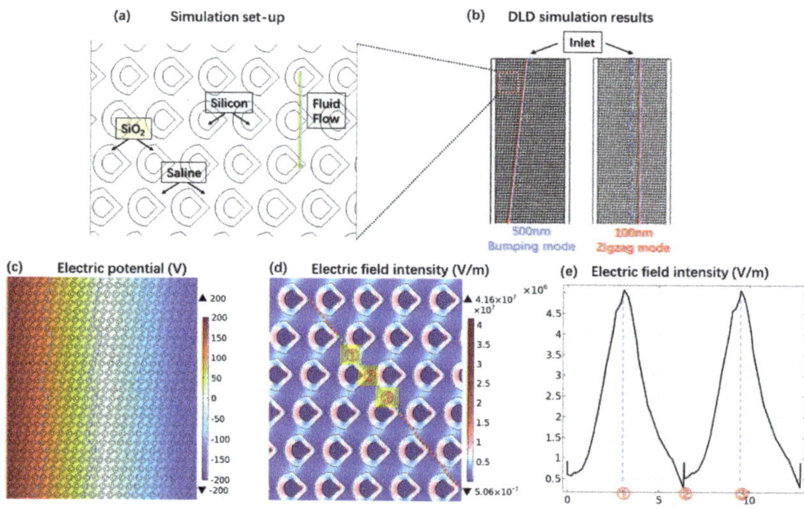

Figure 6. (**a**) Structure and components of the droplet-shaped columns synergistically employing deterministic lateral displacement and dieletrophoresis in simulation. (**b**) Particle-tracking simulation results of 500 nm polystyrene particles on the left and 100 nm polystyrene particles on the right. (**c**) Electric potential of the whole fluid channel and the wavy pattern longitudinal equipotential lines. (**d**) Electric field intensity distribution. (**e**) The line graph of the electric field intensity values along the green line in (**d**).

By applying electrical excitation across the flow channel's boundaries in the left and right sides, the droplet-shaped pillar arrays within the microfluidic region generate highly non-uniform electric fields that are overall perpendicular to the direction of the fluid flow. Such electric fields enhance the device's overall ability to separate exosomes (ranging from 30 to 150 nm) from particles of similar sizes, such as lipoproteins (5 to 200 nm) and retroviruses (80 to 150 nm). Despite their size similarities, these nanoparticles exhibit significant variance in their biochemical compositions and structures [33], resulting in disparate electrical properties, as indicated in Equations (6)–(10).

By changing the frequency of the applied electrical signal, it is feasible to manipulate the complex permittivity of exosomes, similarly sized nanoparticles, and the surrounding medium. Our strategy is to align the complex permittivity of exosomes to zero, whereas that of similar nanoparticles deviates from zero. Under such conditions, exosomes are exempt from the influence of DEP forces, while other nanoparticles are experiencing effective DEP forces. This differential application of DEP forces enables the refined separation of exosomes from particles with overlapping size ranges but divergent biochemical identities. In determining the frequency, we plotted the real part of the Clausius–Mossotti factor (f_{CM}) as a function of the frequency of exosomes and BSA (similar to lipoprotein, as an impurity) based on Equation (10), as shown in Figure S3a (in Supplementary Material). We found that within the 100–1000 kHz frequency range, the $Re[f_{CM}]$ for exosome is close to 0, while for BSA it is 1. This frequency range is well-suited to our needs. Furthermore, we noticed that the $Re[f_{CM}]$ function plots for gold nanoparticles and polystyrene nanoparticles closely resemble those for exosome and BSA, shown in Figure S3b (in Supplementary Material). This suggests that we could use gold and polystyrene nanoparticles for preliminary non-biological sample experiments in future studies.

We initially employed a particle tracking simulation to investigate whether this novel droplet-shaped structure possesses DLD functionality. The simulation setup is illustrated in Figure 6a, where the spaces between the shapes are filled with saline and the fluid characteristic is a laminar flow at a velocity of 300 μm/s. The gap between the droplet-shaped structures is set to 800 nm, with a shift fraction of 0.1. Particles 500 nm and 100 nm in diameter were injected into the middle of the microfluidic channel entrance at intervals of every 0.01 s for a duration of 24 s. The simulation results, shown in Figure 6b, clearly demonstrate that the 500 nm particles exhibit obvious lateral displacement, aligning with the DLD bumping mode, while the trajectory of the 100 nm particles is consistent with the zigzag mode. This indicates that our droplet-shaped array retains the functionality of DLD.

We further verified the DEP performance of this droplet-shaped structure. Our approach integrates the thermally oxidized, non-uniform DLD arrays with DEP functionality by configuring the individual pillar structure into two distinct layers, as Figure 6a shows. The core layer is composed of silicon, whereas the exterior layer is fabricated from SiO_2. The space within the structure is defined as saline solution to simulate the real bodily fluid. A voltage of 400 Vpp at a frequency of 100 kHz was applied across the microfluidic flow channel. Figure 6c displays the two-dimensional electric potential distribution within the flow channel, revealing that the longitudinal equipotential lines exhibit a wavy pattern closely resembling the inclined arrangement of columns within the channel. The significantly lower relative dielectric constant of SiO_2 (3.9) compared to saline solution (80) and silicon (11.7) results in the highest electric field strength being generated within the silicon dioxide layer, as depicted by the two-dimensional electric field distribution in Figure 6d. Consequently, the voltage drop is primarily concentrated within the silicon dioxide layer. The inclined arrangement of the SiO_2 pillars alters the uniform field strength distribution typically observed between parallel plate capacitors, leading to the formation of such wavy equipotential lines. Figure 6e displays the electric field distribution along the green diagonal line in Figure 6d. As is evident from the figure, a periodic electric field with a substantial gradient is generated along the diagonal. This phenomenon is precisely facilitated by the sharp tip of the droplet-shaped design we have implemented. The electric field lines densely converge at the sharp point marked as position 1, creating a region of high electric field intensity. In contrast, the transformation of the original circular shape into a droplet shape significantly enlarges the space above and below position 2, causing the longitudinal distribution of electric field lines to become sparse, thereby resulting in a region of lower electric field intensity. Consider a particle experiencing a positive DEP force. When positioned at location 1, a particle proceeds to travel with the fluid flow, gliding along the inclined sidewall toward location 2. Upon reaching location 2, the particle is drawn towards regions of higher electric field intensity, prompting its displacement to location 3. As this cyclic process persists, the particle is progressively displaced towards the right, a movement mainly governed by the DEP effect.

4. Discussion

In this study, we propose a novel strategy employing thermally oxidized non-uniform DLD arrays with tapered shapes to improve the separation performance of nano-sized particles. Although thermal oxidation has been utilized in the fabrication process of DLD arrays in previous work, the impact of the resultant tapered structures has often been overlooked. This oversight led to discrepancies between the device's real separation threshold and the expected one [25], thus limiting the separation efficiency of the device. In our work, we precisely control the linewidth of the upper and lower layers on the same DLD pillars, thus enabling the upper layer to apply the DLD effect on impurity particles while ensuring that neither layer affects exosomal small particles. In experiments involving fluorescent nanoparticles and biological samples with exosomes, we achieved purities for target particles exceeding 90%. This technique preserves the benefits of conventional DLD methods, being label-free, labor-efficient, and field-free, while minimizing damage to exosomes. Additionally, it circumvents the need for high-precision lithography, thereby substantially lowering the fabrication costs associated with traditional DLD approaches. Furthermore, we propose a synergistic approach combining DEP and DLD to overcome the intrinsic size-based sorting limitation of DLD, which leads to its inability to separate exosomes from nanoparticles of similar sizes (e.g., lipoproteins, retroviruses, and exoparticles). This design not only retains the functional characteristics of DLD but also helps to create a highly non-uniform electric field between the columns after applying alternating voltage on both sides of the array, thereby generating strong DEP forces. By modulating the frequency of the alternating voltage, it is possible to adjust the DEP effects on particles with diameters similar to exosomes to a direction opposite to that of DLD while not exerting force on the exosomes themselves. Thus, the device is capable of separating exosomes based on particle size through DLD and further distinguishing impurities with dimensions similar to exosomes based on the particles' intrinsic electrical properties through DEP. While our device demonstrates potential, there is room for improvement. The throughput and the fluid flow rate of our DLD device are at a low level, leading to long separation times for samples. In future iterations, we plan to implement multi-channel parallel injection and widen and deepen the flow channels to increase the overall device throughput. We have realized the fabrication of the critical pointed structures of the droplet-shaped arrays essential for the effective implementation of the DEP effect. We will soon validate the simulation results of our DEP design through experimental verification, using the outcomes to further refine our device design.

5. Conclusions

In summary, we demonstrated a novel microfluidic exosome separation device featuring thermally oxidized non-uniform deterministic lateral displacement (DLD) arrays. By precisely adjusting the linewidth of the upper and lower layers of the tapered structures formed after thermal oxidation, we have achieved efficient exosome separation with a purity exceeding 90% while significantly reducing the fabrication costs of the device by circumventing the need for high-precision lithography. Additionally, the incorporation of DEP offers a potential strategy for differentiating exosomes from particles of similar dimensions yet distinct biochemical compositions. This work establishes a robust platform for exosome isolation, thereby facilitating subsequent detection of nucleic acids, proteins, and other exosomal constituents. Moreover, we aim to transfer the DEP simulation outcomes into experimental tests to validate the DEP functionality in our device. Further endeavors will also explore the possibility of topological optimization of the microfluidic design, with aspirations to integrate it with a downstream biosensing apparatus in pursuit of the development of a comprehensive exosome analysis platform.

Supplementary Materials: The following supporting information can be downloaded at: https://www.mdpi.com/article/10.3390/bios14040174/s1, Figure S1: The process flow of the PDMS fabrication; Figure S2: Illustration of forming a tapered structure from the arrays with vertical sidewalls through thermal oxidation; Table S1: The parameters of the silicon-on-insulator wafer; Table S2: The parameters of the deep reactive ion etching (DRIE) process; Video S1: A dynamic video showcasing the movement of particles.

Author Contributions: Conceptualization, Y.L., J.Z. and D.W.; data curation, D.W., S.Y., N.W. and H.G.; formal analysis, J.Z., Y.L., D.W. and N.W.; funding acquisition, S.F., Y.L. and J.Z.; investigation, Y.L., D.W., J.Z. and H.G.; methodology, D.W., S.Y. and N.W.; project administration, S.F., Y.L. and J.Z.; resources, S.F., Y.L. and J.Z.; software, D.W., N.W., S.Y. and H.G.; supervision, S.F., Y.L. and J.Z.; validation, D.W., S.Y. and H.G.; visualization, D.W., J.Z. and Y.L.; writing—original draft, D.W.; writing—review and editing, S.F., J.Z. and Y.L. All authors have read and agreed to the published version of the manuscript.

Funding: This work was supported by the National Key Research and Development Program of China (2021YFF1200300, 2023YFB3208200, 2023YFA0915200), the Research Equipment Development Program from Chinese Academy of Sciences (ZDKYYQ20220002), the National Natural Science Foundation of China (No. 62374170), and the Science and Technology Commission of Shanghai Municipality (No. 23J21900200, 22xtcx00100, XTCX-KJ-2022-32).

Institutional Review Board Statement: Not applicable.

Informed Consent Statement: Not applicable.

Data Availability Statement: The data that support the findings of this study are available from the corresponding author upon reasonable request.

Acknowledgments: The authors want to thank the Institutional Center for Shared Technologies and Facilities of Shanghai Institute of Microsystem and Information Technology, CAS for their support throughout this work.

Conflicts of Interest: The authors declare no conflicts of interest. The funders had no role in the design of the study; in the collection, analyses, or interpretation of data; in the writing of the manuscript; or in the decision to publish the results.

References

1. Kalluri, R.; LeBleu, V.S. The biology, function, and biomedical applications of exosomes. *Science* **2020**, *367*, eaau6977. [CrossRef]
2. Pegtel, D.M.; Gould, S.J. Exosomes. *Annu. Rev. Biochem.* **2019**, *88*, 487–514. [CrossRef]
3. Xu, R.; Rai, A.; Chen, M.; Suwakulsiri, W.; Greening, D.W.; Simpson, R.J. Extracellular vesicles in cancer—Implications for future improvements in cancer care. *Nat. Rev. Clin. Oncol.* **2018**, *15*, 617–638. [CrossRef]
4. Tkach, M.; Théry, C. Communication by extracellular vesicles: Where we are and where we need to go. *Cell* **2016**, *164*, 1226–1232. [CrossRef]
5. Valadi, H.; Ekström, K.; Bossios, A.; Sjöstrand, M.; Lee, J.J.; Lötvall, J.O. Exosome-mediated transfer of mRNAs and microRNAs is a novel mechanism of genetic exchange between cells. *Nat. Cell Biol.* **2007**, *9*, 654–659. [CrossRef]
6. Théry, C.; Zitvogel, L.; Amigorena, S. Exosomes: Composition, biogenesis and function. *Nat. Rev. Immunol.* **2002**, *2*, 569–579. [CrossRef]
7. Mathivanan, S.; Fahner, C.J.; Reid, G.E.; Simpson, R.J. ExoCarta 2012: Database of exosomal proteins, RNA and lipids. *Nucleic Acids Res.* **2012**, *40*, D1241–D1244. [CrossRef]
8. Colombo, M.; Raposo, G.; Théry, C. Biogenesis, secretion, and intercellular interactions of exosomes and other extracellular vesicles. *Annu. Rev. Cell Dev. Biol.* **2014**, *30*, 255–289. [CrossRef]
9. Théry, C.; Amigorena, S.; Raposo, G.; Clayton, A. Isolation and characterization of exosomes from cell culture supernatants and biological fluids. *Curr. Protoc. Cell Biol.* **2006**, *30*, 3.22.1–3.22.29. [CrossRef]
10. Théry, C.; Witwer, K.W.; Aikawa, E.; Alcaraz, M.J.; Anderson, J.D.; Andriantsitohaina, R.; Antoniou, A.; Arab, T.; Archer, F.; Atkin-Smith, G.K.; et al. Minimal information for studies of extracellular vesicles 2018 (MISEV2018): A position statement of the International Society for Extracellular Vesicles and update of the MISEV2014 guidelines. *J. Extracell. Vesicles* **2018**, *7*, 1535750. [CrossRef]
11. Zech, D.; Rana, S.; Büchler, M.W.; Zöller, M. Tumor-exosomes and leukocyte activation: An ambivalent crosstalk. *Cell Commun. Signal.* **2012**, *10*, 37. [CrossRef] [PubMed]
12. Davies, R.T.; Kim, J.; Jang, S.C.; Choi, E.J.; Gho, Y.S.; Park, J. Microfluidic filtration system to isolate extracellular vesicles from blood. *Lab Chip* **2012**, *12*, 5202–5210. [CrossRef] [PubMed]

13. Guo, S.C.; Tao, S.C.; Dawn, H. Microfluidics-based on-a-chip systems for isolating and analysing extracellular vesicles. *J. Extracell. Vesicles* **2018**, *7*, 1508271. [CrossRef] [PubMed]
14. Su, W.; Li, H.; Chen, W.; Qin, J. Microfluidic strategies for label-free exosomes isolation and analysis. *TrAC Trends Anal. Chem.* **2019**, *118*, 686–698. [CrossRef]
15. Yuan, D.; Zhao, Q.; Yan, S.; Tang, S.Y.; Alici, G.; Zhang, J.; Li, W. Recent progress of particle migration in viscoelastic fluids. *Lab Chip* **2018**, *18*, 551–567. [CrossRef] [PubMed]
16. Leshansky, A.M.; Bransky, A.; Korin, N.; Dinnar, U. Tunable nonlinear viscoelastic "focusing" in a microfluidic device. *Phys. Rev. Lett.* **2007**, *98*, 234501. [CrossRef] [PubMed]
17. Shao, H.; Im, H.; Castro, C.M.; Breakefield, X.; Weissleder, R.; Lee, H. New technologies for analysis of extracellular vesicles. *Chem. Rev.* **2018**, *118*, 1917–1950. [CrossRef] [PubMed]
18. Ding, L.; Yang, X.; Gao, Z.; Effah, C.Y.; Zhang, X.; Wu, Y.; Qu, L. A holistic review of the state-of-the-art microfluidics for exosome separation: An overview of the current status, existing obstacles, and future outlook. *Small* **2021**, *17*, 2007174. [CrossRef]
19. Huang, L.R.; Cox, E.C.; Austin, R.H.; Sturm, J.C. Continuous particle separation through deterministic lateral displacement. *Science* **2004**, *304*, 987–990. [CrossRef]
20. Davis, J.A.; Inglis, D.W.; Morton, K.J.; Lawrence, D.A.; Huang, L.R.; Chou, S.Y.; Sturm, J.C.; Austin, R.H. Deterministic hydrodynamics: Taking blood apart. *Proc. Natl. Acad. Sci. USA* **2006**, *103*, 14779–14784. [CrossRef]
21. McGrath, J.; Jimenez, M.; Bridle, H. Deterministic lateral displacement for particle separation: A review. *Lab Chip* **2014**, *14*, 4139–4158. [CrossRef] [PubMed]
22. Voldman, J. Electrical forces for microscale cell manipulation. *Annu. Rev. Biomed. Eng.* **2006**, *8*, 425–454. [CrossRef] [PubMed]
23. Shafiee, H.; Sano, M.B.; Henslee, E.A.; Caldwell, J.L.; Davalos, R.V. Selective isolation of live/dead cells using contactless dielectrophoresis (cDEP). *Lab Chip* **2010**, *10*, 438–445. [CrossRef]
24. Hettiarachchi, S.; Cha, H.; Ouyang, L.; Mudugamuwa, A.; An, H.; Kijanka, G.; Kashaninejad, N.; Nguyen, N.T.; Zhang, J. Recent microfluidic advances in submicron to nanoparticle manipulation and separation. *Lab Chip* **2023**, *23*, 982–1010. [CrossRef]
25. Wunsch, B.H.; Smith, J.T.; Gifford, S.M.; Wang, C.; Brink, M.; Bruce, R.L.; Austin, R.H.; Stolovitzky, G.; Astier, Y. Nanoscale lateral displacement arrays for the separation of exosomes and colloids down to 20 nm. *Nat. Nanotechnol.* **2016**, *11*, 936–940. [CrossRef] [PubMed]
26. Davis, J.A. Microfluidic Separation of Blood Components through Deterministic Lateral Displacement. Ph.D. Thesis, Princeton University, Princeton, NJ, USA, 2008.
27. Yang, J.; Huang, Y.; Wang, X.B.; Becker, F.F.; Gascoyne, P.R. Cell separation on microfabricated electrodes using dielectrophoretic/gravitational field-flow fractionation. *Anal. Chem.* **1999**, *71*, 911–918. [CrossRef]
28. Hughes, M.P. Strategies for dielectrophoretic separation in laboratory-on-a-chip systems. *Electrophoresis* **2002**, *23*, 2569–2582. [CrossRef] [PubMed]
29. Kim, D.; Sonker, M.; Ros, A. Dielectrophoresis: From molecular to micrometer-scale analytes. *Anal. Chem.* **2018**, *91*, 277–295. [CrossRef]
30. Xing, X.; Yobas, L. Dielectrophoretic isolation of cells using 3D microelectrodes featuring castellated blocks. *Analyst* **2015**, *140*, 3397–3405. [CrossRef]
31. di Toma, A.; Brunetti, G.; Chiriacò, M.S.; Ferrara, F.; Ciminelli, C. A Novel Hybrid Platform for Live/Dead Bacteria Accurate Sorting by On-Chip DEP Device. *Int. J. Mol. Sci.* **2023**, *24*, 7077. [CrossRef]
32. Moore, J.H.; Varhue, W.B.; Su, Y.H.; Linton, S.S.; Farmehini, V.; Fox, T.E.; Matters, G.L.; Kester, M.; Swami, N.S. Conductance-based biophysical distinction and microfluidic enrichment of nanovesicles derived from pancreatic tumor cells of varying invasiveness. *Anal. Chem.* **2019**, *91*, 10424–10431. [CrossRef] [PubMed]
33. Van der Pol, E.; Böing, A.N.; Harrison, P.; Sturk, A.; Nieuwland, R. Classification, functions, and clinical relevance of extracellular vesicles. *Pharmacol. Rev.* **2012**, *64*, 676–705. [CrossRef] [PubMed]

biosensors

MDPI

Article

A 3D Capillary-Driven Multi-Micropore Membrane-Based Trigger Valve for Multi-Step Biochemical Reaction

Yijun Zhang [1,2], Yuang Li [1,2], Xiaofeng Luan [1,2], Xin Li [1,2], Jiahong Jiang [1], Yuanyuan Fan [1], Mingxiao Li [1], Chengjun Huang [1,2], Lingqian Zhang [1,*] and Yang Zhao [1,*]

1 Institute of Microelectronics of the Chinese Academy of Sciences, Beijing 100029, China
2 University of Chinese Academy of Sciences, Beijing 100049, China
* Correspondence: zhanglingqian@ime.ac.cn (L.Z.); zhaoyang@ime.ac.cn (Y.Z.)

Abstract: Point-of-care testing (POCT) techniques based on microfluidic devices enabled rapid and accurate tests on-site, playing an increasingly important role in public health. As the critical component of capillary-driven microfluidic devices for POCT use, the capillary microfluidic valve could schedule multi-step biochemical operations, potentially being used for broader complex POCT tasks. However, owing to the reciprocal relationship between the capillary force and aperture in single-pore microchannels, it was challenging to achieve a high gating threshold and high operable liquid volume simultaneously with existing 2D capillary trigger valves. This paper proposed a 3D capillary-driven multi-microporous membrane-based trigger valve to address the issue. Taking advantage of the high gating threshold determined by micropores and the self-driven capillary channel, a 3D trigger valve composed of a microporous membrane for valving and a wedge-shaped capillary channel for flow pumping was implemented. Utilizing the capillary pinning effect of the multi-micropore membrane, the liquid above the membrane could be triggered by putting the drainage agent into the wedge-shaped capillary channel to wet the underside of the membrane, and it could also be cut off by taking away the agent. After theoretical analysis and performance characterizations, the 3D trigger valve performed a high gating threshold (above 1000 Pa) and high trigger efficiency with an operable liquid volume above 150 μL and a trigger-to-drain time below 6 s. Furthermore, the retention and trigger states of the valve could be switched for repeatable triggering for three cycles within 5 min. Finally, the microbead-based immunoreaction and live cell staining applications verified the valve's ability to perform multi-step operations. The above results showed that the proposed 3D trigger valve could be expected to play a part in wide-ranging POCT application scenarios.

Keywords: 3D trigger valve; microporous membrane; capillary-driven; gating threshold; POCT

1. Introduction

Microfluidic techniques have shown great potential for point-of-care testing (POCT) applications owing to their advantages of high integration, low reagent usage, and rapid detection [1]. In the past decades, microfluidic devices have been widely used for on-site detection of molecules, cells, pathogens, and microorganisms in different scenarios, such as pregnancy detection, food hygiene detection, and COVID-19 virus detection [2–8]. In conventional microfluidic systems, various microfluidic handlings of liquid samples, e.g., pumping, stopping, mixing, and washing, are usually required for different purposes, in which external or internal liquid control units, such as mechanical (electro, magnetic, etc.) pumps and valves, were usually required [9]. These units dramatically increased the complexity and cost of the microfluidic system. In order to overcome these disadvantages of conventional microfluidic systems in POCT applications, capillary microfluidic chips were introduced in which liquid sample was handled by natural capillary forces and gravity, without other external driving forces [10]. As one of the typical capillary chips, the paper-based strips [11,12] have been successfully used on various POCT applications, such as pregnancy and glucose tests. However, their lack of accurate control limited the application in multi-step complex reactions.

In order to realize a more controllable capillary flow, various capillary valve control units have been reported, including a capillary trigger valve, capillary retention valve, capillary retention burst valve, and delay valve [13–16]. These capillary valves were designed based on the Laplace pressure difference generated by the changing curvature radius of the liquid meniscus, depending on surface tension, contact angle, liquid consistency, and flow channel shape in 2D microfluidic devices. In these 2D devices, capillary channels and valves mainly included burst pressure valves and trigger valves (Figure 1a) with the triggering agent. Different analytical and numerical models have been reported to study the influence of channel shape, surface characteristics, and fluid characteristics on them [17–21]. However, due to the 2D limitation, the holding time of these valves was usually less than 5 min [22], restricting the time for the user to pre-load the liquid sample and their operational applicability. In order to improve the applicability of the 2D valves, the stair-step liquid-triggered valve has been proposed [23–26]. The precise channel with a high aspect ratio (i.e., height-to-width ratio) was etched in horizontal and vertical directions to improve valve reliability. The retention time increased to 30 min, attributed to the geometry optimization, but the gating threshold and operating liquid volume were still low (~443 Pa, ~6 μL). Meanwhile, this method had stringent requirements on the aspect ratio of valves [27]. Attempts, such as a high aspect ratio channel or semi-3D valve [28–30] design, were proposed to improve the performance of trigger valves for POCT use. However, single-pore microchannel-based trigger valves were limited by the reciprocal relationship between the capillary force and the aperture of the microchannel [31]. To sum up, existing 2D valves had the defects of a low gating threshold, low operable fluid volume, long triggering time, and short retention time. To overcome the drawbacks, we needed a new form of the trigger valve to break the deadlock on the trade-off between high fluid switching ability and high operable liquid volume.

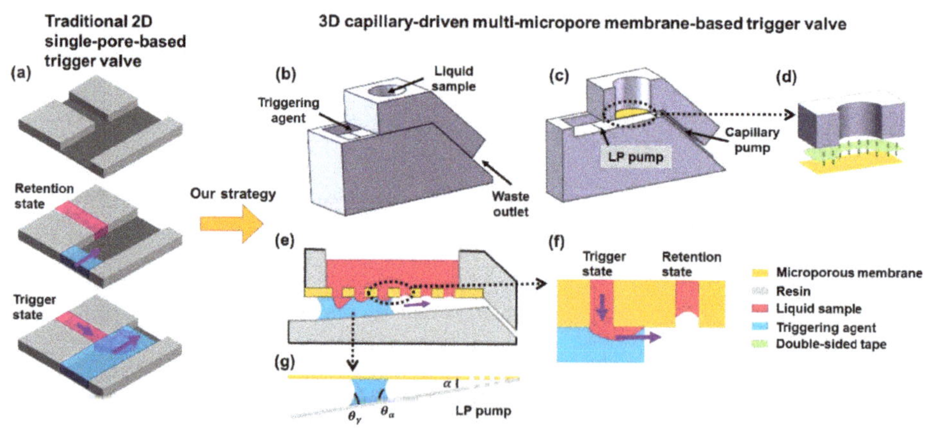

Figure 1. Schematic diagram showing the mechanics and mechanism of the 3D capillary-driven multi-micropore membrane-based trigger valve. (**a**) 3D Schematic and trigger process of traditional 2D single-pore-based trigger valve. (**b**) 3D Schematic of the 3D capillary-driven multi-micropore membrane-based trigger valve. (**c**) Vertical cross-section view of the 3D valve. (**d**) Device assembly schematic. (**e**) Working mechanism of the 3D valve. (**f**) Two states of micropores. (**g**) Laplace pump.

In this paper, we proposed a new 3D capillary-driven multi-micropore membrane-based trigger valve to simultaneously improve the trigger efficiency and operable liquid volume while triggering the valve. With similar trigger structures at the openings of micropores, microporous membranes could achieve a significantly different gating threshold by tuning the dry and wet states. When in the dry state, the gating threshold of the valve could increase to the kPa scale due to the capillary pinning effect of microporous membranes [32], whereas in the wet state, the liquid could drain from millions of micropores on the membrane. Moreover, combining the advantages of the self-driven capillary channel under the membrane for rapidly switching the dry and wet state of the membrane, we proposed the new 3D capillary-driven multi-micropore membrane-based trigger valve, in which the pressure barrier was up to 3500 Pa (5 μm micropores, for instance). Meanwhile, the device was triggered within 6 s (duration from trigger to drain of 150 μL liquid, given 5 μm micropores, for instance) and was implemented in multi-step bio-particle immunostaining chain reactions. Furthermore, in the following sections, we characterized the 3D trigger valve's static performances in different switch states and repeatable performances for cycling switches, respectively, and demonstrated its potential for POCT use in two short applications.

2. Design and Experimental Section

2.1. Working Principle of the 3D Trigger Valve

The 3D trigger valve was composed of a chamber (loading liquid samples), a microporous membrane (trigger valve), and a wedge-shaped and straight capillary channel (pumps), as shown in Figure 1. The working principle of the valve was described as follows. The triggering agent was first dripped at the entrance of the wedge-shaped capillary channel and was self-driven by capillary force into the channel to infiltrate the microporous membrane. Then, the pore valve was triggered, creating sufficient pressure to overcome the flow resistance. The traction produced by the Laplace pressure pump (LP pump) and capillary pump [33,34] continued driving the liquid and triggering subsequent pore valves.

The microporous membrane was the key element that determined the valve status of open or closed. As shown in Figure 1f, in the retention state, the diagram illustrates the liquid state in the valve. The liquid-gas interface tension exerted an upward force on the liquid column in the micropore due to the surface tension, preventing the liquid sample from fully filling the micropores (it failed to break through the gating threshold of a pore [31]). Conversely, when the triggering agent eliminated the transmembrane pressure of the microporous membrane in the triggering state, as shown in Figure 1f, the liquid-liquid interface formed an outward tension for liquid samples due to the intermolecular force [35]. To improve the trigger efficiency, we designed a wedge-shaped capillary channel under the microporous membrane based on the feeding process of the shorebird effect [36]. When $\theta_a < \frac{\pi}{2} + \frac{\alpha}{2}$ and $\alpha \geq \theta_a - \theta_r$ (where θ_a, θ_r, and α were the advancing angle, the receding angle, and the wedge angle, respectively, as shown in Figure 1g), the liquid would fill the corner of the wedge-shaped capillary channel and flow to the slope capillary channel eventually [37,38]. Then, the wedge-shaped capillary channel and microporous membrane returned to the dry state, and liquid samples were retained again. The chamber reverted to a no-leaking vessel and started a new round of reciprocating switching.

With the advantage of the triggering ability of the multi-micropore membrane-based valve and the self-driven ability of the wedge-shaped capillary channel, the combined 3D trigger valve could gain the advantages of befitting the gating threshold and fast response and is suitable for POCT scenarios.

2.2. Device Design, Fabrication, and Surface Modification

The overall design idea of the 3D trigger valve followed the basic principles of the mesoscopic scale, which aimed to manipulate liquid samples with a volume of 5 to 200 μL, as is often used in traditional biomedical experiments using 96- or 384-well plates (174927, Thermo Fisher Scientific, Waltham, MA, USA). For that reason, the main structure of the device was designed at the millimeter geometry size. Meanwhile, the feature size of the wedge-shaped capillary channel was 1.5 mm, referring to the capillary length of the water, which is about 2.7 mm [37]. Finally, considering the feature sizes of the single cells and biological microspheres, the 3, 5, and 8 μm micropore diameters were preferred.

The 3D trigger valve was designed as a general-purpose component to achieve multi-step biochemical reaction operations in various POCT scenarios. The requirements of operational capability and duration were referred to the commonly performed biomedical experiments using 96- or 384-well plates. Since the feature sizes of the traditional 2D valves were in micrometers, they were only suitable for the operation and reaction of liquid samples ranging from nanoliter to sub-microliter scale. To meet the requirements in those experiments, we could identify the operable liquid volume of the 3D trigger valve to tens of microliters and the operation duration to tens of seconds per cycle. As shown in Figure 1b, the diameter and height of the cylindrical chamber for the liquid sample were 6 mm and 5 mm, respectively. The wedge-shaped capillary channel's length, width, and entrance height were 10, 8, and 1.5 mm, respectively. The wedge angle was 3.4°, which had the highest droplet speed because of the balance between the magnitude of the Laplace pressure and its direction [39]. The height and length of the slope capillary channel were designed to be 0.7 mm and 10 mm, respectively. For comparative study, the micropore sizes were chosen as 3 μm, 5 μm, or 8 μm, respectively. Correspondingly, the porosity of the three microporous membranes was 7.1%, 11.8%, and 15.1%, and their thickness was 9 μm, 10 μm, and 15 μm, respectively.

The 3D trigger valve was manufactured by a 3D printer (Lite 300, UnionTech, Shanghai, China) using transparent resin (11122, Mohou, Beijing, China). The polymeric microporous membranes with different pore sizes and porosities were purchased from Lanzhou Heavy Ion Research Facility (HIRFL) lab, CAS of China, where the membranes were fabricated on commercial polycarbonate (PC) membrane using a high-speed heavy ion bombardment and sodium hydroxide etching manufacturing processes.

SEM graphics of microporous membranes with different apertures were taken. We measured the diameter of multiple micropores at different positions, and the average value was obtained (i.e., the pore size of microporous membranes). The number of micropores per unit area was defined as the density of the microporous membrane (ρ), and the porosity (φ) of the microporous membrane was calculated using the formula ($\varphi = \rho\pi R^2$).

To improve the hydrophilicity and consistency of the surface characteristics of microporous membranes, we treated them with oxygen plasma for 2 min. The water contact angle of the above-modified surfaces was photographed and measured at room temperature.

2.3. *Device Performance Evaluation*

2.3.1. Evaluation of the Gating Threshold, Seepage Velocity, and Response Time

The whole device evaluation equipment consisted of a horizontal stage, a high-power microscope (HY-300X) with adjustable magnification in both vertical and horizontal directions, a camera (SONY CMOS, 1080P@60fps), a light source, and the 3D trigger valve.

As shown in Figure S1 in the Supplementary Materials, after the device was placed vertically on a horizontal table, the deionized water added a drop of red ink to facilitate the observation used in the characterization experiments. Microporous membranes with diameters of 3 μm, 5 μm, and 8 μm were attached to the side of the transparent tube with a length of 45 cm and an inner diameter of 6 mm. The stained deionized water was slowly injected from the upper side, and the gating thresholds of membranes were calculated by recording the height of the liquid before it broke through the membrane. The seepage velocity of microporous membranes was evaluated by measuring the time for the 1.4 mL liquid to seep through the infiltrated membrane. For the gating threshold, both the untreated and oxygen-plasma-treated microporous membranes were measured. The seepage velocity was measured with the treated microporous membrane. The trigger efficiency of the valve was also accessed by measuring the entire draining time of 150 μL of deionized water. Each experiment with different conditions was repeated three times.

In the transparent tube, the liquid pressure formed by the gravity of the liquid column could be equivalent to the gating threshold of the microporous membranes, which was equal to the product of density, depth, and gravity constant ($P = \rho g h$). The gating threshold of the microporous membranes could be calculated by substituting the height of the liquid column obtained by the experiment. The porous media seepage theory, dominated by surface molecular forces, could be used to calculate the seepage velocity of microporous membranes. When infiltrating the bottom of microporous membranes, the time (t) required for the liquid height to drop from 8 cm to 3 cm (liquid volume $V = \pi R^2 h$) was recorded to calculate the average seepage velocity ($u = \frac{V}{t}$).

2.3.2. Reopen and Reclose Performance Evaluation

To evaluate the 3D trigger valve's performance for the cycling switch, we repeatedly added liquid samples and triggering agents to reopen and reclose the valve and observed the volume of the liquid above the membrane. The pore size of the microporous membrane was 5 μm, and the liquid samples were three groups of 150 μL of deionized water (red ink). We recorded the change of the liquid height using a horizontally positioned microscope to assess the time response of the 3D valve, which showed how quickly the valve was turned on and off. The experiment was repeated three times.

2.3.3. Demonstration of Multi-Step Microfluidic Operations for Bio-Particle Immunostaining

In order to evaluate the capability of multi-step operations, which are usually required in most bio-detection applications, demonstrations of immunostaining of polystyrene microspheres and cells in the proposed 3D trigger valve were performed. The immunostaining process of polystyrene microspheres (PS microspheres, Tianjin BaseLine Chrom Tech Research Centre, Tianjin, China) was as follows. Two 3D microfluidic valves with 5 μm micropores were used as the experimental and control groups. The reaction reagents were 80 μL 150 μg/mL antibody solution (F9512, Sigma-Aldrich, Saint Louis, MO, USA) and 20 μL PS microspheres with a diameter of 7 μm. The difference between the experimental group and the control group was that the former used PS microspheres sealed with the 100 μg/mL IgG antigen (I4506, Sigma-Aldrich, Saint Louis, MO, USA), while the latter used PS microspheres sealed with Bovine Serum Albumin (BSA) (Thermo Fisher, Waltham, MA, USA). After 15 min for the reaction, the Phosphate Buffer Saline (PBS) (GIBCO, Life Technologies Corp., New York, NY, USA) was added to the valve as the liquid sample and triggering agent to wash repeatedly.

The immunostaining process of A549 cells (China Infrastructure of Cell Line Resource) was as follows. First, the 2.5 × 10^5/mL 50 μL A549 cell suspension was added into the chamber of valves. Then, we added a 2% 50 μL calcein AM solution (Thermo Fisher, Waltham, MA, USA) to the experimental group and 50 μL PBS to the control group. Similarly, we washed with PBS after 15 min.

After the immunostaining reaction, the PS microspheres and A549 cells were observed with the fluorescence microscope and CCD camera. The obtained images under the bright and fluorescence fields were then merged for collaborative observation.

3. Results and Discussions

3.1. Microfluidic Trigger Valve Geometry and Surface Characterization

The fabricated 3D trigger valve is characterized in Figure 2, which shows the photograph of the device and the microscopic images of micropores and capillary channels. Figure 2a,b shows the 3D trigger valve in the assembled state and parted state, which consists of a cylindrical chamber for sample loading (filled with red ink), a waste outlet, and different channels for self-driving. Combined with the operable liquid volume commonly used in the laboratory and the size of the 96-well plate, the chamber was designed with a total volume of 150 μL. The 3D valve can handle a 25-fold higher volume of liquid sample than 2D valves [27]. Figure 2c shows a typical SEM image of the microporous membranes with a 5 μm pore size. The microporous membrane was closely bonded to the device structure with the thin double-sided adhesive tape (3M55236, Shanghai, China) (Figure 1d). The thin double-sided tape was cut to the annulus with an inner diameter equal to the diameter of the cylindrical chamber and pasted at the bottom of the cylindrical chamber. The device base was stably placed on a horizontal operating platform, and the bottom of the upper chamber was glued to the device base. Figure 2d–f shows the slope capillary and wedge-shaped capillary channels with a wedge angle of around 3.4° to ensure sufficient Laplacian pressure for the liquid to be completely drained.

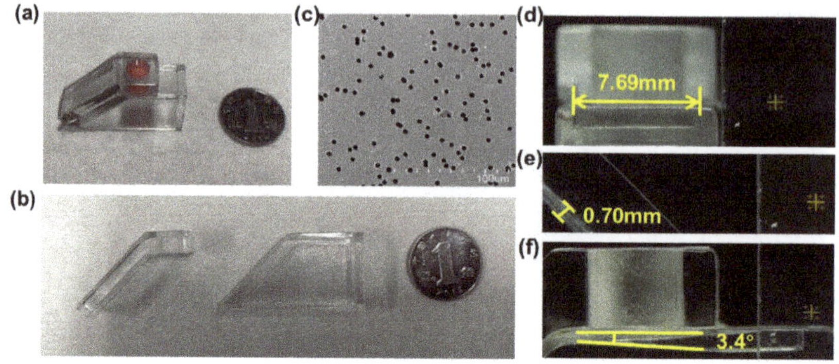

Figure 2. Fabrication of the 3D trigger valve. (**a**) Photograph of the fabricated valve in the assembled state. (**b**) Photograph of the fabricated valve in the disassembled state. (**c**) SEM image of the microporous membrane (pore size = 5 µm, porosity = 11.8%, thickness = 10 µm). (**d**) Microscopic image of the wedge-shaped capillary channel with ~8 mm in width. (**e**) Microscopic image of the slope capillary channel with 0.7 mm in height. (**f**) Microscopic image of the wedge-shaped capillary channel with 3.4° in the wedge angle.

Figure 3a–c shows the microscopic images of microporous membranes and the corresponding SEM images of a single micropore, indicating that microporous membranes' pore sizes were etched uniformly to 3 µm, 5 µm, and 8 µm, respectively. Figure 3d,e shows that the contact angle of the untreated and oxygen-plasma-treated membranes was 70° and 35°, respectively. Oxygen plasma treatment could reduce the contact angle by two times and significantly improve the hydrophilicity of microporous membranes.

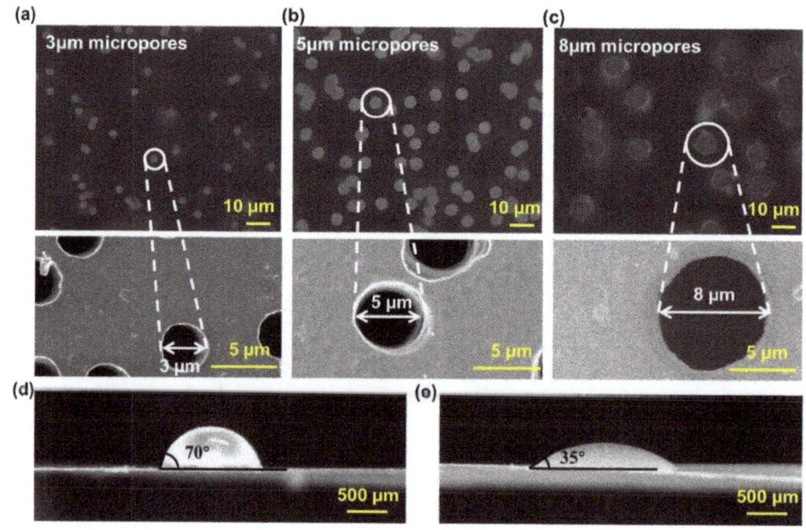

Figure 3. Morphology characterization of microporous membranes. (**a**) The micropores of 3 µm (porosity = 7.1%, thickness = 9 µm). (**b**) The micropores of 5 µm (porosity = 11.8%, thickness = 10 µm).(**c**) The micropores of 8 µm (porosity = 15.1%, thickness = 15 µm). (**d**) The 70° contact angle of the deionized water on the surface of untreated microporous membranes. (**e**) The 35° contact angle of the deionized water on the surface of pretreated microporous membranes.

3.2. Retention and Conducting Performance Evaluation

Since the pore size and porosity would affect the retention performance of the valve, the gating thresholds of the untreated and treated microporous membranes with 3 μm, 5 μm, and 8 μm pore sizes were evaluated. Figure 4a–c shows the tests of untreated membranes, with the gating threshold of 3959.2 Pa for 3 μm pores, 3482.3 Pa for 5 μm pores, and 2080.9 Pa for 8 μm pores. The obtained results showed that with the decrease in the pore size and porosity, the gating threshold increased significantly to require more pressure to trigger, as shown in Figure 3d. According to previous reports, the gating threshold of deionized water in the 2D valve with a 300 μm radius was 420 Pa [31]. However, the gating threshold of the 3D trigger valve was about 3500 Pa due to the dramatic increase in the gating threshold of a single micropore. These advantages ensured the margin and reliability of the 3D trigger valve.

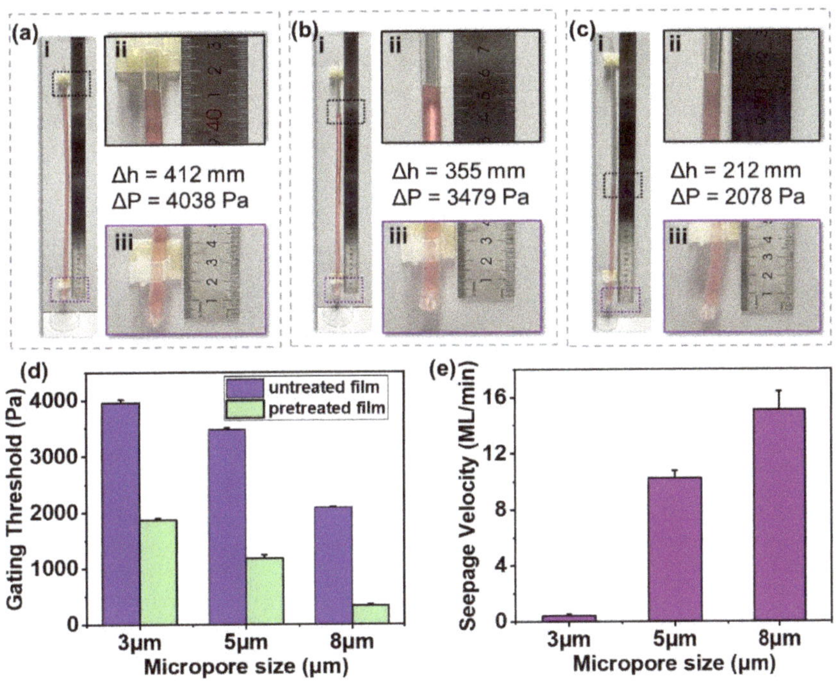

Figure 4. Retention and conducting performances of microporous membranes with the high gating threshold. (**a**) The retention performance of microporous membranes with the 3 μm pore size. (**b**) The retention performance of the microporous membrane with the 5 μm pore size. (**c**) The retention performance of the microporous membrane with the 8 μm pore size. (**d**) Gating thresholds of untreated and pretreated microporous membranes with different pore sizes. (**e**) Seepage velocity of pretreated microporous membranes with different pore sizes.

On the other hand, the contact angle would also affect the retention performance of the valve. As shown in Figure 4d, the gating threshold of the untreated microporous membrane with a 5 μm pore size was 3482.3 Pa, while that of the oxygen-plasma-treated microporous membrane with a 5 μm pore size was 1179.0 Pa. Hydrophilic treatment could reduce the gating threshold of microporous membranes by a factor of 3.

Figure 4e shows the tests of oxygen-plasma-treated microporous membranes with different pore sizes, with the seepage velocity of 0.40 mL/min for 3 μm pores, 10.21 mL/min for 5 μm pores, and 15.08 mL/min for 8 μm pores. The seepage velocity was positively correlated with the pore size and porosity. Compared to 2D single-pore trigger valves depending only on the capillary force, the 3D trigger valve had high-density micropores (~$1 \times 10^6/\text{cm}^2$), which dramatically improved the conducting performance.

3.3. Trigger Performance Evaluation

To evaluate the trigger efficiency of the 3D trigger valve, we recorded the trigger times (total draining time of 150 μL of deionized water added with a drop of red ink) with different pore sizes and illustrated them in Figure 5. Trigger efficiency referred to the duration for draining the entire liquid sample, which embodied the conductivity of the 3D valve. In this paper, the cylindrical chamber volume of the device was taken as the standard to evaluate and compare the time required to drain the 150 μL of deionized water totally. The shorter the trigger time, the higher the trigger efficiency. The valve could be triggered for pretreated microporous membranes with a pore size of 5 μm or above within 6 s (MOV S1). However, it was necessary to continuously supply a large volume of trigger agent to achieve complete drainage for microporous membranes with a small pore size (≤3 μm) due to the decrease in the seepage velocity. For untreated microporous membranes, the valve with 3 μm micropores could not be triggered, and the valve with 5 μm micropores increased the trigger time to 13 s, while the valve with 8 μm micropores was unaffected. Therefore, to ensure a sufficient gating threshold and improve the trigger efficiency as much as possible, the oxygen-plasma-treated microporous membrane with a 5 μm aperture was used in the subsequent experiments.

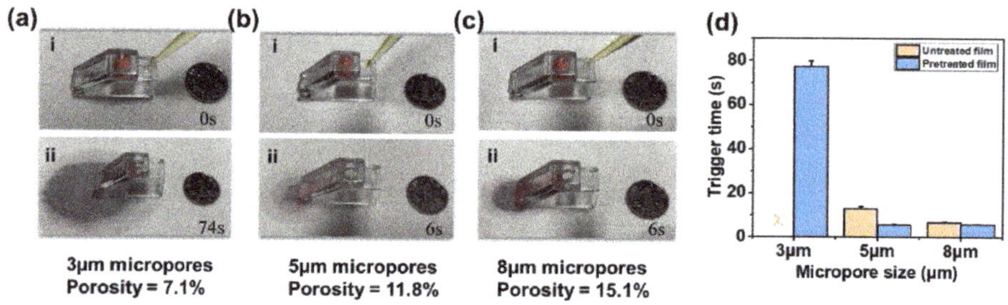

Figure 5. Trigger performances of the 3D trigger valve with different gating thresholds. (**a**) The trigger performance of the valve with the 3 μm pore size. (**b**) The trigger performance of the valve with the 5 μm pore size. (**c**) The trigger performance of the valve with the 8 μm pore size. (**d**) Trigger performance of untreated and pretreated microporous membranes with different pore sizes.

3.4. Reopen and Reclosed Evaluation of the Valve

The reopening and reclosing ability of the 3D trigger valve with a 5 μm microporous membrane was evaluated, as shown in Figure 6a. The liquid height in the chamber changed in cycles, as shown in Figure 6b and MOV S2. The test results showed that the repetition could be at least three times. In the repeatable retention-trigger cycle, the repetition trigger time was 6 s, and the valve returned to the retention state (drained state) within 60 s, as shown in Figure 6c. Additionally, the 3D trigger valve could be retained without the triggering agent for at least 30 min unless affected by evaporation (Figure S2), while the 2D valve was only cut off for a maximum of 5 min [22]. The results indicated that the 3D trigger valve had the potential for repeatable triggering for multi-step operations.

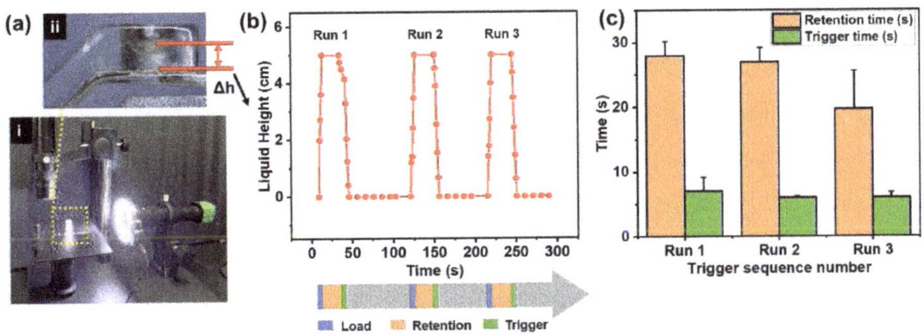

Figure 6. Performance characterization of the 3D trigger valve with repeatable retention-trigger cycle. (**a**) Setup diagram. (**b**) Liquid height variation in three times repeatable retention-trigger cycle. (**c**) The time required for three repetitions to retain and trigger.

3.5. Triggering Ability for Multi-Step Microfluidic Operations in Bio-Particle Immunostaining Applications

Based on the advantage of repeatable triggering for multi-step operations in the device, we carried out multi-step microfluidic operations and POCT applications. Bio-particle immunostaining refers to the common particle fluorescence phenomenon in bio-detection applications, which often requires multi-cycling washing operations to reduce the interference of background fluorescence. In this paper, the immunostaining experiments of polystyrene microspheres and cells were taken as instances to demonstrate multi-cycling staining and washing operations on the 3D valve. Figure 7a shows the procedure of multi-step microfluidic operations (e.g., sample loading, keeping and reacting, trigger to drainage, and washing), which were performed to mimic the major procedures in a typical biochemical reaction. Two 3D trigger valves with 5 μm micropores were each used as the experimental and the control groups (Figure 7b). The major procedures of PS microsphere and cell line immunostaining are shown in Figure 7c,e. After each step, the bright field microscopic images and fluorescent images were captured and are shown in Figure 7d–f. As shown in Figure 7d, the PS microspheres modified by antigen had no fluorescence phenomenon (Figure 7d-ii). No fluorescence was observed when the antibody was reacted with PS microspheres modified with BSA (Figure 7d-v). The fluorescence phenomenon was only observed in the case of the reaction of antibodies and antigen-modified PS microspheres due to the specific binding (Figure 7d-viii). All antigen-modified PS microspheres in the field showed specific binding.

Figure 7e shows the multi-step procedure of live cell staining. A549 cells had no fluorescence under normal conditions (Figure 7f-ii), while they showed green fluorescence after calcein AM staining (Figure 7f-v). The merged image shown in Figure 7f-vi illustrates that all cells in the field were alive, verifying the precise control and nontoxicity of the valve to live cells. Through the staining experiments of PS microspheres and live cells, we found that the device could accurately manipulate the liquid. Multi-step operations are common procedures for many biological detections, such as pregnancy detection and COVID-19 virus detection, which include the steps of sample loading, reaction on micro-particles, and washing [40]. We demonstrated the applicability of the 3D trigger valve to multi-step operations in microfluidic experiments of micro-particles (within 20 min), which had the potential to be applied to POCT.

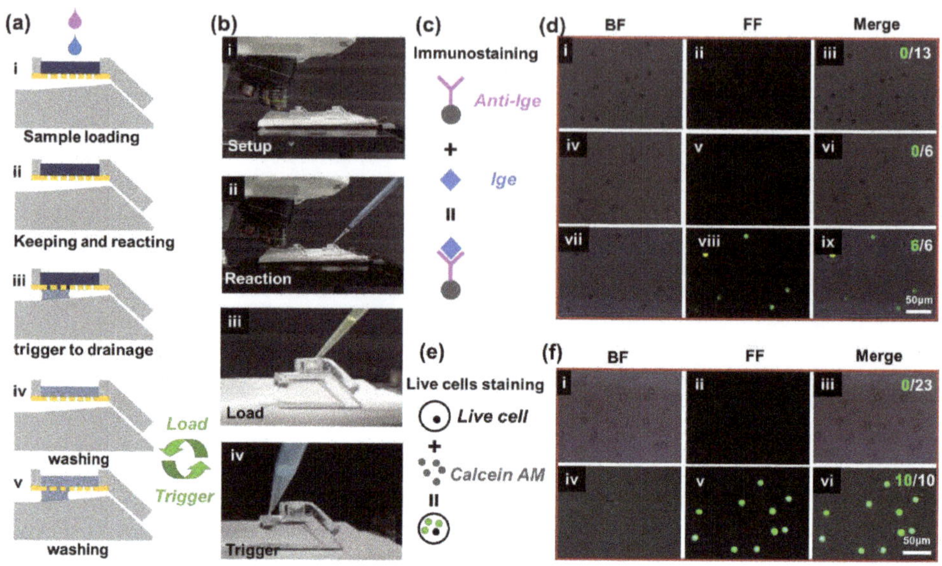

Figure 7. The results of biochemical reactions with multi-step operations. (**a**) The procedure of multi-step chain biochemical reactions. (**b**) Experimental equipment and operation flow chart. (**c**) The principle of immunostaining. (**d**) Staining reaction results of immunostaining. (**e**) The principle of live cell staining. (**f**) Staining reaction results of live cells.

Compared with typical 2D trigger valves, the advantages of the 3D valve we propose could be briefly summarized as follows: (1) High gating threshold: 2D valves have a lower gating threshold of 443 Pa [27], while the 3D valve we propose has a higher gating threshold of 3500 Pa (5 μm pore size). The 10-fold higher gating threshold indicates that the device has a higher margin and makes it hard for liquids to leak. (2) High operable liquid volume: 2D valves can only handle 6 μL or an even smaller amount of liquid sample [27], while the 3D valve we proposed can handle a 150 μL liquid sample. The 25-fold larger volume indicates that the device can accommodate more liquid samples. (3) High conductivity: The conductivity of 2D valves is only 10–15 μL/min [1], while the 3D valve we propose can drain the 150 μL of liquid within 6 s after triggering, and the trigger performance is stable, which indicates that the device can handle more liquid samples. (4) Long retention time: 2D valve can only be cut off for a maximum of 5 min [22], while the 3D valve we proposed can keep the cut-off for hours (without considering atmospheric evaporation). It indicates that the device has the capability for temporary sample storage. (5) Multi-cycling reopening and reclosing capability: The 3D valve we proposed can be switched on and off repeatedly more than three times within 5 min, while this parameter was rarely reported in traditional 2D valves. It indicates that the device can be used for multi-step reactions.

4. Conclusions

In this study, a new 3D trigger valve was proposed and evaluated based on the theory of Laplace pressure. The valve showed a high gating threshold (~1000 Pa), high trigger efficiency (~6 s), and a large operable liquid volume. The device also showed the capability for repeatable triggering for multi-step operations. Furthermore, the valve was implemented in the staining experiments of PS microspheres and live cells. The capillary-driven 3D trigger valve was implemented for multi-step biochemical reactions, and the obtained results indicated that the device had the potential to be used for POCT applications without an external driving force.

Supplementary Materials: The following supporting information can be downloaded at: https://www.mdpi.com/article/10.3390/bios13010026/s1, Figure S1 shows the device for evaluating the gating threshold. Figure S2 shows the retention performance of the 3D trigger valve. MOV S1 showed the trigger process of the valve with the 5 μm pore size. MOV S2 showed the repeatable retention-trigger cycle of the 3D trigger valve.

Author Contributions: Conceptualization, Y.Z. (Yijun Zhang) and Y.Z. (Yang Zhao); Data curation, Y.Z. (Yijun Zhang) and Y.Z. (Yang Zhao); Formal analysis, Y.Z. (Yijun Zhang); Funding acquisition, Y.F., C.H., and Y.Z. (Yang Zhao); Investigation, Y.Z. (Yijun Zhang); Methodology, Y.Z. (Yijun Zhang); Project administration, C.H. and Y.Z. (Yang Zhao); Resources, Y.L. and X.L. (Xiaofeng Luan); Software, Y.L., X.L. (Xiaofeng Luan), X.L. (Xin Li), J.J. and M.L.; Supervision, C.H. and Y.Z. (Yang Zhao); Validation, Y.Z. (Yijun Zhang); Visualization, Y.Z. (Yijun Zhang); Writing—original draft, Y.Z. (Yijun Zhang); Writing—review & editing, Y.Z. (Yijun Zhang), C.H., L.Z. and Y.Z. (Yang Zhao). All authors have read and agreed to the published version of the manuscript.

Funding: This research was funded by the National Key Research and Development Program of China, grant number 2018YFC2001100, the National Natural Science Foundation of China, grant numbers 62171441 and 61904198, the Scientific Research and Equipment Development Project of CAS, grant number YJKYYQ20210031, and the Youth Innovation Promotion Association of Chinese Academy of Sciences.

Institutional Review Board Statement: Not applicable.

Informed Consent Statement: Not applicable.

Data Availability Statement: Not applicable.

Acknowledgments: This work was supported by the National Key Research and Development Program of China (No. 2018YFC2001100), the National Natural Science Foundation of China (No. 62171441, No. 61904198), the Scientific Research and Equipment Development Project of CAS (YJKYYQ20210031), and the Youth Innovation Promotion Association of Chinese Academy of Sciences.

Conflicts of Interest: The authors declare no competing financial interests.

References

1. Yafia, M.; Ymbern, O.; Olanrewaju, A.O.; Parandakh, A.; Sohrabi Kashani, A.; Renault, J.; Jin, Z.; Kim, G.; Ng, A.; Juncker, D. Microfluidic chain reaction of structurally programmed capillary flow events. *Nature* **2022**, *605*, 464–469. [CrossRef]
2. Adams, A.A.; Charles, P.T.; Veitch, S.P.; Hanson, A.; Deschamps, J.R.; Kusterbeck, A.W. REMUS100 AUV with an integrated microfluidic system for explosives detection. *Anal. Bioanal. Chem.* **2013**, *405*, 5171–5178. [CrossRef] [PubMed]
3. Gervais, L.; Hitzbleck, M.; Delamarche, E. Capillary-driven multiparametric microfluidic chips for one-step immunoassays. *Biosens. Bioelectron.* **2011**, *27*, 64–70. [CrossRef] [PubMed]
4. Li, Y.; Jiao, X.; Du, X.; Wang, F.; Wei, Q.; Wen, Y.; Zhang, X. Wettability alteration in a functional capillary tube for visual quantitative point of care testing. *Analyst* **2018**, *143*, 3001–3005. [CrossRef] [PubMed]
5. Scanlon, T.C.; Dostal, S.M.; Griswold, K.E. Erratum for "A high-throughput screen for antibiotic drug discovery" (Volume 111, issue 2, pp. 232–243). *Biotechnol. Bioeng.* **2019**, *116*, 475. [CrossRef] [PubMed]
6. Tachibana, H.; Saito, M.; Shibuya, S.; Tsuji, K.; Miyagawa, N.; Yamanaka, K.; Tamiya, E. On-chip quantitative detection of pathogen genes by autonomous microfluidic PCR platform. *Biosens. Bioelectron.* **2015**, *74*, 725–730. [CrossRef] [PubMed]
7. Xu, R.H.; Wei, W.; Krawczyk, M.; Wang, W.; Luo, H.; Flagg, K.; Yi, S.; Shi, W.; Quan, Q.; Li, K.; et al. Circulating tumour DNA methylation markers for diagnosis and prognosis of hepatocellular carcinoma. *Nat. Mater.* **2017**, *16*, 1155–1161. [CrossRef]
8. Zhao, M.; Nelson, W.C.; Wei, B.; Schiro, P.G.; Hakimi, B.M.; Johnson, E.S.; Anand, R.K.; Gyurkey, G.S.; White, L.M.; Whiting, S.H.; et al. New generation of ensemble-decision aliquot ranking based on simplified microfluidic components for large-capacity trapping of circulating tumor cells. *Anal. Chem.* **2013**, *85*, 9671–9677. [CrossRef]
9. Zhang, C.; Xing, D.; Li, Y. Micropumps, microvalves, and micromixers within PCR microfluidic chips: Advances and trends. *Biotechnol. Adv.* **2007**, *25*, 483–514. [CrossRef] [PubMed]
10. Achille, C.; Parra-Cabrera, C.; Dochy, R.; Ordutowski, H.; Piovesan, A.; Piron, P.; Van Looy, L.; Kushwaha, S.; Reynaerts, D.; Verboven, P.; et al. 3D Printing of Monolithic Capillarity-Driven Microfluidic Devices for Diagnostics. *Adv. Mater.* **2021**, *33*, e2008712. [CrossRef]
11. Patil, Y.; Dotseth, K.; Shapiro, T.; Pushparajan, D.; Binderup, S.; Horn, J.R.; Korampally, V. Modular design of paper based switches for autonomous lab-on paper micro devices. *Biomed. Microdevices* **2020**, *23*, 1. [CrossRef] [PubMed]

12. Soum, V.; Park, S.; Brilian, A.I.; Kwon, O.S.; Shin, K. Programmable Paper-Based Microfluidic Devices for Biomarker Detections. *Micromachines* **2019**, *10*, 516. [CrossRef] [PubMed]
13. Lai, C.C.; Chung, C.K. Numerical analysis and experiment of high-efficiency long-term PDMS open-surface mixing chip. *J. Micromech. Microeng.* **2019**, *29*, 075003. [CrossRef]
14. Menges, J.; Meffan, C.; Dolamore, F.; Fee, C.; Dobson, R.; Nock, V. New flow control systems in capillarics: Off valves. *Lab Chip* **2021**, *21*, 205–214. [CrossRef]
15. Safavieh, R.; Juncker, D. Capillarics: Pre-programmed, self-powered microfluidic circuits built from capillary elements. *Lab Chip* **2013**, *13*, 4180–4189. [CrossRef]
16. Zhang, Y.; Chen, Y.; Huang, J.; Liu, Y.; Peng, J.; Chen, S.; Song, K.; Ouyang, X.; Cheng, H.; Wang, X. Skin-interfaced microfluidic devices with one-opening chambers and hydrophobic valves for sweat collection and analysis. *Lab Chip* **2020**, *20*, 2635–2645. [CrossRef]
17. Man, P.F.; Mastrangelo, C.H.; Burns, M.A.; Burke, D.T. Microfabricated capillarity-driven stop valve and sample injector. In Proceedings of the MEMS 98. IEEE. Eleventh Annual International Workshop on Micro Electro Mechanical Systems. An Investigation of Micro Structures, Sensors, Actuators, Machines and Systems (Cat. No.98CH36176), Heidelberg, Germany, 25–29 January 1998; pp. 45–50.
18. Barman, U.; Lagae, L.; Jones, B. Capillary stop valve actuation by thermo-pneumatic- pressure for lab-on-chip systems. *Microsyst. Technol.* **2020**, *27*, 681–692. [CrossRef]
19. Chu, K.C.; Tsao, H.K.; Sheng, Y.J. Pressure-gated capillary nanovalves based on liquid nanofilms. *J. Colloid Interface Sci.* **2020**, *560*, 485–491. [CrossRef]
20. Guo, W.; Hansson, J.; van der Wijngaart, W. Capillary pumping independent of the liquid surface energy and viscosity. *Microsyst. Nanoeng.* **2018**, *4*, 2. [CrossRef] [PubMed]
21. Zhang, H.; Tran, H.H.; Chung, B.H.; Lee, N.Y. Solid-phase based on-chip DNA purification through a valve-free stepwise injection of multiple reagents employing centrifugal force combined with a hydrophobic capillary barrier pressure. *Analyst* **2013**, *138*, 1750–1757. [CrossRef]
22. Zimmermann, M.; Hunziker, P.; Delamarche, E. Valves for autonomous capillary systems. *Microfluid. Nanofluid.* **2008**, *5*, 395–402. [CrossRef]
23. Chen, X.; Chen, S.; Zhang, Y.; Yang, H. Study on Functionality and Surface Modification of a Stair-Step Liquid-Triggered Valve for On-Chip Flow Control. *Micromachines* **2020**, *11*, 690. [CrossRef] [PubMed]
24. Ghosh, S.; Aggarwal, K.; Vinitha, T.U.; Nguyen, T.; Han, J.; Ahn, C.H. A new microchannel capillary flow assay (MCFA) platform with lyophilized chemiluminescence reagents for a smartphone-based POCT detecting malaria. *Microsyst. Nanoeng.* **2020**, *6*, 5. [CrossRef]
25. Lai, C.C.; Chung, C.K. Facile design and fabrication of capillary valve for mixing using two-step PDMS moulding. *Micro. Nano. Lett.* **2018**, *13*, 1408–1411. [CrossRef]
26. Zhang, L.; Jones, B.; Majeed, B.; Nishiyama, Y.; Okumura, Y.; Stakenborg, T. Study on stair-step liquid triggered capillary valve for microfluidic systems. *J. Micromech. Microeng.* **2018**, *28*, 065005. [CrossRef]
27. Olanrewaju, A.O.; Robillard, A.; Dagher, M.; Juncker, D. Autonomous microfluidic capillaric circuits replicated from 3D-printed molds. *Lab Chip* **2016**, *16*, 3804–3814. [CrossRef]
28. Bhargava, K.C.; Thompson, B.; Malmstadt, N. Discrete elements for 3D microfluidics. *Proc. Natl. Acad. Sci. USA* **2014**, *111*, 15013–15018. [CrossRef]
29. Olanrewaju, A.; Beaugrand, M.; Yafia, M.; Juncker, D. Capillary microfluidics in microchannels: From microfluidic networks to capillaric circuits. *Lab Chip* **2018**, *18*, 2323–2347. [CrossRef]
30. Papadimitriou, V.A.; Segerink, L.I.; van den Berg, A.; Eijkel, J.C.T. 3D capillary stop valves for versatile patterning inside microfluidic chips. *Anal. Chim. Acta.* **2018**, *1000*, 232–238. [CrossRef]
31. Chen, J.M.; Chen, C.-Y.; Liu, C.-H. Pressure Barrier in an Axisymmetric Capillary Microchannel with Sudden Expansion. *Jpn. J. Appl. Phys.* **2008**, *47*, 1683–1689. [CrossRef]
32. Hou, X.; Hu, Y.; Grinthal, A.; Khan, M.; Aizenberg, J. Liquid-based gating mechanism with tunable multiphase selectivity and antifouling behaviour. *Nature* **2015**, *519*, 70–73. [CrossRef] [PubMed]
33. Mahlberg, L.; Hermann, M.; Ramsay, H.; Salomons, T.; Stamplecoskie, K.; Oleschuk, R.D. Portable microfluidic platform employing Young–Laplace pumping enabling flowrate controlled applications. *Microfluid. Nanofluid.* **2021**, *25*, 48. [CrossRef]
34. Vasilakis, N.; Papadimitriou, K.I.; Morgan, H.; Prodromakis, T. High-performance PCB-based capillary pumps for affordable point-of-care diagnostics. *Microfluid. Nanofluid.* **2017**, *21*, 103. [CrossRef]
35. Biot, M.A. General Theory of Three-Dimensional Consolidation. *J. Appl. Phys.* **1941**, *12*, 155–164. [CrossRef]
36. Prakash, M.; Quere, D.; Bush, J.W. Surface tension transport of prey by feeding shorebirds: The capillary ratchet. *Science* **2008**, *320*, 931–934. [CrossRef] [PubMed]
37. Luo, C.; Heng, X.; Xiang, M. Behavior of a liquid drop between two nonparallel plates. *Langmuir* **2014**, *30*, 8373–8380. [CrossRef]
38. Zheng, Y.; Bai, H.; Huang, Z.; Tian, X.; Nie, F.Q.; Zhao, Y.; Zhai, J.; Jiang, L. Directional water collection on wetted spider silk. *Nature* **2010**, *463*, 640–643. [CrossRef] [PubMed]

39. Son, J.; Bae, G.Y.; Lee, S.; Lee, G.; Kim, S.W.; Kim, D.; Chung, S.; Cho, K. Cactus-Spine-Inspired Sweat-Collecting Patch for Fast and Continuous Monitoring of Sweat. *Adv. Mater.* **2021**, *33*, e2102740. [CrossRef] [PubMed]
40. Stokes, W.; Berenger, B.M.; Venner, A.A.; Deslandes, V.; Shaw, J.L.V. Point of care molecular and antigen detection tests for COVID-19: Current status and future prospects. *Expert Rev. Mol. Diagn.* **2022**, *22*, 797–809. [CrossRef]

biosensors

MDPI

Article

Development of an Open Microfluidic Platform for Oocyte One-Stop Vitrification with Cryotop Method

Shu Miao [1,†], Chenxi Guo [2,†], Ze Jiang [3,†], Hao-Xiang Wei [3], Xin Jiang [1,*], Jingkai Gu [2], Zhuo Hai [2], Tianren Wang [2,*] and Yun-Hui Liu [3]

1 School of Mechanical Engineering and Automation, Harbin Institute of Technology, Shenzhen 518055, China
2 Shenzhen Key Laboratory of Fertility Regulation, Reproductive Medicine Center, The University of Hong Kong-Shenzhen Hospital, Shenzhen 518005, China
3 The T Stone Robotics Institute, Department of Mechanical and Automation Engineering, The Chinese University of Hong Kong, Hong Kong SAR, China
* Correspondence: x.jiang@ieee.org (X.J.); wtrcmu@126.com (T.W.)
† These authors contributed equally to this work.

Abstract: Oocyte vitrification technology is widely used for assisted reproduction and fertility preservation, which requires precise washing sequences and timings of cryoprotectant agents (CPAs) treatment to relieve the osmotic shock to cells. The gold standard Cryotop method is extensively used in oocyte vitrification and is currently the most commonly used method in reproductive centers. However, the Cryotop method requires precise and complex manual manipulation by an embryologist, whose proficiency directly determines the effect of vitrification. Therefore, in this study, an automatic microfluidic system consisting of a novel open microfluidic chip and a set of automatic devices was established as a standardized operating protocol to facilitate the conventional manual Cryotop method and minimize the osmotic shock applied to the oocyte. The proposed open microfluidic system could smoothly change the CPA concentration around the oocyte during vitrification pretreatment, and transferred the treated oocyte to the Cryotop with a tiny droplet. The system better conformed to the operating habits of embryologists, whereas the integration of commercialized Cryotop facilitates the subsequent freezing and thawing processes. With standardized operating procedures, our system provides consistent treatment effects for each operation, leading to comparable survival rate, mitochondrial membrane potential (MMP) and reactive oxygen species (ROS) level of oocytes to the manual Cryotop operations. The vitrification platform is the first reported microfluidic system integrating the function of cells transfer from the processing chip, which avoids the risk of cell loss or damage in a manual operation and ensures the sufficient cooling rate during liquid nitrogen (LN_2) freezing. Our study demonstrates significant potential of the automatic microfluidic approach to serve as a facile and universal solution for the vitrification of various precious cells.

Keywords: open microfluidic chip; cell manipulation; vitrification

1. Introduction

Oocyte cryopreservation is an effective technique widely used in human-assisted reproductive technology and fertility preservation [1]. Through collecting and preserving oocytes at a young age, the cryopreservation technique preserves the dramatically declined ability of oocytes to fertilize and develop with age and allows women to delay pregnancy until they wish to have a child [1–3]. Since Chen et al. reported the first successful cryopreservation of human oocytes in 1986 [4], considerable efforts have been paid to address the challenges of mature oocytes cryopreservation, including the large cell volume, small relative surface area, high water content, special intracellular contents and the presence of meiotic spindles [1,5–8].

Oocyte cryopreservation technology can be categorized into slow-rate freezing and vitrification according to different freezing principles [9–12]. In a vitrification process, the cells are rapidly cooled down by direct or indirect contact with liquid nitrogen (LN_2), leading to a glass-like solidification of biological samples in the absence of ice crystals formation [13]. Currently, vitrification is the most commonly used method for the cryopreservation of human oocytes due to improved survival and pregnancy rates. The most commonly used vitrification method in clinical practice is the Cryotop method [5,14], as shown in Figure 1. This operation mode has the advantages of high freezing efficiency and convenient operation, and is accepted by most embryologists as the gold standard. During Cryotop method for vitrification, the washing sequences and timings for each CPA (including equilibrium solution (ES) and vitrification freezing solution (VS)) treatment should be precisely controlled to avoid the overexposure of cells to CPAs, which will cause irreversible osmotic damage to oocytes. The clinical operation of the Cryotop method is manually performed by skilled embryologists within 15–20 min; however, the failure of manual vitrification still occurs due to cell loss, mechanical and osmotic damage etc. [15]. It is necessary to develop a standardized protocol for oocyte vitrification based on the Cryotop method, which is more reliable and minimizes the influence of operators.

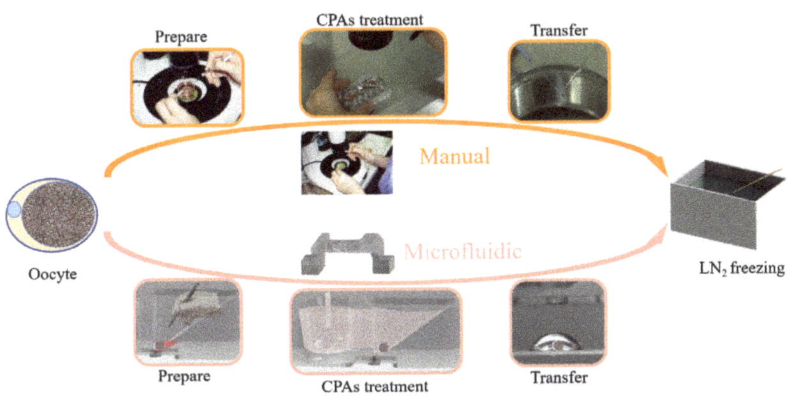

Figure 1. Schematic showing manual and microfluidic vitrification approaches. Vitrification involves multiple steps of cell pick-and-place before freezing in liquid nitrogen.

Microfluidic systems have been demonstrated as promising solutions for various applications in assisted reproduction, such as cumulus–oocyte complexes (COCs) removal, sperm screening, in vitro fertilization and embryo culture [16–19]. Through precisely manipulating the fluids at the microscale, microfluidic systems can realize complex fluid behaviors such as co-flow diffusion, droplet generation, and concentration gradient generation using different channel designs [20–24]. For cryopreservation of oocytes and embryos [6,25–32], two categories of microfluidic platforms—confinement devices and free-flow devices—have been developed to simplify the process of CPAs loading and unloading [33].

Confinement devices use various mechanisms (e.g., physical obstruction, pressure difference or electrowetting) to manipulate oocytes/embryos [6,25–30,32] and provide a temporal concentration gradient of CPAs around the cells to complete vitrification pretreatment. Guo et al. designed a curved-channel and microcolumn array microfluidic device to immobilize the oocyte and linearly load CPAs under diffusion, which reduced the permeability damage caused by sudden changes in CPAs concentration [6]. Tirgar et al. fabricated a microchannel with a width smaller than the embryo size to limit the movement of the embryo, and integrated a capillary pump to introduce the CPAs instead of the microsyringe pump, which simplified the operation process [28]. Pyne et al. developed the EWOD platform and used the digital microfluidic method to complete the vitrification pretreatment of embryos [26]. In a confinement device, the CPAs surrounding the oocyte can be controlled throughout the loading and unloading process through diffusion, enabling operators to perform experiments with concentration profiles, which can be hardly achieved with manual vitrification.

Nevertheless, the fine confinement structure causes considerable difficulties in chip fabrication and packaging. In contrast to confinement devices, free-flowing devices do not immobilize cells, and instead construct multiple parallel channels with adjacent channels and control the environment of CPAs around cells through diffusion under laminar flow [31]. These devices introduce oocytes/embryos and CPAs into parallel channels and allow CPAs to freely diffuse into cells through long serpentine channels. Despite the simpler fabrication process and more stable concentration change compared with confinement devices, it is difficult to correct the position of the oocyte/embryo in the channel of free-flowing devices and the operation period is much longer. It should be noted that most of the existing confinement and free-flow devices are closed systems, which means that the oocyte/embryo will be put into LN_2, accompanying the bulky chips. As a result, the cooling rate of the cells will be dramatically decreased and may not reach the requirement for vitrification. Therefore, it is critical to construct a microfluidic platform that enables transfer of the cell with droplet volume to Cryotop accurately with sufficiently high cooling rate.

Herein, we report a convenient, universal, user-friendly and high-efficiency microfluidic system for oocytes vitrification, which consists of an open microfluidic chip and a companion device. The design of the open microfluidic chip is based on the Cryotop method, using a confinement structure without packaging and integrating the transfer channel and capillary valve, which reduces the difficulty of fabrication and operation. Compared to the closed systems, our open chip (similar to a Petri dish) is more convenient and more consistent with the operating habits of embryologists. This system provides a simplified and standardized vitrification process in which the operator needs only to load the oocyte into the designated area of the chip and operate the companion device to obtain the same vitrification pretreatment effect as the Cryotop method. Compared with manual operation, the open microfluidic system avoids the risk of cell loss or damage caused by frequent oocyte movement, the loading of CPAs in a concentration gradient reduces osmotic damage and the standardized process eliminates uncertainty of manual operation. In this study, mouse oocytes were selected as experimental candidates and vitrified through the manual and microfluidic protocols, which were then examined after a standard thawing process. The results showed no significant difference in survival rate and quality (characterized by MMP and ROS level) among thawed oocytes treated by the two methods, demonstrating the validity of our system.

In short, to achieve the platform for oocyte vitrification, the contributions of this research work can be summarized as follows: (1) a newly designed prototype based on the clinical vitrification Cryotop method, featuring an embryologist-centered configuration; (2) a novel open microfluidic chip based on convection–diffusion and surface tension, validated through numerical simulation and experiments; (3) evaluation of system reliability based on the survival rate, MMP and ROS level of mouse oocytes.

2. Materials and Methods

2.1. Device Design and Fabrication

During oocyte vitrification, the main challenge is to ensure the quality of the CPAs exchange and the manipulation of the oocyte. The high-efficiency CPAs exchange directly affects the survival rate and development rate of oocytes. Compared with the traditional closed microfluidic, the open microfluidic is more user-friendly and has a simpler manufacturing process. In this work, an open microfluidic chip with the companion device was developed to achieve oocyte vitrification pretreatment and transfer the oocyte to the Cryotop after pretreatment (Figure 2). The open microfluidic chip (Figure 3) consisted of two independent chambers (left: solution exchange chamber, right: oocyte chamber), a capillary gap, a transfer channel and a capillary valve. The solution exchange chamber was a cylindrical shape with a radius of 1.5 mm and a depth of 2.2 mm, designed for the loading and unloading of CPAs. The oocyte chamber was a 2.5 mm depth square conical shape, with 4 mm (length) × 1.2 mm (width) on the upper surface and 0.4 mm (length) × 0.3 mm (width) on the bottom, designed for positioning and placing the oocyte. The capillary gap, having dimensions of 40 μm (width) × 560 μm (length) × 2.2 mm (depth), as a bridge, connected the chambers on both sides to achieve the solution exchange by spontaneous capillary flow (SCF) and prevent the oocyte (diameters in the range of 70 to 100 μm) from being lost in the process of treatment [34,35]. The oocyte chamber (2.5 mm) was deeper than the capillary gap (2.2 mm) and ensured that a certain volume of liquid would be contained around the oocyte to prevent osmotic damage caused by evaporation after the solution unloading. The transfer channel with a radius of 100 μm was designed to connect the oocyte chamber, and the capillary valve (upper part: radius 100 μm, depth 300 μm; bottom part: radius 150 μm, depth 200 μm). In microfluidic chips, the influence of gravity and inertia is reduced at the microscale, and the influence of surface tension becomes significant and even dominates under certain conditions [36]. A capillary valve was designed to temporarily block the transfer channel, in which the liquid was simply stopped by the surface tension caused by a sudden expansion of the microchannel cross-section. After vitrification pretreatment of the oocyte, the capillary valve was opened by the introduction of positive pressure.

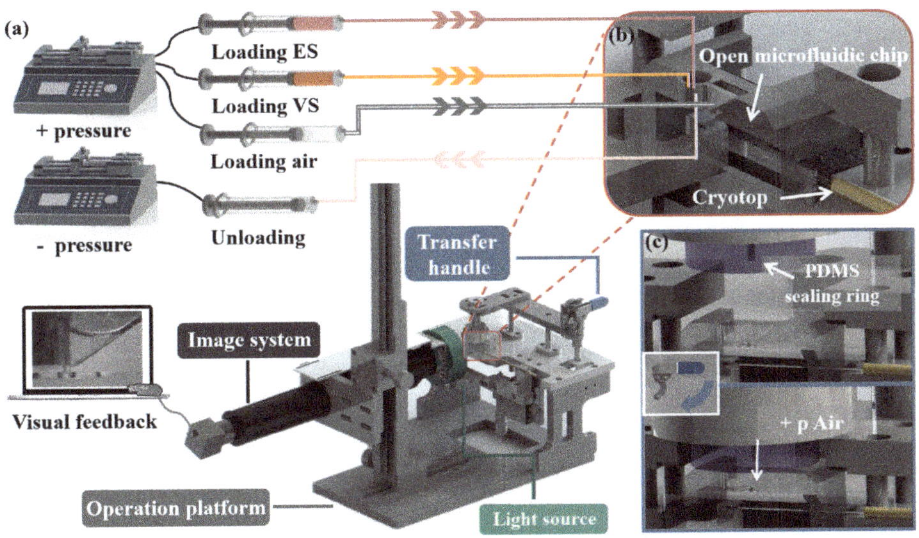

Figure 2. The structure diagram of the vitrification system and working principle of system. (**a**) The vitrification system composed of the open microfluidic chip and the companion system (syringe pumps, operation platform, transfer system and the imaging system). The operator manual placed the oocyte into the chip, completed the vitrification pretreatment and transferred with the assistance of the companion system. The real-time internal conditions of the chip on the whole operation process were presented to the operator by visual feedback. (**b**) Above all, Cryotop and the open microfluidic chip containing the oocyte were fixed on the operation platform. The syringes used for loading (ES, VS and air) and unloading were, respectively, connected to the chambers on both sides of the chip and PDMS sealing ring with soft silicone tubes. (**c**) Schematic diagram of the transfer process. After vitrification pretreatment, the operator pressed the transfer handle, and the PDMS sealing ring was pressed on the upper surface of the open microfluidic chip to form a hermetic space. The oocyte was transferred to Cryotop by external pressure.

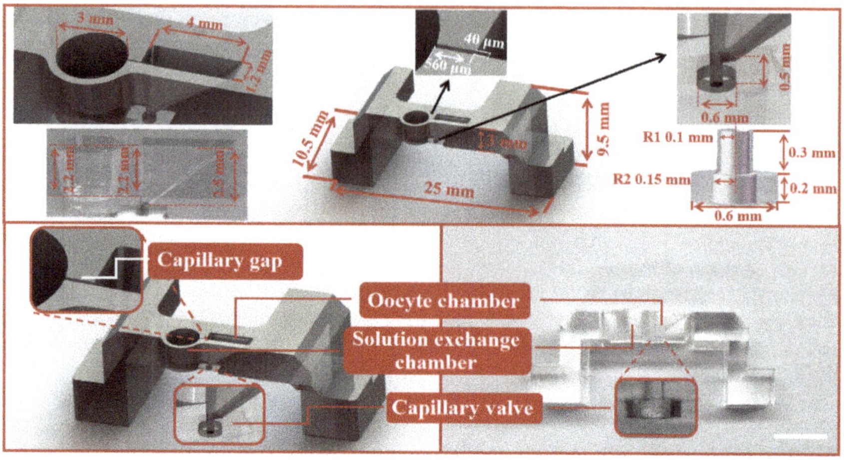

Figure 3. The dimension detail, structure diagram and physical diagram of the open microfluidic chip, composed of capillary gap, oocyte chamber, solution exchange chamber and capillary valve. Scale bars are 2 mm.

The companion device consisted of two syringe pumps (LSP02, LONGER, Baoding, China), the operation platform, the transfer system (includes PDMS sealing ring and a transfer handle) and the imaging system (Figure 2a). One of the syringe pumps provided positive pressure (to load CPAs and transfer the oocyte) and the other provided negative pressure (to unload CPAs). During oocyte vitrification pretreatment, CPAs were loaded and unloaded by syringe pump. During the transferring process (Figure 2c), the operator pressed the transfer handle (PDMS sealing ring was pressed above the chip) and the syringe pump provided positive pressure for opening the capillary valve; then, the oocyte was transferred to the Cryotop. The imaging system, including a CCD camera (MER-503-36U3M, DAHENG IMAGING, Beijing, China) and a light source (CR-6390-R, DAHENG IMAGING, Beijing, China), was used to track the oocyte in the chip. Considering that the oocyte would suffer damage when exposed to a light source, we chose a red LED (Peak wavelength: 623 nm), which could minimize the damage to the oocyte. The chip and the companion device were fabricated by acrylic and aluminum alloy, respectively, through computer numerical control (CNC) machining (SuZhou Wenhao Co, Ltd, Suzhou, China).

2.2. System Operation

The CPAs (ES and VS) were prepared in HEPES-buffered tissue culture medium (Invitrogen, Waltham, MA, USA) supplemented with 10% *v/v* fetal bovine serum (FBS; Invitrogen, Waltham, MA, USA) based on the typical protocols. Briefly, the ES contained 7.5% *v/v* dimethyl sulfoxide (DMSO; Sigma-Aldrich, Darmstadt, Germany) and 7.5% *v/v* ethylene glycol (EG; Sigma-Aldrich, Darmstadt, Germany), and the VS contained 15% *v/v* DMSO, 15% *v/v* EG and 0.4 mol/L sucrose (Sigma-Aldrich, Darmstadt, Germany).

Above all, for the operator needed to prepare 4 disposable syringes (5 mL range, BD, Franklin Lakes, NJ, USA), three of them (loading ES, VS and air, respectively) were placed in the syringe pump that provided positive pressure, and the other one (no loading) was placed in the syringe pump that provided negative pressure. The syringes (loading ES, VS and no loading) were connected with the solution exchange chamber of the open microfluidic chip with soft silicone tubes, while the loading and unloading of CPAs were controlled by the syringe pump, respectively. Finally, a soft silicone tube was used to connect the syringe (loading air) to the gas port on the PDMS sealing ring, and Cryotop was placed in the designated position. After the preparations were completed, the overall vitrification protocol (Figure 4) was divided into the following steps:

- Step 0: The operator manually places the oocyte at the bottom of the oocyte chamber with a small amount of medium through the stereo microscope.
- Step 1: Load 20 μL ES (50 μL/min) into the solution exchange chamber. Part of the ES will be introduced into the oocyte chamber through the spontaneous aspirating of the capillary gap. Wait for 9–12 min for equilibration of the oocyte in ES.
- Step 2: Unload ES (100 μL/min) from the solution exchange chamber until the liquid level of residual solution around oocyte is below the capillary gap.
- Step 3: Load 20 μL VS (50 μL/min) into the solution exchange chamber; part of the VS will be introduced into the oocyte chamber, in which the oocyte is treated for 30 s. Note that due to the high concentration of cryoprotectant in VS, excessive operation time will have adverse effects on the oocyte. Steps 3–5 should be completed within 1 min.
- Step 4: Unload VS (100 μL/min) from the solution exchange chamber until the liquid level of residual solution around oocyte is below the capillary gap, which is similar to Step 2.
- Step 5: After the CPAs exchange is completed, the operator presses the transfer handle to transfer the oocyte from the channel to the Cryotop by syringe pump. Finally, the operator puts the Cryotop into LN_2 to complete the vitrification.

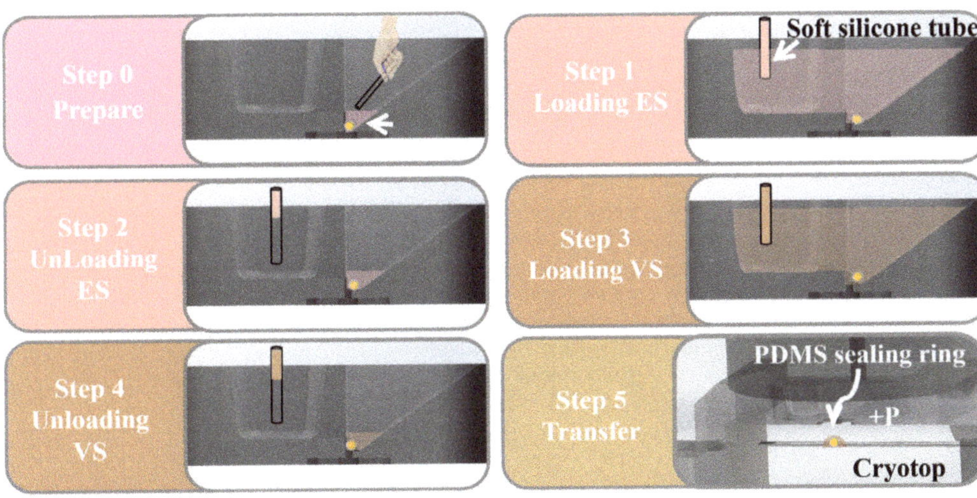

Figure 4. The operation steps of vitrification on the open microfluidic chip: preparation, loading and unloading ES/VS and transfer.

2.3. Fluid Simulation

The critical designs of our microfluidic chip are the variation rate of CPA concentration in the oocyte chamber and the reliability of the capillary valve. Therefore, to ensure that the designed chip can meet the requirements, numerical simulations of the fluid behavior in the abovementioned key structures were performed using COMSOL Multiphysics 5.6.

In the process of CPAs treatment, CPAs were introduced into the oocyte chamber through the capillary gap, and the CPAs concentration around the oocyte was changed by diffusion. The gradual change of CPAs concentration avoided the osmotic pressure shock caused by traditional manual operations. However, the cytotoxic solutes in the CPAs might cause damage to oocytes if the treatment time was too long due to the relatively slower diffusion process. To ensure that the concentration of CPAs around the oocyte could reach the concentration required for pretreatment, we used numerical simulation to predict the time-dependent changes in the concentration in the right chamber. Since the concentration of solute in CPAs was low compared with the solvent, it could be assumed that solute molecules only interact with water molecules in the solvent. Thus, the diffusion of CPAs concentration around oocytes could be described by Fick's law:

$$-\nabla \cdot (-D\nabla c + c\mathbf{u}) = 0 \tag{1}$$

where c represents the concentration and $\mathbf{u}$ is the fluid velocity field. In the process of CPAs processing, VS processing played a vital role because the time and concentration needed to be strictly controlled to prevent the damage to the oocyte. The solutes within VS were mainly EG, DMSO and sucrose, among which sucrose had the lowest concentration and diffusion coefficient. Therefore, we selected sucrose, as a candidate, and numerically simulated its transient concentration in the oocyte chamber.

Another critical part of our chip is the capillary valve, which was a passive valve and used surface tension to control fluid flow. When the external driving pressure was less than the critical value, the valve was closed. Once the external driving pressure exceeded the critical value, the valve would be opened and transfer the oocyte to the Cryotop. In order to ensure the reliability of the capillary valve, it was necessary to determine the critical burst value of the capillary valve. In the capillary valve we designed (Figure 5), there were two microchannel sections with sudden expansion (the first: the radius increases from 100 μm to 150 μm; the second: the radius 150 μm to the outside). When a liquid column was trapped in a vertical capillary of radius R, the capillary force should be equal to the downward force F acting on the liquid column to reach equilibrium [37,38]. The capillary force was generated by the meniscus on the lower surfaces of the liquid column. The external force F might originate from the pressure difference between the upper and lower surfaces and the gravity of the liquid column itself. Therefore, under ideal conditions:

$$-2\pi R\gamma_{lg}\cos\theta = F \tag{2}$$

where γ_{lg} is the liquid–gas surface tension coefficient and θ represents the contact angle of the meniscus. In our chip (Figure 5a), the external force F consisted of the weight of the liquid column and the external pressure over the cross-sectional area. When the liquid reached the capillary valve, the meniscus stopped at the edge of the microchannel outlet (Figure 5b). Under the action of external driving pressure, the meniscus deformed until its contact angle with the wall reached the critical advancing angle value θ_a (Figure 5c). Then, the valve opened and the meniscus continued to move forward (Figure 5d). The analytical expression of burst pressure (P_b) could be obtained according to Equation (2):

$$P_b = -2\gamma_{lg}\frac{\cos(\theta_a + \beta)}{r} \tag{3}$$

where β is the channel expansion angle and the r is its radius. When $\theta_a + \beta$ is equal to π the burst pressure P_b will reach its maximum value. The burst pressure was mainly determined by the surface tension coefficient and the channel radius. The radius at the second sudden expansion of the microchannel cross-section (150 μm) was greater than that at the first (100 μm); therefore, the burst pressure of the second was less than that of the first. The second structure was only used to protect the liquid column from contacting the outside as much as possible.

At the interface between the liquid phase and the gas phase, the level set method was used to describe the behavior between the two phases on the interface. In order to predict the burst pressure value, a numerical model was established to simulate the change of the meniscus shape under different driving pressures.

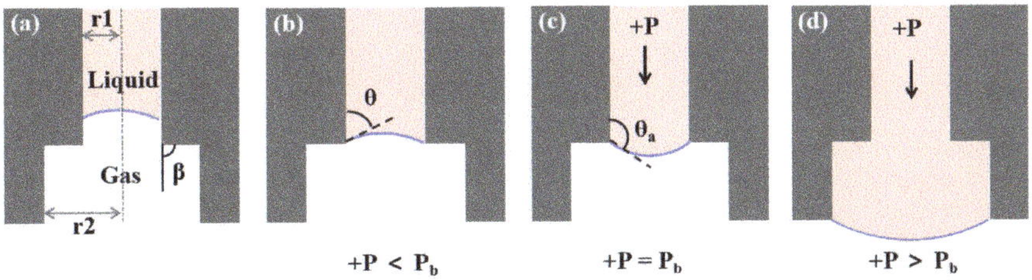

Figure 5. Different situations of liquid flow in capillary valve. (**a**) The internal structure diagram of the capillary valve; the inner radius (r1: 100 μm, r2: 150 μm) had two sudden expansions and the channel expansion angle β is 90°. (**b**) The liquid reached the capillary valve, without external pressure, and the meniscus stopped at the edge of the microchannel outlet. (**c**) The external pressure gradually increased, and the convexity of the meniscus and contact angle became larger. (**d**) When the external pressure exceeded the critical burst pressure, the capillary valve was opened and droplet was transferred to Cryotop.

2.4. Characterization of the Open Microfluidic Performance

The concentration of CPAs in the oocyte chamber was experimentally characterized using fluorescent tracer. We used sodium fluorescein (Sigma-Aldrich, Darmstadt, Germany), having a molecular weight (MW: 332.3 g/mol) close to sucrose (MW: 342.3 g/mol), as a tracer. The fluorescence intensity of the culture medium containing different concentrations of tracer was quantified at 490 nm, leading to the correlation between tracer concentration and fluorescence intensity. Then, we added the culture medium (0.4 mol/L) to the solution exchange chamber and recorded the changes in the fluorescence intensity of the oocyte chamber in real time.

To obtain the burst pressure of the capillary valve, a constant pressure pump (PC01, Fluidiclab, Shanghai, China) was used to provide additional pressure to the interior of the oocyte chamber, which was prefilled with VS. The variation of the meniscus at the capillary valve was characterized under different pressures.

The performance of the chip to realize the CPAs exchange and oocyte transfer was evaluated by an imaging system. Since the real oocyte was almost transparent and hard to observe, opaque polystyrene (PS) microparticles (Rigor Science, Wuxi, China) with a similar diameter to oocytes (80 μm) were used to verify the effectiveness of the chip design. After the microparticle was placed into the oocyte chamber, the operator conducted the abovementioned vitrification procedures while the imaging system monitored the liquid level and the particle position inside the chip.

2.5. Mouse Maintenance and Oocytes Collection

Female C57BL/6J mice (The Jackson Laboratory, Bar Harbor, ME, USA) between 6 and 10 weeks of age were used in this study. The animals were housed under a controlled environment with free access to water and food, and with lights switched on between 6:00 and 18:00. All experimental protocols were approved by the regional ethics committee of the University of Hong Kong–Shenzhen Hospital. To collect fully grown germinal vesicle (GV) oocytes, mice were super-stimulated with the injection of 10 IU pregnant mare serum gonadotropin (PMSG; Cat. G4527, Sigma, St. Louis, MO, USA).The COCs were obtained by manually rupturing antral ovarian follicles from mice ovaries 48 h after injection of PMSG. Cumulus cells were removed by mechanically repeated pipetting via a 75 μm (in diameter) tip (Cat. MXL3-IND-75, The STRIPPER® Tips, ORIGIO, CooperSurgical, Målov, Denmark) before collecting ready-to-use GV oocytes. To collect MII oocytes, mice were injected with 10 IU PMSG first; after 48 h, the mice were injected again with 10 IU hCG (Cat. HOR-250, Chorionic Gonadotropin Human, PROSPEC, Kenilworth, NJ, USA). The MII oocytes were obtained by punctuation of tubal enlargement after 16 h of hCG injection. The GV and MII oocytes were cultured in M2 medium (Cat. M7167, Sigma, St. Louis, MO, USA) at 37 °C before the cryopreservation.

2.6. Manual Vitrification Procedure

Manual vitrification of the oocyte was accomplished using our previously configured ES and VS solution and Cryotop. Specifically, in the vitrification process, the operator used an embryo transfer tube to transfer the oocytes into ES and VS sequentially, and the processing time was 9–12 min and 30 s, respectively. The oocyte was placed on the Cryotop; then, the Cryotop was inserted into LN_2 to complete vitrification.

2.7. Thawing Procedure

The oocytes vitrified by manual operation and microfluidic system were manually thawed using the Cryotop Kit (Kitazato BioPharma, Shizuoka, Japan). During the thawing process, the Cryotop with vitrified oocyte was removed from the LN_2, and the tip of Cryotop was quickly immersed into a prewarmed (37 °C) thawing solution (TS) for 1 min. Then, the oocyte was transferred into a dilution solution (DS), washing solution (WS1) and washing solution 2 (WS2) in sequence, with processing times of 3 min, 5 min and 5 min, respectively. Finally, the oocyte was transferred into the culture medium at 37 °C in a humidified atmosphere of 5% CO_2.

2.8. Evaluation of Success Rate and Oocyte Survival Rate

An operation was considered successful when the system achieved all steps of the vitrification protocol and successfully transferred the processed oocyte to the Cryotop. Manual vitrification experiments were also performed by embryologist. The post-freezing survival rate of thawed oocytes was measured to further quantify the performance of the system. Survivability was measured directly by examining the morphology of oocytes after thawing. According to commonly used criteria to judge the survivability of oocyte cryopreservation, an oocyte was considered healthy/alive when it had no abnormal shape, membrane damage or degeneration or fragmentation of cytoplasm.

2.9. JC-1 Staining and ROS Level Measurement of the Oocytes

The MMP was evaluated using the JC-1 probe (Cat. T3168, Invitrogen, Waltham, MA, USA). JC-1 dye exhibits potential-dependent accumulation in mitochondria, as indicated by an emission shift in the fluorescence from green (529 nm) to red (590 nm). Thus, mitochondrial depolarization could be indicated by a decrease in the red/green fluorescence intensity ratio. Thawed oocytes were cultured in M2 medium containing 2 mM JC-1 for 30 min at 37 °C, after which they were washed 3 times with PBS. The samples were then immediately imaged by confocal microscopy. To measure ROS level in thawed oocytes, carboxy-H2DCFD (Cat. C400, Invitrogen, Waltham, MA, USA), a fluorescent oxidative stress indicator, was applied in this assay. Oocytes were pretreated with M2 medium containing 10 mM H_2O_2 for 5 min. They were then washed and incubated with M2 medium containing 10 μM carboxy-H2DCFD for 30 min at 37 °C, after which they were washed 3 times with PBS and then immediately imaged by confocal microscopy.

2.10. Confocal Imaging

Images of the fluorescently labeled oocytes were acquired using a ZEISS LSM 900 with an Airyscan 2 laser scanning confocal microscope and Hybrid Detectors (HyD). Images were captured with a HC PL APO ×20/0.7 NA CS2 dry objective lens. Alexa Fluor 488, Alexa Fluor 546 and DAPI fluorescence was captured with an argon-ion laser operating at 488 nm, a HeNe laser operating at 561 nm, or a diode-pumped solid-state laser operating at 405 nm using 488 nm excitation/519 nm detection, 552 excitation/575 nm detection, and 405 nm excitation/461 nm detection, respectively.

2.11. Statistical Analysis

The data were analyzed with Minitab version 18 (Minitab Inc., State College, PA, USA) using the student's t-test. p-values less than 0.05 were considered to be statistically significant. Graphs were generated using Microsoft Excel and figures were prepared with CorelDraw version X8 (Corel Corp., Ottawa, ON, Canada).

3. Results and Discussion

3.1. Fluid Simulation and Performance Characterization of the Open Microfluidic Chip

The rate of change of CPAs concentration in the oocyte chamber had a significant effect on vitrification. We selected sucrose and numerically simulated its transient concentration in the oocyte chamber. Once the VS was introduced into the oocyte chamber, the sucrose concentration changed due to diffusion between the newly introduced VS and the residual ES, according to Equation (1); the snapshots of the simulated diffusion process are shown in Figure 6a. The average concentration of CPAs in the bottom region of the oocyte chamber was calculated, and its variation with respect to time is shown in Figure 6b. The sucrose concentration around the oocyte increased exponentially within 0–1000 ms and reached a plateau after 2000 ms. The constant sucrose concentration after 2000 ms indicated that the diffusion of VS was complete. Therefore, the time for VS to reach the desired concentration by diffusion was estimated to be 2 s, which was negligible compared to the processing time (30 s), validating the rationality of our designed system.

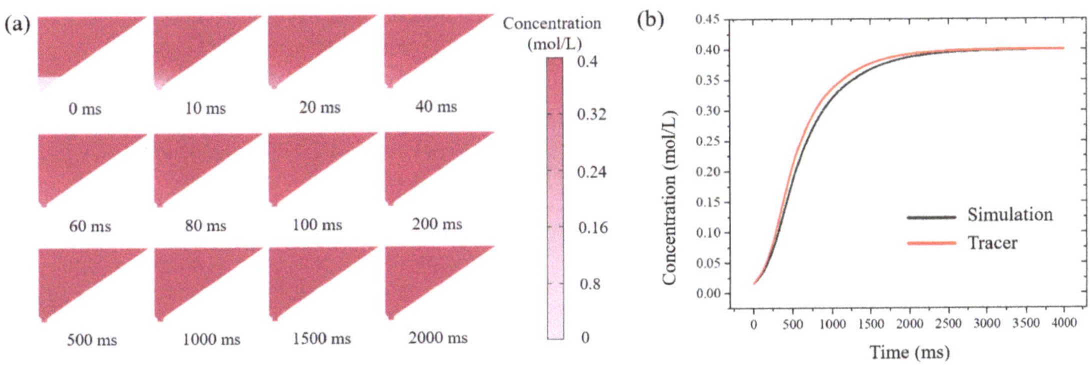

Figure 6. Theoretical and experimental VS concentration around the oocyte with respect to time. (**a**) Simulated time-dependent sucrose concentration in the oocyte chamber under diffusion condition (initial concentration upper: 0.4 mol/L; bottom: 0 mol/L). (**b**) Gray: the numerically predicted sucrose concentration in the oocyte chamber over time; Red: the experimentally measured intensity of the fluorescent tracer in the oocyte chamber.

Moreover, we analyzed the diffusion of tracer in the chip and the changes in fluorescence intensity at the bottom of the oocyte chamber at different time points using image J@(NIH, USA), shown in Figure 6b. The bottom fluorescence increased exponentially within 1000 ms and reached a plateau at 2000 ms, which was consistent with the simulation results. Tracer experiments and numerical simulations showed that, on our designed open microfluidic chip, CPAs could be introduced in the form of a concentration gradient, which would reduce the risk of osmotic shock due to CPAs mutation compared with the manual Cryotop method.

In order to simulate the critical burst pressure of the capillary valve, the VS with higher density and greater gravity was selected as the candidate, and the contact angle of the PMMA chip was 95°. Under different external pressures, the change of the meniscus of the capillary valve is shown in the Figure 7. Obviously, the protruding degree of the lower meniscus increased with the increase in external pressure, and the contact angle also increased. When the external pressure reached 975 Pa, the capillary valve was opened, and the liquid flowed to the outside. Therefore, 975 Pa was considered as the critical burst pressure of the capillary valve, which agreed well with the theoretical prediction (950 Pa) of Equation (3) above.

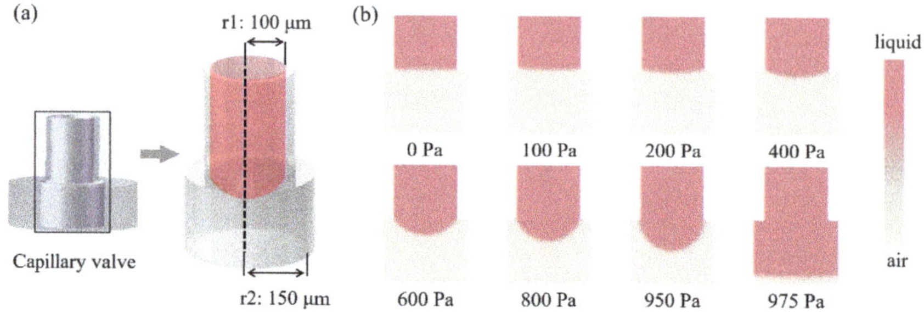

Figure 7. Numerical simulation of critical burst pressure of capillary valve. (**a**) The capillary valve modeling. (**b**) Deformation of the lower meniscus under different external pressures (liquid: pink; air: gray).

To further evaluate the reliability of the capillary valve, a constant pressure pump was used to test the critical burst pressure of the valve.The valve was opened when the pressure reached 900 Pa. The slightly lower critical pressure than the calculated values could be attributed to the output fluctuation of the pump. The hydraulic pressure of VS in the fully filled oocyte chamber was about 30 Pa, which was far less than the critical burst pressure. Therefore, we believed that the designed capillary valve could reliably block the liquid during CPAs treatment and allowed oocytes to pass through under exerted pressure.

After verifying the reliability of two key parts of the chip, the overall workflow of the chip was characterized. The PS microparticle within the chip was tracked with the imaging system and the different processing steps were recorded, as shown in Figure 8. The operator first placed the microparticle in the oocyte chamber (as shown in Figure 8a), and then performed the abovementioned (Figure 4) operation process. Steps 1–4 were the CPAs treatment process (as shown in Figure 8b–e), in which steps 2 and 4 were the removal of CPAs. The immobilization of the microparticle in the oocyte chamber indicated that the capillary gap had the effect of blocking the passage of microparticle/oocyte, and the different depths of the chambers on both sides could ensure that there was a certain amount of liquid remaining after the unloading of CPAs to avoid osmotic pressure damage caused by the rapid evaporation of the liquid. There was no leakage of fluid in the oocyte chamber prior to step 5, demonstrating the reliability of the capillary valve. When the vitrification pretreatment was completed, the operator pressed the handle to introduce external pressure. Figure 8f shows that the external pressure reached the critical burst pressure of the capillary valve, and the droplet containing the microparticle was about to be transferred to the Cryotop. Figure 8g shows the transfer of the droplet to the Cryotop with almost no liquid residue in the chip, and Figure 8h is the microparticle transferred to the Cryotop after processing. So far, the system has completed the entire vitrification pretreatment and transfer.

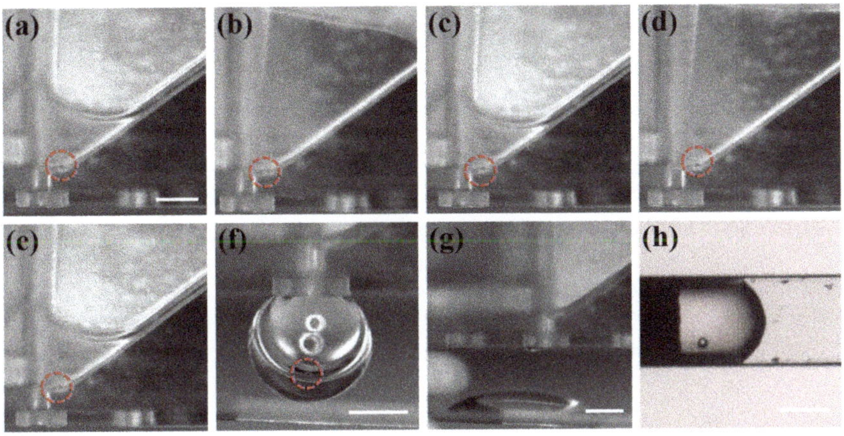

Figure 8. The microparticle was tracked by the imaging system over the entire process of system vitrification. (**a**) The operator placed the microparticle in the chip. (**b–g**) The microparticle completed vitrification pretreatment and was transferred to Cryotop. (**h**) The droplet on Cryotop contained microparticle. Scale bars in (**a**,**f**,**g**,**h**) are 400 μm.

3.2. Success Rate and Oocyte Survival Rate

As summarized in Tables 1 and 2, the system achieved a similar success rate compared with manual operation in the GV stage (100% versus 97.3%) and MII stage (100% versus 97.9%). Among the system vitrification of 45 oocytes in the GV stage and 39 oocytes in the MII stage, all of them were successfully vitrified. During manual operation, the human operator might occasionally fail to locate the oocyte within the limited time of the protocol and fail to aspirate the oocyte. The survival rates of thawed oocytes were calculated to further quantify the system performance. Compared with the manual Cryotop operation, the survival rate of the system was slightly higher (GV stage: 97.4% versus 94.4% , MII stage: 97.8% versus 97.8%), which was due to the different loading methods of CPAs and the operational reproducibility. On the one hand, the CPAs concentration changed abruptly under manual operation, while the CPAs concentration in the microfluidic chip changed more gently. On the other hand, in experiments involving complex manual operations, it was difficult for the operator to ensure the consistency of the experiment. Compared with the microfluidic system proposed by Jiang et al. [25], our system had a higher survival rate (97.8% versus 75.0%). This might be attributed to the employment of commercial Cryotop in our platform, which had a higher cooling rate than the independently designed carrier. With the precise control of volume and treatment time in CPAs, the system reduced the risk of overexposure and osmotic shock, and achieved a desirable oocyte survival rate as manual operation.

Table 1. Experimental results of mouse GV stage oocytes vitrification.

	GV Stage		
	Number	**Success Rate (%)**	**Survival Rate (%)**
Manual	37	97.3	94.4
Microfluidic	39	100	97.4

Table 2. Experimental results of mouse MII stage oocytes vitrification.

	MII Stage		
	Number	Success Rate (%)	Survival Rate (%)
Manual	47	97.9	97.8
Microfluidic	45	100	97.8
Jiang et al. [25]	20	/	75

3.3. No Difference between Manual Manipulation and Microfluidics System for Oocytes Cryopreservation

The GV oocytes and MII oocytes were cryopreserved via manual vitrification and microfluidics vitrification, respectively. We firstly demonstrated that there was no big difference on morphological changes between both groups of GV or MII oocytes. No obvious degeneration or fragmentation of cytoplasm was detected in thawed oocytes (Figure 9), which suggested that the oocytes could survive vitrification and thawing procedures performed either via traditional manual operations or the microfluidics system.

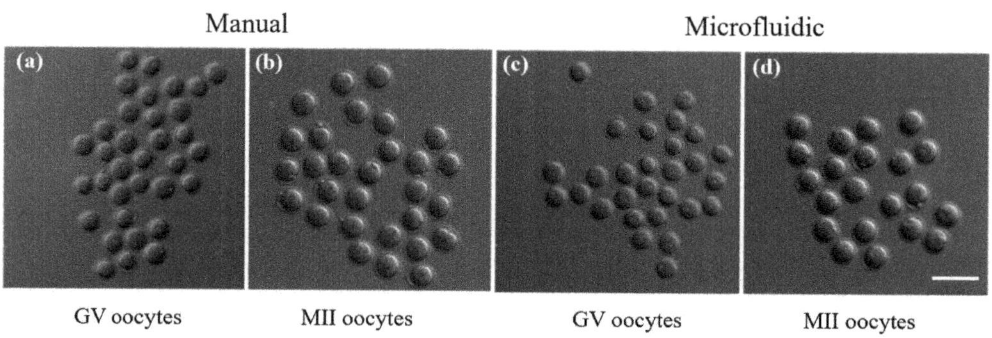

Figure 9. Evaluation of oocytes thawed from microfluidics vitrification via comparing with traditional manual vitrification method. (**a**,**c**) GV oocytes and (**b**,**d**) MII oocytes (with polar body extruded) were thawed from manual vitrification and microfluidics vitrification. Scale bar is 100 μm.

To further evaluate the quality of vitrification-thawed oocytes from these two methods, we next performed several biomolecular experiments to test whether the functional changes occurred in these oocytes. JC-1 live staining experiments were performed on thawed MII oocytes to check the MMP level, which is a typical biomarker to assess the physiologic function of mitochondrion in oocytes. The red fluorescent signals labeled JC-1 aggregates and green fluorescent signals labeled JC-1 monomers. The ratio of red/green fluorescence represented the level of MMP. According to this red/green fluorescence ratio, the representative images captured via a confocal microscope demonstrated a slightly better MMP level in the microfluidics group compared with that in the manual group. However, there was no significant difference after statistical analysis between both groups (Figure 10a,b).

Moreover, the ROS levels from cytoplasm of vitrification-thawed MII oocytes were measured as one of the criteria to evaluate the biological functions of these cryopreserved oocytes. The representative images showed that the thawed oocytes in both groups have a comparable ROS level after the recovery from H_2O_2 pretreatment. The low expression of green fluorescence revealed the normal ROS levels and healthy conditions in these thawed oocytes, whereas the high ROS levels usually represent a high-cell-stress condition or the preliminary phase of apoptosis [39]. The quantification of green fluorescence intensity demonstrated no significant difference between both groups, suggesting that the microfluidics vitrification method conducted similar results with respect to the traditional vitrification method (Figure 10c,d). Overall, based on the results collected from above biomolecular experiments, we conclude that no morphological changes or functional changes were detected from the cryopreserved oocytes via both systems.

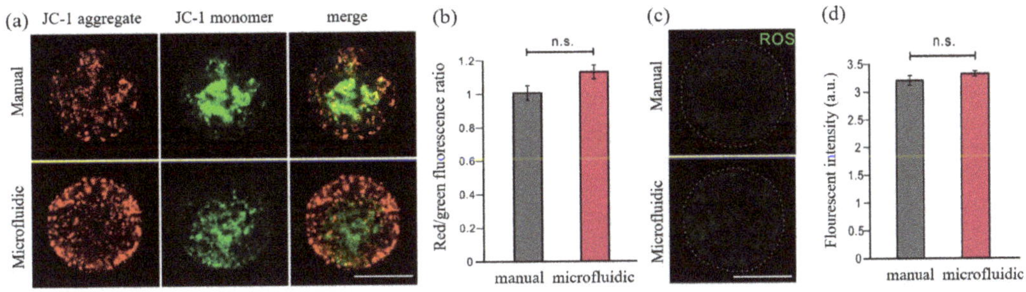

Figure 10. JC-1 staining and ROS level measurement of the oocytes. (**a**) The MMP level of thawed MII oocytes was measured by JC-1 staining. (**b**) The red/green fluorescence ratio of oocytes was quantified between manual and microfluidic groups. (**c**) The ROS level of thawed GV oocytes between manual and microfluidic group. (**d**) The ROS levels were measured and quantified in two groups. The students' *t*-test was used to compare the difference in two groups. n.s. means no significant difference between two compared groups. Scale bar is 50 μm.

4. Conclusions

In summary, we demonstrate a microfluidic platform for achieving standardized vitrification of oocytes, which is compatible with the commonly used Cryotop method. Theoretically, our platform is applicable to all embryos or oocytes following the standard vitrification process. With an open microfluidic chip, the system realized the gentle loading of CPAs, the positioning of the oocyte and the transfer of oocyte to a Cryotop. The flow behavior inside the chip, i.e., the variation of CPAs concentration around the oocyte and the reliability of the capillary valve, was systematically investigated through theoretical and experimental methods. Mouse GV and MII oocytes were then selected as biological samples to verify the vitrification effect of the open microfluidic chip platform. The results demonstrate the considerable stability and reliability of the designed chip to complete the vitrification pretreatment and transfer of the oocyte within the specified operating time, with a success rate of 100%. Evaluating the quality of thawed mouse GV/MII stage oocytes vitrified by two methods, our system showed a higher survival rate (97.6% versus 94.0%) and consistent MMP and ROS levels compared with the manual group conducted by the senior embryologist. To the best of our knowledge, this is the first report of the microfluidic system based on the Cryotop gold standard to complete the vitrification procedure without any subsequent processing, making it accessible to embryologists who are not familiar with microfluidic manipulation and also much lower in cost than commercial automated vitrification platforms (3000 USD versus 75,000 USD) [40]. The estimated cost of 3000 USD for our platform includes the production of microfluidic chips in addition to machining of mechanical structures and the affiliated sensors and actuators. Considering the tiny size and preciousness of human oocytes, the manual operation should be very precise and reliable, which improves the difficulty of training a qualified operator. Facilitated by our platform, the vitrification process is effectively simplified and standardized, which reduces the requirement for operators. It normally takes 15–20 min for a skillful human operator to manually conduct the treatment and freezing processes for an oocyte. With the help of our platform, this period can be controlled to within 15 min. It is worth noting that our platform can effectively avoid misoperation during manual operation and eliminate the differences caused by individual operators. Finally, we believe that the developed vitrification microfluidic system will greatly facilitate assisted reproduction and provide a standard solution for germ cell vitrification.

Author Contributions: S.M., C.G. and Z.J. conceived the project, designed the experimental outline and wrote the manuscript. S.M. and Z.J. designed microfluidic chip and performed CFD simulations. C.G., J.G. and Z.H. collected the mouse oocytes and executed mouse oocytes vitrification experiments. H.-X.W. worked on the data processing and analysis. T.W. and Y.-H.L. provided comments on the data and the manuscript. X.J. worked on the experiments design, data processing and analysis and manuscript preparation. All authors have read and agreed to the published version of the manuscript.

Funding: This work was supported by the National Natural Science Foundation of China (Grant No. 61873072), in part by the National Natural Science Foundation of China (Grant No. U1813202), in part by the Shenzhen Science and Technology Program of China (No. RCYX20200714114705073, No. KQTD20190929172749226) and in part by the Hong Kong Centre for Logistics Robotics.

Institutional Review Board Statement: The animal study was reviewed and approved by Ethical Committee of the University of Hong Kong–Shenzhen Hospital. All experiments were performed at the University of Hong Kong–Shenzhen Hospital, China.

Informed Consent Statement: Not applicable.

Data Availability Statement: Not applicable.

Conflicts of Interest: The authors declare no conflict of interest.

References

1. Kuwayama, M.; Vajta, G.; Kato, O.; Leibo, S.P. Highly efficient vitrification method for cryopreservation of human oocytes. *Reprod. Biomed. Online* **2005**, *11*, 300–308. [CrossRef]
2. Lucena, E.; Bernal, D.P.; Lucena, C.; Rojas, A.; Moran, A.; Lucena, A. Successful ongoing pregnancies after vitrification of oocytes. *Fertil. Steril.* **2006**, *85*, 108–111. [CrossRef] [PubMed]
3. Gook, D.A.; Edgar, D.H. Human oocyte cryopreservation. *Hum. Reprod. Update* **2007**, *13*, 591–605. [CrossRef] [PubMed]
4. Chen, C. Pregnancy after human oocyte cryopreservation. *Lancet* **1986**, *327*, 884–886. [CrossRef]
5. Kuwayama, M. Highly efficient vitrification for cryopreservation of human oocytes and embryos: The Cryotop method. *Theriogenology* **2007**, *67*, 73–80. [CrossRef]
6. Guo, Y.; Yang, Y.; Yi, X.; Zhou, X. Microfluidic method reduces osmotic stress injury to oocytes during cryoprotectant addition and removal processes in porcine oocytes. *Cryobiology* **2019**, *90*, 63–70. [CrossRef]
7. Al-Hasani, S.; Diedrich, K.; Van der Ven, H.; Reinecke, A.; Hartje, M.; Krebs, D. Cryopreservation of human oocytes. *Hum. Reprod.* **1987**, *2*, 695–700. [CrossRef]
8. Clark, N.A.; Swain, J.E. Oocyte cryopreservation: Searching for novel improvement strategies. *J. Assist. Reprod. Genet.* **2013**, *30*, 865–875. [CrossRef]
9. Cobo, A.; Bellver, J.; Domingo, J.; Pérez, S.; Crespo, J.; Pellicer, A.; Remohí, J. New options in assisted reproduction technology: The Cryotop method of oocyte vitrification. *Reprod. Biomed. Online* **2008**, *17*, 68–72. [CrossRef]
10. Bromfield, J.; Coticchio, G.; Hutt, K.; Sciajno, R.; Borini, A.; Albertini, D.F. Meiotic spindle dynamics in human oocytes following slow-cooling cryopreservation. *Hum. Reprod.* **2009**, *24*, 2114–2123. [CrossRef]
11. Moussa, M.; Shu, J.; Zhang, X.; Zeng, F. Cryopreservation of mammalian oocytes and embryos: Current problems and future perspectives. *Sci. China Life Sci.* **2014**, *57*, 903–914. [CrossRef] [PubMed]
12. Mandawala, A.A.; Harvey, S.C.; Roy, T.; Fowler, K. Cryopreservation of animal oocytes and embryos: Current progress and future prospects. *Theriogenology* **2016**, *86*, 1637–1644. [CrossRef] [PubMed]
13. Cobo, A.; García-Velasco, J.A.; Coello, A.; Domingo, J.; Pellicer, A.; Remohí, J. Oocyte vitrification as an efficient option for elective fertility preservation. *Fertil. Steril.* **2016**, *105*, 755–764. [CrossRef] [PubMed]
14. Antinori, M.; Licata, E.; Dani, G.; Cerusico, F.; Versaci, C.; Antinori, S. Cryotop vitrification of human oocytes results in high survival rate and healthy deliveries. *Reprod. Biomed. Online* **2007**, *14*, 72–79. [CrossRef]
15. Miao, S.; Chen, D.; Nie, Q.; Jiang, X.; Sun, X.; Dai, J.; Liu, Y.H.; Li, X. Development of a Vision-Based Robotic Manipulation System for Transferring of Oocytes. In Proceedings of the 2021 IEEE/RSJ International Conference on Intelligent Robots and Systems (IROS), Prague, Czech Republic, 27 September–1 October 2021; IEEE: Piscataway, NJ, USA, 2021; pp. 7470–7475.
16. Yetkinel, S.; Kilicdag, E.B.; Aytac, P.C.; Haydardedeoglu, B.; Simsek, E.; Cok, T. Effects of the microfluidic chip technique in sperm selection for intracytoplasmic sperm injection for unexplained infertility: A prospective, randomized controlled trial. *J. Assist. Reprod. Genet.* **2019**, *36*, 403–409. [CrossRef]
17. Weng, L. IVF-on-a-Chip: Recent advances in microfluidics technology for in vitro fertilization. *Slas Technol. Transl. Life Sci. Innov.* **2019**, *24*, 373–385. [CrossRef]
18. Thapa, S.; Heo, Y.S. Microfluidic technology for in vitro fertilization (IVF). *JMST Adv.* **2019**, *1*, 1–11. [CrossRef]
19. Chen, Y.S.; Lo, T.W.; Huang, H.Y.; Li, L.M.; Wang, Y.W.; Yao, D.J.; Hsu, W.S.; Liu, C.H. A microfluidic lab chip for the manipulation and co-culturing of embryos with stromal cells. *Sens. Actuators B Chem.* **2021**, *349*, 130820. [CrossRef]

20. Doufène, K.; Tourné-Péteilh, C.; Etienne, P.; Aubert-Pouëssel, A. Microfluidic systems for droplet generation in aqueous continuous phases: A focus review. *Langmuir* **2019**, *35*, 12597–12612. [CrossRef]
21. Damiati, S.; Kompella, U.B.; Damiati, S.A.; Kodzius, R. Microfluidic devices for drug delivery systems and drug screening. *Genes* **2018**, *9*, 103. [CrossRef]
22. Li, H.; Torab, P.; Mach, K.E.; Surrette, C.; England, M.R.; Craft, D.W.; Thomas, N.J.; Liao, J.C.; Puleo, C.; Wong, P.K. Adaptable microfluidic system for single-cell pathogen classification and antimicrobial susceptibility testing. *Proc. Natl. Acad. Sci. USA* **2019**, *116*, 10270–10279. [CrossRef]
23. Zhang, X.; Wei, X.; Men, X.; Wu, C.X.; Bai, J.J.; Li, W.T.; Yang, T.; Chen, M.L.; Wang, J.H. Dual-Multivalent-Aptamer-Conjugated Nanoprobes for Superefficient Discerning of Single Circulating Tumor Cells in a Microfluidic Chip with Inductively Coupled Plasma Mass Spectrometry Detection. *ACS Appl. Mater. Interfaces* **2021**, *13*, 43668–43675. [CrossRef] [PubMed]
24. Zhang, X.; Wei, X.; Men, X.; Jiang, Z.; Ye, W.Q.; Chen, M.L.; Yang, T.; Xu, Z.R.; Wang, J.H. Inertial-force-assisted, high-throughput, droplet-free, single-cell sampling coupled with ICP-MS for real-time cell analysis. *Anal. Chem.* **2020**, *92*, 6604–6612. [CrossRef] [PubMed]
25. Jiang, B.; Huang, B.; Cai, G.; Chen, Y.; Wu, T. Facile and highly efficient loading and freezing of cryoprotectants for oocyte vitrification based on planar microfluidics. *Microfluid. Nanofluidics* **2021**, *25*, 63. [CrossRef]
26. Pyne, D.G.; Liu, J.; Abdelgawad, M.; Sun, Y. Digital microfluidic processing of mammalian embryos for vitrification. *PLoS ONE* **2014**, *9*, e108128. [CrossRef] [PubMed]
27. Martín-Villalba, T.; Fidalgo, J.S.; Carasa, P.; Garcia, D.; Munuera, C.D.; Cancio-Villalonga, D.; Álvarez-Argüelles, S.; Fernandez, G.; Bajo, E.; Fernandez, A.; et al. Validation of an oocyte vitrification using davitri, a novel automated microfluidics device. *Fertil. Steril.* **2021**, *116*, e200–e201. [CrossRef]
28. Tirgar, P.; Sarmadi, F.; Najafi, M.; Kazemi, P.; AzizMohseni, S.; Fayazi, S.; Zandi, G.; Ziaie, N.; Shoushtari Zadeh Naseri, A.; Ehrlicher, A.; et al. Toward embryo cryopreservation-on-a-chip: A standalone microfluidic platform for gradual loading of cryoprotectants to minimize cryoinjuries. *Biomicrofluidics* **2021**, *15*, 034104. [CrossRef]
29. Lai, D.; Ding, J.; Smith, G.; Smith, G.; Takayama, S. Slow and steady cell shrinkage reduces osmotic stress in bovine and murine oocyte and zygote vitrification. *Hum. Reprod.* **2015**, *30*, 37–45. [CrossRef]
30. Heo, Y.S.; Lee, H.J.; Hassell, B.A.; Irimia, D.; Toth, T.L.; Elmoazzen, H.; Toner, M. Controlled loading of cryoprotectants (CPAs) to oocyte with linear and complex CPA profiles on a microfluidic platform. *Lab chip* **2011**, *11*, 3530–3537. [CrossRef]
31. Song, Y.S.; Moon, S.; Hulli, L.; Hasan, S.K.; Kayaalp, E.; Demirci, U. Microfluidics for cryopreservation. *Lab Chip* **2009**, *9*, 1874–1881. [CrossRef]
32. Miao, S.; Jiang, Z.; Luo, J.; Zhong, F.; Wei, H.; Sun, X.; Jiang, X.; Jiang, M.; Liu, Y. A Robotic System with Embedded Open Microfluidic Chip for Automatic Embryo Vitrification. *IEEE Trans. Biomed. Eng.* **2022**. [CrossRef] [PubMed]
33. Liu, Y.; Cui, K.; Kong, Q.; Zhang, L.; Ge, S.; Yu, J. A self-powered origami paper analytical device with a pop-up structure for dual-mode electrochemical sensing of ATP assisted by glucose oxidase-triggered reaction. *Biosens. Bioelectron.* **2020**, *148*, 111839. [CrossRef]
34. Berthier, J.; Brakke, K.A.; Berthier, E. *Open Microfluidics*; John Wiley & Sons: Hoboken, NJ, USA, 2016.
35. Berthier, J.; Brakke, K.; Furlani, E.; Karampelas, I.; Poher, V.; Gosselin, D.; Cubizolles, M.; Pouteau, P. Whole blood spontaneous capillary flow in narrow V-groove microchannels. *Sens. Actuators B Chem.* **2015**, *206*, 258–267. [CrossRef]
36. Glière, A.; Delattre, C. Modeling and fabrication of capillary stop valves for planar microfluidic systems. *Sens. Actuators A Phys.* **2006**, *130*, 601–608. [CrossRef]
37. De Gennes, P.G.; Brochard-Wyart, F.; Quéré, D. *Capillarity and Wetting Phenomena: Drops, Bubbles, Pearls, Waves*; Springer: Berlin/Heidelberg, Germany, 2004; Volume 315.
38. Wang, Z.; Chang, C.C.; Hong, S.J.; Sheng, Y.J.; Tsao, H.K. Trapped liquid drop in a microchannel: Multiple stable states. *Phys. Rev. E* **2013**, *87*, 062401. [CrossRef] [PubMed]
39. Udono, H.; Nishida, M. Metformin-ROS-Nrf2 connection in the host defense mechanism against oxidative stress, apoptosis, cancers, and ageing. *Biochim. Biophys. Acta (BBA)-Gen. Subj.* **2022**, *1866*, 130171. [CrossRef] [PubMed]
40. Roy, T.K.; Brandi, S.; Tappe, N.M.; Bradley, C.K.; Vom, E.; Henderson, C.; Lewis, C.; Battista, K.; Hobbs, B.; Hobbs, S.; et al. Embryo vitrification using a novel semi-automated closed system yields in vitro outcomes equivalent to the manual Cryotop method. *Hum. Reprod.* **2014**, *29*, 2431–2438. [CrossRef] [PubMed]

biosensors

MDPI

Review

Single-Cell RNA Sequencing in Organ and Cell Transplantation

Roozbeh Abedini-Nassab [1,*], Fatemeh Taheri [2], Ali Emamgholizadeh [1] and Hossein Naderi-Manesh [3,4]

1 Faculty of Mechanical Engineering, Tarbiat Modares University, Tehran P.O. Box 1411944961, Iran
2 Biomedical Engineering Department, University of Neyshabur, Neyshabur P.O. Box 9319774446, Iran
3 Department of Nanobiotechnology, Faculty of Bioscience, Tarbiat Modares University, Tehran P.O. Box 1411944961, Iran; naderman@modares.ac.ir
4 Department of Biophysics, Faculty of Bioscience, Tarbiat Modares University, Tehran P.O. Box 1411944961, Iran
* Correspondence: abedini@modares.ac.ir

Abstract: Single-cell RNA sequencing is a high-throughput novel method that provides transcriptional profiling of individual cells within biological samples. This method typically uses microfluidics systems to uncover the complex intercellular communication networks and biological pathways buried within highly heterogeneous cell populations in tissues. One important application of this technology sits in the fields of organ and stem cell transplantation, where complications such as graft rejection and other post-transplantation life-threatening issues may occur. In this review, we first focus on research in which single-cell RNA sequencing is used to study the transcriptional profile of transplanted tissues. This technology enables the analysis of the donor and recipient cells and identifies cell types and states associated with transplant complications and pathologies. We also review the use of single-cell RNA sequencing in stem cell implantation. This method enables studying the heterogeneity of normal and pathological stem cells and the heterogeneity in cell populations. With their remarkably rapid pace, the single-cell RNA sequencing methodologies will potentially result in breakthroughs in clinical transplantation in the coming years.

Keywords: single-cell RNA sequencing; transplantation; microfluidics; biomarkers

1. Introduction

Single-cell ribonucleic acid sequencing (scRNA-seq) is now considered the state-of-the-art technology for assessing the function of individual cells within a large tissue sample. It is a powerful technique for profiling the expression of genes in individual cells, providing a new way to study biological processes and identify rare cell types. This analytical method has found application across various domains, including cancer biology [1,2], neurobiology [3,4], immunology [5,6], and transplantation [7,8]. In areas such as cancer research, scRNA-seq allows scientists to understand how tumors differ at the cellular level and how they develop over time. Similarly, in neuroscience, scRNA-seq can be used to uncover how different neural circuits are formed and regulated, providing new insight into the development and functioning of the brain. In addition, scRNA-seq can be used to study the diversity of microbial populations within complex environments, such as the human gut [9]. Recently, scRNA-seq has also been used to study the role of graft–host interaction after transplantation and for preclinical evaluation of the grafts and therapeutic stem cells to ensure reliable and safe transplantation [10].

The development of scRNA-seq is partly due to the recent advances in the field of biomicrofluidics. In particular, droplet-based microfluidic chips, in addition to various applications in different fields [11], offer the encapsulation of thousands of single-cells and barcode-carrying beads in droplets, where the genomic data of the single cells can be captured [12]. After being captured with barcoded beads The messenger ribonucleic acids (mRNAs) of individual cells are sequenced. The obtained data are then analyzed to find the gene expression level at the single-cell resolution.

Currently, transplantation remains the primary solution for organ failure; however, associated biological challenges still hinder its success rate [13]. Monitoring the graft health and interaction with the host to detect post-transplant complications and possibly prevent them is an essential need in determining the long-term success of the transplantation and a key component of transplant therapy. While solid biopsy is the gold standard for allograft monitoring, analyzing body fluids, such as blood and urine, is also becoming popular as a less invasive and more patient-friendly method capable of reliably detecting transplant complications. As such, scRNA-seq is being used to evaluate post-transplant graft health by analyzing the gene expression profile and its changes. By applying scRNA-seq on allograft biopsies at various time points, the transcriptomic profiles of cells in rejection, as well as the changes in cell interactions after anti-rejection treatments, can be identified [14]. This type of analysis can be used to assess the degree to which the graft is compatible with the host's tissue and can help to identify potential problems that could lead to graft rejection.

In addition to providing insights into the transplantation quality and the graft–host interactions, scRNA-seq enables the discovery of new biomarkers to predict transplantation outcomes. These biomarkers can then be used to assess the likelihood of a successful outcome prior to the transplantation procedure.

Stem cell transplantation is considered another area where scRNA-seq could have a transformative impact. Stem cells regularly divide to maintain, develop, and repair the organs and tissues [15]. In stem cell transplantation, stem cells from a donor are transplanted into a recipient in order to treat a wide range of diseases and injuries, such as leukemia and spinal cord injuries. Unfortunately, the transplantation process is not always successful. Numerous factors can affect the success of the transplantation, such as the quality of the stem cells, the compatibility of the donor and recipient, and environmental factors (e.g., diet and pollutants).

By applying scRNA-seq to stem cell transplantation, it is possible to gain a better understanding of stem cell biology and investigate how the various factors can influence their behavior. This approach would enable researchers to develop and optimize more efficient and successful transplantation protocols and to improve stem cell selection methods. It also could open the door to a new era of personalized, targeted therapies that are more effective and efficient than ever before.

Another application of scRNA-seq is evaluating transplanted tumors for research purposes. The transplantation of tumor tissue into animal models is an established field for studying the effect of the surrounding microenvironment on tumor behavior, including tumorigenesis and metastasis [16]. The scRNA-seq technology captures this information from the transplanted tumor at single-cell resolution.

In this review, we first discuss advanced available scRNA-seq methods and compare them. Then, we delve into the application of scRNA-seq in solid transplantation, including heart, kidney, lung, and liver. Next, we review the use of scRNA-seq in stem cell transplantation. Then, we discuss selected studies where scRNA-seq addresses questions in tumor transplantation. In each section, essential information and the relevant comparisons are tabulated.

2. Single-Cell RNA Sequencing

As an emerging technology used to measure expression levels of genes within individual cells, scRNA-seq offers an unparalleled level of resolution, allowing scientists to gain a deeper understanding of complex biological processes, such as cellular heterogeneity and cellular interactions. The scRNA-seq approaches are generally classified into two broad categories: plate-based and microfluidic-based methods. Microfluidic tools offer novel sequencing methods with significant advantages, including fast handling, labor saving, and high throughput. Thus, these technologies have been widely used recently. The basic principle of scRNA-seq is to capture and sequence the mRNA molecules present within individual cells. By doing so, it is possible to quantify the expression levels of various genes, providing a snapshot of the transcriptome of each cell. The acquired sequencing data can then be used to infer biological characteristics such as the cell type and its state of activation. In addition, it offers insights into the dynamics and interactions of molecular networks within individual cells.

To obtain the mentioned data from the single cells, single-cell transcriptome analysis typically involves three main steps [17]. The first step is to dissociate and isolate live single cells from tissues and keep them alive. They need to be protected against gene and protein expression variations due to cell dissociation and handling. This step may be performed using mechanical and enzymatic-based methods. The obtained single cells may then be selected based on their membrane protein expressions using fluorescence-activated cell sorting. The second step is to reverse-transcribe mRNA selectively to achieve a cDNA library. This step involves sub-steps of RNA capture, reverse transcription, RNA sequencing, and library construction. The third step is to sequence and analyze the cDNA library obtained from the single cells. Data analysis involves several sub-steps, including quality control, preprocessing, quantification, dimensionality reduction, and visualization. Early errors in data not removed during the preprocessing steps and quality checks may propagate throughout the rest of the analysis. Dead cells with damaged cell membranes, cell doublets, or bead doublets need to be detected and filtered. For example, the leakage of mRNAs from the damaged membranes has a substantial impact on the overall read counts. Also, cDNA amplification may introduce an unwanted bias toward the amplification of cDNA transcripts with specific sizes.

2.1. Single-Cell Isolation Methods

In order to isolate single cells for lysis and analysis, different methods are used. Here, we mention six widely used cell isolation methods, four of which are based on microfluidic tools.

2.1.1. FACS

Fluorescence-activated cell sorting (FACS) systems sort individual particles and cells based on fluorescent signals obtained from them [18,19]. It is possible to use these systems in single-cell analysis methods to isolate fluorescently labeled single cells based on their specific surface molecules. The fluorescent labels are excited with the means of a laser beam, after which the cells are sorted into specific microwells of interest [20] for further steps toward scRNA-seq. The advantages of cell sorting based on FACS include their high accuracy. However, they are not typically suitable for sorting samples with low cell numbers.

2.1.2. Micromanipulation

Mechanical micromanipulation is considered a traditional method to put individual particles into separate chambers to be studied. This technique is based on manual cell micropipetting under a microscope or based on pipetting robots. In these methods, the cell concentration in the cell suspension is lowered to ensure pickup at single-cell resolution. Mechanical micromanipulation is widely used in biological research; however, it needs skilled operators, it is low-throughput, and the cells may be damaged [21–23].

2.1.3. Passive Hydrodynamic Trap-Based Microfluidics

Microfluidic chips provide good control of small volumes of fluids, as well as tiny objects. Hence, cell handling based on these devices is developed as a more modern technique. Hydrodynamic trap-based microfluidic chips are one of these methods that are widely used in the field [24–26]. In this method, the cell suspension is injected into the microfluidic device, where the single cells are captured in the trap sites. By entering a cell into a trap site, the hydrodynamic resistivity of that path increases, limiting the occupation of that trap with the other cells. This passive method based on hydrodynamic forces does not require additional forces to manipulate the particles; however, the unwanted shear stress applied to trapped cells may alter the cell behavior [27]. Also, the cell concentration needs to be controlled to avoid channel clogging. Since the traps are not typically isolated, another challenge is the high chance of cell content contamination. Recently, a method for trapping cell-encapsulating droplets has been introduced, answering most of the mentioned drawbacks of the old systems [28]. In this system, the cells do not experience high shear stresses, and the droplet acts as an isolated microchamber, preventing unwanted cross-contaminations.

2.1.4. Active Cell Manipulation Microfluidics

Microfluidic chips enhanced with active particle manipulation methods use at least one force other than hydrodynamic forces to manipulate cells. Methods based on magnetic [29–34], electric [35–38], acoustic [39–43], and optical forces [44–48], each of which, with their advantages and disadvantages, have been developed. In general, the active methods are rather more complicated than the passive methods; however, they typically offer more precise control of individual cells. Magnetic-based techniques require cell magnetic nanoparticle labeling, while other methods do not. There are cell mutation concerns in optical manipulations [49]. Also, they may have difficulties in opaque environments. The methods based on electric forces (e.g., electrowetting [50,51]) require complex wiring systems. In acoustic systems, manipulating specific single particles is typically not offered. Microfluidics chips employing a combination of different techniques have shown promising results [52,53]. These methods borrow desired specifications from different techniques to offer unique capabilities, while undesired aspects are limited. For example, a recently developed device called the magnetomicrofluidic chip assembles single cells in semi-isolated low-shear microchambers in a microfluidic chip using the combination of hydrodynamic traps and magnetic or electric forces [29,54,55]. In these systems, the undesired shear stress and the cell content contamination seen in hydrodynamic trapping systems are eliminated, while they provide high-throughput precise particle manipulation.

2.1.5. Valve-Based Microfluidics

To control the fluid flow inside the microchannels in a microfluidic chip, microvalves are used [56]. Typically, the microvalves have been developed using a multilayer microfluidic structure [57]. Quake valves that control the flow rate in a fluid-carrying channel by applying a pneumatic pressure on a PDMS membrane on top of that channel to deflect it are considered one of the most common microvalve designs [57]. Based on these valves, hundreds of individually addressable microchambers on a single chip are fabricated [58], where single cells can then be stored for scRNA-seq [59]. The microchambers can be fully isolated using these chambers; however, if the cells are punched under a pressed valve, their contents will be dispersed throughout the microchannels.

2.1.6. Droplet-Based Microfluidics

Among various microfluidic single-cell isolation techniques, droplet-based microfluidics have drawn significant attention with applications in scRNA-seq. In this method, individual cells are encapsulated into droplets acting as tiny microchambers [60–63]. Various microfluidic chip designs for producing microdroplets have been proposed. These designs are generally categorized into three groups: co-flow [64,65], cross-flow [66–68], and flow focusing [69,70].

2.2. scRNA-Seq Methods

Drop-seq is one of the widely used scRNA-seq methods that use droplet-based microfluidic chips. It was developed by Dr. Macosko and was first introduced in 2015 [60]. Figure 1 illustrates a schematic of typical steps in Drop-seq. A flow-focusing droplet-based microfluidic chip is used to encapsulate single cells with single barcode-carrying beads into microdroplets. The microdroplets are formed at a cross-junction, where cell and bead suspension (to form the dispersed phase) and the oil (the continuous phase) meet. The dispersed passes through the junction and expands into the main channel. At the cross-junction, the continuous phase forms a neck in the dispersed phase, which then shrinks until it breaks up and the droplet forms. Based on the ratio of the viscous force to the inertial tension force, various droplet generation regimes may be observed (e.g., squeezing, dripping, or jetting). In Drop-seq and other droplet-based single-cell analysis tools, the dripping regime is commonly used to form highly monodispersed droplets.

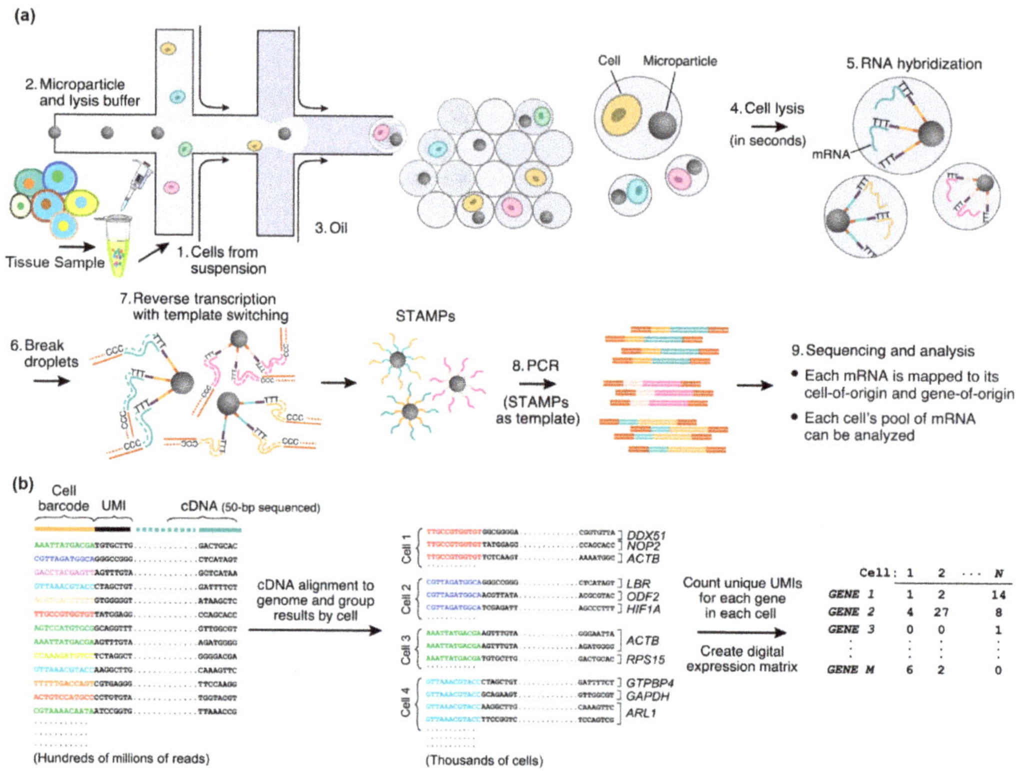

Figure 1. Schematic of the Drop-seq single-cell RNA-sequencing steps is illustrated. (**a**) The library preparation steps are shown. Single cells are separated from a tissue to form the single-cell suspension, which then forms the first aqueous phase to be injected into the microfluidic chip. The other aqueous phase carries the barcoded beads suspended in a lysis buffer. The droplet-based microfluidic chip in a T-junction joins these two aqueous flows, where they form the discrete phase (i.e., droplets encapsulating the required materials) in an oil continuous phase. In the formed microdroplets, the single cells are lysed, and their mRNAs are captured with the primers on the microparticles. Then, the droplets are broken, and the collected mRNAs are reverse-transcribed, forming STAMPs. After PCR, NGS and analysis are performed. (**b**) The analysis steps after sequencing are shown. Sequencing reads are aligned to a reference genome to find the gene of origin of the cDNA. Also, they are organized based on their cell barcodes to count UMIs for each gene in each cell, based on which the expression matrix is extracted. Reprinted from [60], with permission from Elsevier Copyright 2015 Elsevier Inc.

After droplet formation, the lysis buffer breaks the cell membrane, and their mRNAs, which are then captured with the barcode-carrying beads, are released. These single beads in each droplet contain oligonucleotides and unique molecular identifiers (UMIs). They contain primers for the reverse transcription of the mRNAs captured from single cells, and the resulting cDNA is sequenced to uncover the gene expression profile. The resulting sequencing data are then used to study the expression of thousands of genes within cells at the single-cell level. Compared with the old protocols, Drop-seq offers numerous advantages, including high throughput and reasonable cost per cell [71]. These features allow researchers to analyze tens of thousands of single cells in an experiment. This capability can be utilized to study cell heterogeneity, signaling pathways, and developmental trajectories. Drop-seq has been widely used in the biology scientific community and has fundamentally contributed to our understanding of single-cell biology and various diseases.

The data obtained from the sequencing need to be computationally analyzed. Because of the complexity of the information generated by scRNA-seq, a dimensional reduction step is needed before visualizing it. Some popular dimensional reduction methods are principal component analysis (PCA), t-Stochastic Neighbor Embedding (t-SNE), and Uniform Manifold Approximation and Projection (UMAP). The goal of these methods is to keep the most influential features and omit the ones with less impact. After recusing the dimension of the data, a unique coordinate is typically assigned to each cell such that the ones with similar genetic information are placed in close proximity to be clustered based on methods such as the k-nearest neighbor [17].

In Drop-seq, in order to prevent more than a single bead from being in a droplet, its concentration is lowered. However, numerous droplets then lack beads, which may not be a major problem in normal analysis. However, when the cell numbers in samples are limited, losing cells in droplets with no bead becomes challenging. To overcome this problem, InDrop technology is proposed [72]. This method is similar to Drop-seq, but it uses barcoded hydrogels as a more complex barcoding system compared with the beads, which allows for higher throughput and better cell type identification. In this method, high concentrations of hydrogels can be injected into the microfluidic chip, while their size and flexibility prevent having more than a single hydrogel in a droplet.

The gel beads in emulsion (GEMs) utilized in InDrop find analogous applications in the 10× Genomics Chromium system, which is another well-known tool used for scRNA-seq. Similar to Drop-seq, this method uses a microfluidic system to encapsulate individual cells with barcode-carrying beads and perform reverse transcription and cDNA amplification. The system is high-throughput and relatively easy to use, with a straightforward sample preparation protocol and data analysis workflows. Moreover, the system is compatible with a variety of sample types, including both fresh and frozen tissues, as well as cultured cells. The 10× Genomics Chromium system has been used in a variety of applications, including the identification of rare cell types, the characterization of complex tissues and organs, and the study of developmental processes and disease progression [12,73]. Drop-seq, InDrop, and 10× Genomics are widely considered the most successful scRNA-seq technologies in the field.

Seq-well is another scRNA-seq technology that was introduced in 2017 [74]. The Seq-well technology uses a microfluidic chip with thousands of individual wells. Each well is designed to capture a single cell and a single bead, where the cell is then lysed, and its RNA is captured with the bead and barcoded using UMIs. Similar to the methods mentioned above, the resulting cDNA is then amplified and sequenced using next-generation sequencing technology. Since this method does not require any specialized equipment, it is accessible to many researchers. However, because the microfluidic chip is loaded with cells and beads using pipets, no control over individual particles has been offered. As a result, the microwells may be filled with more than one cell or bead, which may be challenging.

In some scRNA-seq methods, as opposed to the microfluidic systems, microplates are used. Smart-seq is one such example that captures the full-length RNAs of single cells and amplifies them for high-throughput sequencing [75,76]. In this method, the cells are lysed in a hypotonic solution, and RNAs are converted to full-length cDNAs. After PCR amplification, the full-length cDNAs are used to construct standard Illumina sequencing libraries. Its updated version, called Smart-seq2, was developed in the lab of Dr. Sandberg and was first published in 2014 [77,78]. In Smart-seq2, reverse transcription (RT), template switching, and preamplification were optimized to offer an increased cDNA yield from single cells, better sensitivity, and enhanced repeatability.

Smart-seq3 combines full-length transcriptome coverage with a 5′ unique molecular identifier RNA counting strategy that enables the in silico reconstruction of thousands of RNA molecules per cell [79]. This technique is suitable for analyzing samples with limited cell numbers. In 2020, Smart-seq3, which combines full-length transcriptome coverage with a 5′ unique molecular identifier RNA counting method, came out. With Smart-seq3, more transcripts per cell compared with Smart-seq2 can be captured. All Smart-seq methods can be enhanced by providing more precise control over individual cells using microfluidic systems such as fLuidigm C1.

Another plate-based scRNA-seq method, in which FACS sorts single cells into 96- or 384-well plates, is the single-cell combinatorial indexed RNA sequencing (sci-RNA-seq) technique. It uses combinatorial cell barcoding and UMIs to enable the high-throughput analysis of large numbers of cells with high sensitivity and accuracy [80]. By barcoding individual cells with a unique combination of DNA tags, the method allows for the identification and quantification of each transcript in each cell. One key advantage of sci-RNA-seq is its ability to offer full-length transcript coverage. This feature allows the identification of novel splice variants and the accurate quantification of transcript isoforms. The method also offers high sensitivity and the detection of low-abundance transcripts and rare cell types.

MATQ-seq (Microbial RNA-Seq for Meta-Transcriptomics and -Transcriptome Analysis) is another high-throughput sequencing-based method, which is mostly used in a mixed microbial community [81]. This technique is based on pipetting single cells into PCR tubes and uses selective hybridization methods to capture mRNA, followed by cDNA synthesis, amplification, and sequencing steps. The PCR amplification step is replaced by in vitro transcription (IVT) in a method called CEL-Seq to increase efficiency [82]. Then, Cell-seq2, which has been optimized for higher sensitivity and less hands-on time, was introduced [83]. Although Cell-seq uses single-cell pipetting, Cell-seq2 works based on either robotic liquid handles or a Fluidigm C1 microfluidic chip.

Nanopore sequencing is one of the most promising techniques that offers long reads. In this approach, while the biomolecule zips through a nanopore in a membrane, it changes a characteristic identifying the biomolecule sequences [84]. The MinION (Oxford Nanopore Technologies, Oxford, UK) is one of the earliest commercialized real-time, portable nanopore sequencers to have been widely used. Single-cell RNA sequencing based on nanopore devices has also been presented [85,86]. Thus, this tool has the potential to be used in studying transcriptomes in transplanted organs.

Each of the mentioned methods has its own advantages and disadvantages, and the choice of method depends on the application of interest, the biology research question, and the available resources. Scientists may choose a method based on parameters such as cost, throughput, sensitivity, full-length transcript coverage, and compatibility with their samples. Table 1 lists the advantages and disadvantages of some of the important scRNA-seq methods.

Table 1. Comparison of some of the common scRNA-seq methods.

Methods	Advantages	Disadvantages	References
Smart-seq Smart-seq 2 Smart-seq 3	Full-length transcript coverage, high sensitivity, low technical noise	Low throughput, requires manual cell isolation, high cost per cell	[75,78,79]
Drop-seq	High throughput, low cost per cell, large-scale parallel processing, droplet-based microfluidics	Limited coverage of full-length transcripts, low sensitivity, high technical noise	[60]

Table 1. *Cont.*

Methods	Advantages	Disadvantages	References
In-Drop	High throughput, low cost per cell, large-scale parallel processing, droplet-based microfluidics, efficient for analyzing limited cell numbers compared with Drop-seq	Limited coverage of full-length transcripts, low sensitivity, high technical noise	[72]
10× Genomics Chromium	High throughput, easy to use, compatible with a wide range of samples, droplet-based microfluidics	Limited coverage of full-length transcripts (if paired with long reads technologies such as nanopore sequencing [87,88]), lower sensitivity compared with Smart-seq3	[89]
Seq-well	High throughput, low cost, easy scalability, ability to multiplex samples, based on microfluidics	Limited coverage of full-length transcripts, lower sensitivity compared with Smart-seq3, Needs manual pipetting.	[74]
sci-RNA-seq	High throughput, high sensitivity, full-length transcript coverage	More technically challenging than some other methods, requires specialized equipment	[80]
MATQ-seq	High throughput, low technical noise, high sensitivity, full-length transcript coverage	More technically challenging than some other methods. Needs manual pipetting.	[81]
Nanopore Sequencing	Long reads, portable	Relatively higher error rates	[87,90]

3. Single-Cell RNA Sequencing in Organ Transplantation

In this section, various organ transplantations in which scRNA-seq is used are discussed.

3.1. Heart Transplantation

By identifying the cellular composition of the cardiac system and the cell states before and after transplantation, crucial information about the graft can be obtained. This idea is investigated by using the scRNA-seq technique to find the single-cell transcriptomic atlas of human cardiac arteries and identify the cellular compositions in various cardiac arteries [91]. In these experiments, various cell populations, including vascular smooth muscle cells, fibroblasts, myofibroblasts, macrophages, T cells, B cells, endothelial cells, natural killer cells, mast cells, and oligodendrocytes are detected. These achievements are useful as a reference to find disease-associated cell populations in vascular and heart diseases. As shown in Figure 2, they have combined scRNA-seq (n = 7) with single-nucleus RNA-sequencing (snRNA-seq) (n = 38) to obtain data from heart samples (left ventricular (LV) tissues from 27 healthy donors and 18 patients with dilated (nonischemic) cardiomyopathy).

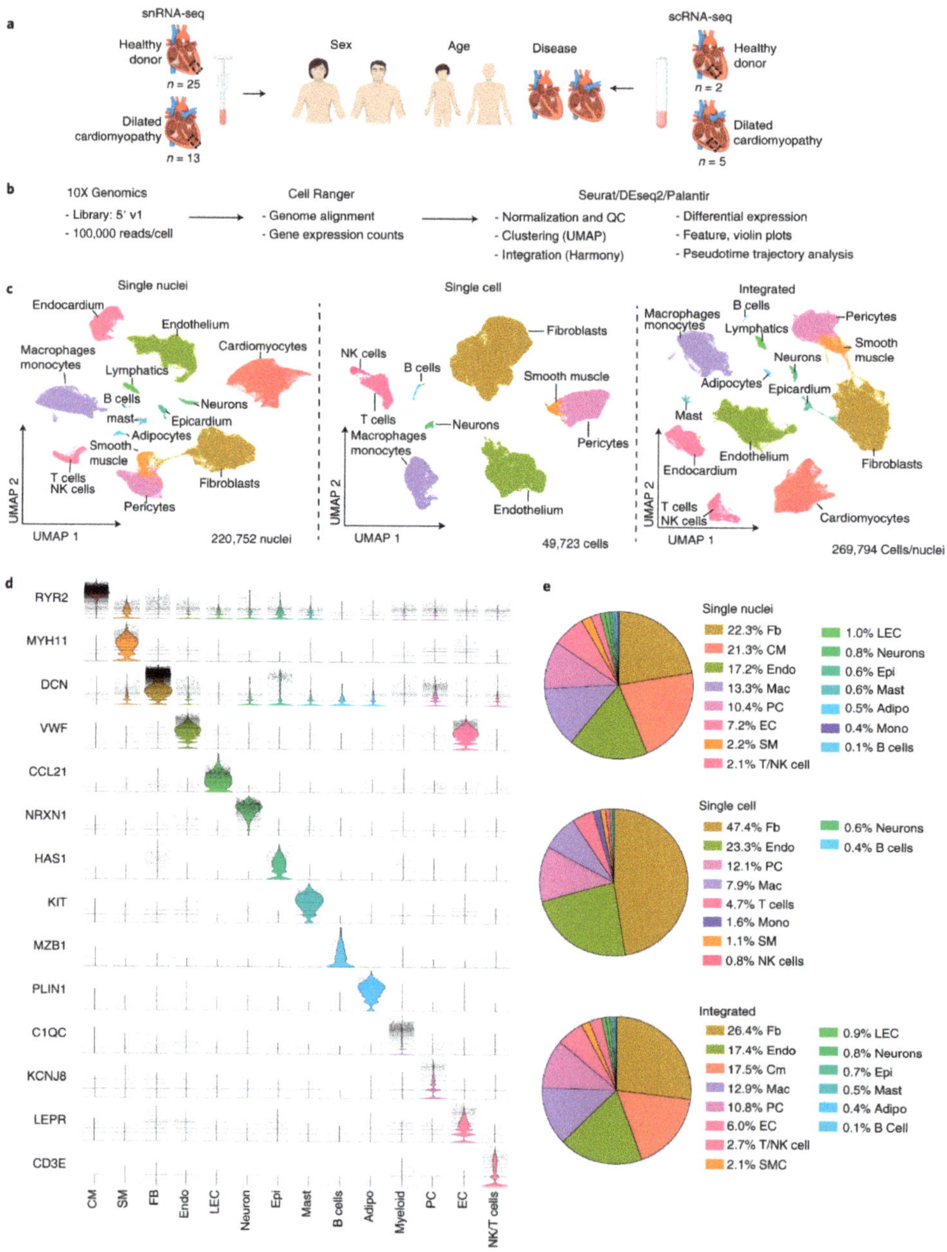

Figure 2. (**a**) Schematic representing the single-nucleus RNA-sequencing and single-cell RNA-sequencing experiments of heart tissues. (**b**) The analysis pipeline is depicted. It contains tissue preprocessing, library generation, alignment (Cell Ranger), and further data analysis. (**c**) Clustering. (**d**) Violin plots based on analyzing the integrated dataset showing characteristic biomarkers of identified cell populations. (**e**) Pie chart presenting the cell proportion. The figure is taken from [91] with permission under a Creative Commons Attribution 4.0 International License (http://creativecommons.org/licenses/by/4.0/, accessed on 1 February 2024).

In a similar study, researchers used scRNA-seq technology to identify intra-graft cell heterogeneity in acute heart allograft rejection in mice [7]. They employed the 10× Chromium platform, based on which about 2000 genes in almost 20,000 cells from two allogeneic heart grafts and about 20,000 cells from two syngeneic heart grafts were detected. They used this technique to identify the cell types (i.e., 21 distinct cell populations) and states associated with acute rejection and introduced the potential predictive biomarkers. They reported five cell clusters, including two resident macrophage groups, two infiltrating macrophage groups (predominantly from allogeneic grafts, one in an active state and one silent), and one dendritic cell-like monocyte group. The authors re-clustered endothelial cells into five subclusters, one of which was from allogeneic grafts. This cell population showed expression of Ubiquitin D, which they claimed was upregulated in heart and kidney rejection. Endothelial cells have been found by other researchers to show tissue-specific identities and unique transcriptomic signatures as well [92].

In another work, researchers used scRNA-seq to study transplant arteriosclerosis as a key challenge in long-term transplantation survival [93]. They used unbiased clustering analyses on a mouse model to identify 21 cell clusters at various disease stages and some novel subpopulations enriched in the allografts. They also reported the local formation of tertiary lymphoid tissues and possible intra-graft alloimmune response modulation. They reported the potential role of Ccl21a and Cxcr3 in regulating early chemotaxis and immune cell infiltration. They also used single-cell analysis to compare the immune response in mouse allograft and the atherosclerosis models. Researchers claim that their results depict both similarities and differences in atherosclerosis models and allograft-induced transplant arteriosclerosis. The innate and adaptive immune responses exist in the two models; however, distinct cell subpopulations may mediate the responses.

Based on the scRNA-seq results in a preprint [94], donor and recipient macrophage populations co-exist within the heart allograft. They claim that donor CCR2+ macrophages play a key role in allograft rejection, and MYD88 signaling inhibition in donor macrophages suppresses the allograft rejection. These results suggest that the signals from the transplanted graft from the donor and not the signals from the recipient macrophage populations define the destiny of the patient.

Researchers found cell-specific transcriptional signatures that are associated with age and heart failure [91]. They realized that cardiomyocytes go to common disease-associated cell states; however, fibroblasts and myeloid cells become diverse. They also found that endothelial cells and pericytes show transcriptional shifts.

Researchers compared cellular composition in various arteries in heart-transplanted patients and realized that some artery-specific vascular smooth muscle cell and fibroblast subpopulations exist [95]. In healthy conditions, the communication between vascular smooth muscle cells and fibroblast is reported to be dominant. They reported the enrichment of atherosclerosis-associated genes in endothelial cells and macrophages. They also reported that intercellular communication between endothelial cells and immune cells may increase during atherosclerosis. Based on this study, they believe that interactions between ICAM1/VCAM1 (EC1) and ITGB2 (immune cells, especially inflammatory macrophages) are important factors in the pathogenesis of atherosclerosis.

Various studies have shown the power of scRNA-seq in identifying previously unknown cell types and gene expression profiles in heart transplantation. In these experiments, different donor and receptor species have been considered. In some studies, human tissues are transplanted, and in other animal models of heart transplantation, they are demonstrated. In Table 2, some examples of these studies with their key findings are tabulated.

Table 2. Examples of key findings in applying scRNA-Seq on heart transplantation.

Key Findings	Methods/Technologies	Donor/Recipient Species	References
Cellular diversity: The scRNA-seq technique has been used to reveal the cellular heterogeneity in the heart, including immune cells, fibroblasts, endothelial cells, macrophages, and cardiomyocytes.	10× Genomics Chromium	Mice, human, pig	[91,96,97]
Immune cell populations in rejection: The scRNA-seq technique has identified various immune cell subsets involved in graft rejection, such as T cells, B cells, natural killer cells, and macrophages. Transcriptional profiles and functional states of these cells during rejection are considered key findings.	inDrop Microfluidics, 10× Genomics Chromium	Mice, human	[7,98,99]
Gene expression profile variations during rejection: Specific gene expression changes during rejection in various cell types (e.g., interferon-stimulated genes upregulation in T cells, proinflammatory pathways activation in macrophages, and upregulation of extracellular matrix genes in fibroblasts) are identified.	10× Genomics Chromium	Mice, human	[7,100–102]
Potential therapeutic target recognition: The scRNA-seq technique has been used to identify novel potential targets for therapeutic purposes in heart transplantation (e.g., IL-18 signaling and Hif1a inhibiting in T cells and CXCR6 in natural killer cells and T cells).	10× Genomics Chromium	Mice, human	[91,103,104]
Biomarker discovery: The scRNA-seq has been used to find gene expression signatures that can be considered biomarkers for predicting organ rejection or checking the responses to immunosuppressive therapy.	10× Genomics Chromium	Mice, human	[91,105]

3.2. Kidney Transplantation

The scRNA-seq technique has fundamentally leveraged our knowledge of renal cell identities and their genomic biomarkers [106]. These achievements identify the pathophysiology of kidney conditions, including early diabetic nephropathy [107], tumor compositions [108], and allograft rejection [5].

Multiple cell populations and subpopulations that are available in the kidney and scRNA-seq can identify this cellular heterogeneity at single-cell resolution. This identification can be enhanced by the integration of single-cell transcriptome and chromatin accessibility datasets. snRNA-seq and single nucleus assay for transposase-accessible chromatin using sequencing are used to generate cell-type-specific chromatin accessibility and transcriptional profiles of the kidney. Researchers have shown that most of the accessible chromatin regions are closely associated with the expressed genes. This multi-omics method allows the detection of unique cell states within the cellular population in the kidney [109].

In a recent study, a mouse kidney allograft rejection model and scRNA-seq were used to study CD45+ leukocytes in allografts on days seven and fifteen after transplantation [110]. Researchers detected 20 immune cell types (See Figure 3) and found that macrophages and CD8+ T cells made the main cell populations at both time points. They reported that in the transition from acute rejection toward chronic rejection, the proportion of CD8+ T cells dropped. However, the proportion of macrophages and dendritic cells highly increased, with $Ly6c^{lo}Mrc1^{+}$ and $Ly6c^{lo}Ear2^{+}$ macrophages being the main subgroups. They concluded that the drop in CD8+ T cells, B cells, and neutrophils and the rise in $Ly6c^{lo}Ear2^{+}$ and $Ly6c^{lo}Mrc1^{+}$ macrophages contribute to the transition from acute rejection to chronic rejection. Clonal CD8+ T cell responses have been reported to show important roles in rejection [111].

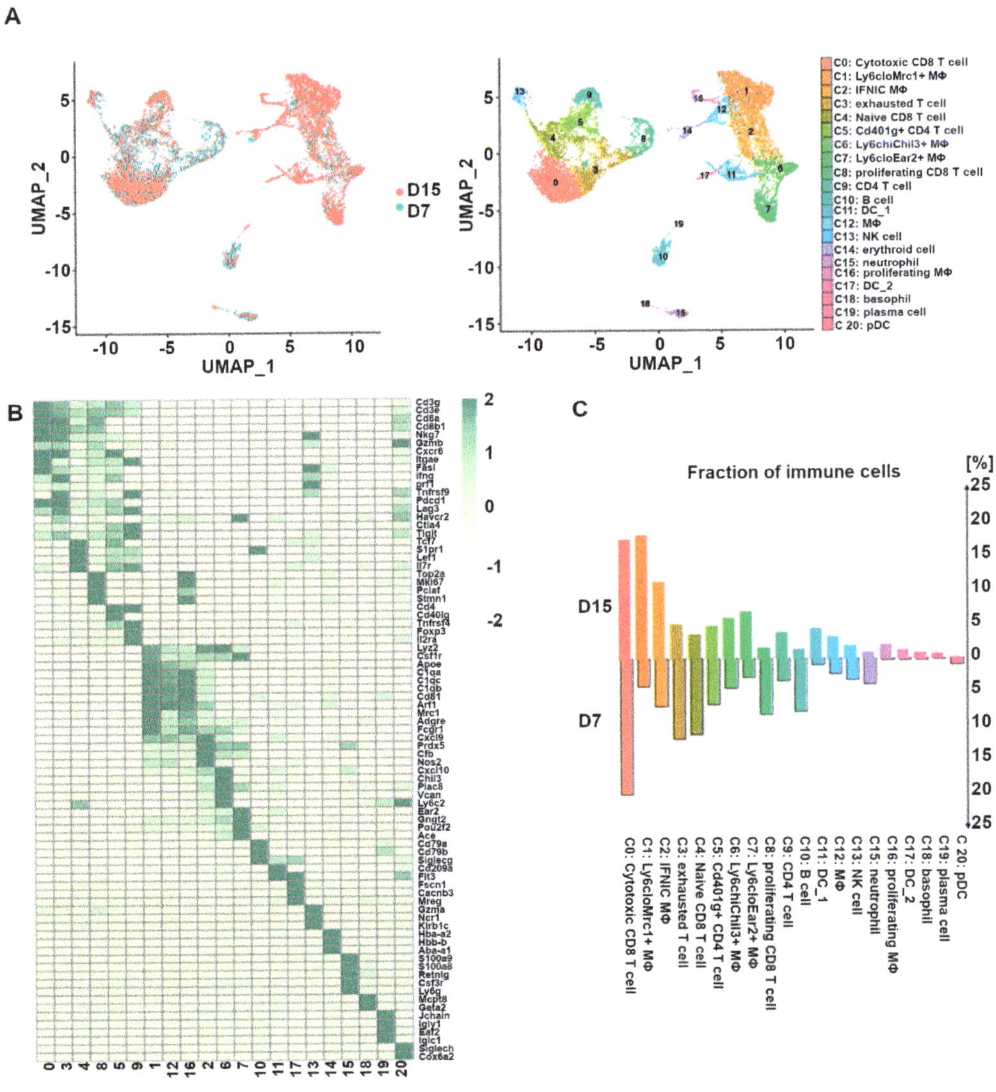

Figure 3. The presence of 20 CD45+ immune cell clusters is detected based on the scRNA-seq of mice kidney allograft. (**A**) UMAP plots of cell clusters identified based on biomarkers. Each dot represents a single cell, with its color demonstrating the cluster designation. (**B**) Heatmap representing the biomarkers of each cluster of kidney graft immune cells. (**C**) Bar plot demonstrating the proportions of the 20 identified immune cell populations in the kidney allografts obtained 7 days and 15 days after transplantation, respectively. The colors are chosen according to clusters in (**A**). UMAP, Mφ, DC, NK, and pDC stand for Uniform Manifold Approximation and Projection, macrophage, dendritic cell, natural killer, and plasmacytoid dendritic cell, respectively. The figure is taken and reproduced with permission from [110] under the terms of the Creative Commons Attribution License (CC BY) (http://creativecommons.org/licenses/by/4.0/, accessed on 1 February 2024).

Another research group employed kidney transplant biopsies and combined germline DNA sequencing with scRNA-seq to analyze the transcriptional profiles of donor- and recipient-derived leukocytes in acute antibody-mediated rejection and non-rejecting states [112]. They identified the major kidney cell types, as well as lymphocytes and macrophages. The ratio of leukocyte donor/recipient can be utilized as a rejection status indicator. They found that macrophages and T cells have distinct transcriptional profiles between the donor and recipient groups. They also claim that donor macrophages persist for years after graft transplantation.

Another group used scRNA-seq to identify 16 cell types and states in a human kidney biopsy specimen [5]. By comparing the results from a healthy adult kidney and a kidney transplant biopsy, they identified rejection-related, segment-specific proinflammatory responses. Of the three endothelial cell subclusters they found, two groups were active, one of which expressed the Fc receptor pathway and Ig internalization genes, which stands for antibody-mediated rejection.

Similarly, scRNA-seq has been applied to three healthy human kidneys and two chronic kidney transplant rejection (CKTR) biopsies [113]. Based on unsupervised clustering analysis of the obtained data, they identified 15 cell types (e.g., five natural killer T cell subclasses, CD4+ T cells, CD8+ T cells, cytotoxic T lymphocytes, regulatory T cells, natural killer cells, two memory B cell subtypes, CD14+ and CD16+ monocyte groups, and a novel subpopulation (myofibroblasts) in fibroblasts expressing collagen and extracellular matrix components). They also distinguished the CKTR group by the higher numbers of immune cells and myofibroblasts. The identification of the functional differences between the cell subpopulations and between samples from healthy and graft-rejected patients was based on the single-sample gene set enrichment (ssGSEA) algorithm. Other groups found a B cell subset ($CD19^{+}IGLC_{3}^{low}IGKC^{high}TCL_{1}A\text{-}CD_{127}^{+}$) to be much lower in renal allograft recipients with accommodation (i.e., allograft recipients that are neither rejected nor infected after immunosuppression) than that in healthy people [114].

B cell subsets have attracted the attention of other researchers as well. Clark and coworkers showed that in kidney rejection, infiltrating B cells contributes to specific innate signaling pathways, which may be related to inflammation [115]. In particular, they showed that B-innate cells generate inflammation-specific antibodies and drive local inflammation in transplanted kidney rejection.

Researchers believe that myeloid cells play a key role in transplant rejection [116,117]. They used scRNA-seq to study the murine kidney transplantation model to study the contribution of these cell subsets and their signaling pathways in graft rejection [118]. They showed that kidney allograft-infiltrating myeloid cells differentiate from monocytes to proinflammatory macrophages. They also identified Axl as a key gene in the differentiation of proinflammatory macrophages in transplanted kidneys, which also promotes the differentiation and proliferation of donor-specific T cells. scRNA-seq analysis has also been used to detect and study the type and status of monocytes/macrophages in kidney transplantation [119]. They form two different subpopulations: resident and infiltrating monocytes/macrophages. The number of resident macrophages decreases during kidney rejection. In these macrophages, the relevant genes during phagocytosis are upregulated.

The scRNA-seq technique has also been used in kidney organoid transplantation [120]. Human-induced pluripotent stem cell (iPSC)-derived kidney organoids have multiple nephron-like structures that show some renal functions [121]. Although these organoids have attracted much attention in disease modeling and drug screening applications, their reproducibility and reduction in off-target cell generation is a challenge. Researchers have used scRNA-seq to answer this need [120] and showed that cell proportion variations exist between different iPSC lines mainly because of off-target cells. They also analyzed transplanted organoids in mice and realized that off-target cells diminish after transplantation.

In addition to the transcriptomic profile of the cell types within the allograft, spatial transcriptomics in acute kidney injury is important and affects the cells heterogeneously. Researchers identified the spatial transcriptomic signature of ischemia/reperfusion injury and cecal ligation puncture as two murine acute kidney injury models [122]. They realized that spatial reduced expression is associated with the injury pathways. They used scRNA-seq to find epithelial cell–immune dialogue in kidney injury spatially.

Single-cell RNA sequencing has also been applied to kidney solid biopsies and CD14^{+} peripheral blood mononuclear cells (PBMCs) obtained from Immunoglobulin A nephropathy (IgAN), and the results have been compared to sequences of normal samples [123]. This analysis is used to study the molecular events in IgAN progression that can be used in disease treatment. In IgAN, JCHAIN upregulation is related to the in situ dimerization and deposition of IgA1. The pathological mesangium also indicates high cell–cell communication between renal parenchymal cells and immune cells. Also, the expression of genes specific to kidney-resident macrophages and CD8+ T cells depict abnormal regulation related to proliferation and inflammation.

Although analyzing kidney biopsy samples has resulted in valuable information, biopsy-associated complications, biopsy specimen scoring variability, invasiveness, and costs are considered concerns in the field. Also, studying the kidney behavior at multiple time points after transplantation to understand its dynamic behavior is an important need; however, repeated biopsies are not convenient and safe, if practically possible [124]. Urinary and PBMCs from kidney transplant recipients are other sources of organ injury biomarkers [125]. Transcriptomic analysis and clustering of these cells are considered noninvasive analysis methods on which researchers are working. Muthukumar, from Weill Cornell Medical College, and his coworkers performed scRNA-seq on urinary cells obtained from kidney transplant recipients, with biopsies classified as acute T cell-mediated rejection, chronic active antibody-mediated rejection, and normal conditions [126]. They claimed that urine samples matched to the acute T cell-mediated rejection biopsy showed increased macrophages, dendritic cells, T cells, and NK cells, while the one matched to normal biopsy displayed dominant kidney tubular epithelial cells. This approach is an innovative method for uncovering the complex cellular landscape of kidney allograft rejection at the single-cell resolution.

Researchers have generated the profiles of various PBMC cell types and their gene expression using scRNA-seq in chronic antibody-mediated rejection patients [127]. Based on their results, MT-ND6, CCL4L2, CXCR4, NFKBIZ, DUSP1, and ZFP36 were upregulated in these patients. They also reported that MAPK and NFκB signaling pathways were activated. They claimed that single-cell sequencing is a potential strategy for understanding the details of the peripheral blood lymphocyte in chronic antibody-mediated rejection patients.

It has been shown that the integration of spatial and single-cell transcriptomics can localize the cell–cell communication between the epithelial and immune cells [128] in kidney injury. This method is used to spatially map the transcriptomic signature of acute kidney injury in murine models, which can also be applied to human kidney tissue [122]. In Table 3, some key findings of applying scRNA-seq on kidney transplantation are listed. Also, some genes play the biomarker role for transplant problems. In Table 4, some biomarkers in different kidney transplantation studies are tabulated.

Table 3. Examples of key findings in applying scRNA-Seq on kidney transplantation.

Key Findings	Methods/Technologies	Donor/Recipient Species	References
Cellular diversity: The scRNA-seq technique has been used to reveal the cellular heterogeneity in the kidney, including immune cells, macrophages, IFNg, myeloid, and T cell subclusters. These heterogeneities represent distinct signatures that have different roles in allograft loss.	10× Genomics	Mice, human	[5,112,113,118,129,130]

Table 3. *Cont.*

Key Findings	Methods/Technologies	Donor/Recipient Species	References
Immune cell populations in rejection: The scRNA-seq technique has identified various immune cell subsets involved in graft rejection, such as T cells, B cells, neutrophils, myeloid cells, dendritic cells, stromal cells, and macrophages. Transcriptional profiles and functional states of these cells during rejection are considered key findings.	10× Genomics	Mice, human	[113,115,116,129]
Gene expression profile variations during rejection: Altering myeloid cell differentiation and their behavior based on upregulating expressions of ribosomal protein genes may affect the implant.	10× Genomics	Mice, human	[112,116,118,127]
Potential therapeutic target recognition: The scRNA-seq technique has been used to identify novel potential targets for therapeutic purposes in kidney transplantation.	10× Genomics	Mice, human	[5,111,116,127,131]
Biomarker discovery: The scRNA-seq technique has been used to find gene expression signatures that can be considered biomarkers for predicting organ rejection or checking the responses to immunosuppressive therapy.	10× Genomics,	Mice	[17,132,133]
Uncover novel cell types: The scRNA-seq technique assists in finding novel cell types and statuses without any bias or RNA degradation.	10× Genomics,	Human, mice	[113,129,131,134]
Cells (e.g., some glomerular endothelial cells) from the recipient or donor (e.g., the renal architecture) may represent endothelial chimerism.	10× Genomics	Human	[116]
The scRNA-seq technique shows that leukocyte populations mostly express sex-linked genes from recipients, which may be linked to immune cell infiltration. For example, natural killer cells and monocytes are involved in kidney rejection,	10× Genomics	Human	[116]

Table 4. Examples of important biomarkers of kidney transplantation complications.

Gene Biomarkers	Methods/Technologies	Donor/Receptor	References
IFNg, GSVA, and DEGs	10× Genomics	Mice	[110]
RTK and Axl	10× Genomics	Mice	[118]
PDGF, ECM, and TGF-β	10× Genomics	Human	[113]
Nphs2CremT/mG, SclCremT/mG, Cdh16CremT/mG, AQP3, and HSD11B2	Droplet-based	Mice	[131]
CXCL10	10× Genomics	Human	[116]
TRDC, CD4, CD8A, KLRK1, ITGAX, CD19, and CD14	10× Genomics	Human	[111]
PGs, *GGT5*, and *EMILIN1*	10× Genomics, Drop-seq	Human	[129]

Table 4. *Cont.*

Gene Biomarkers	Methods/Technologies	Donor/Receptor	References
CD3E, MS4A1, SDC1 (CD138), and TPSAB1	Droplet Microfluidics	Human	[112]
ALDOB, GATM, GPX35, JUN, VIM, HSP, ALDOB, GPX3, GATM, CTGF, CXCL12. CAV1, COL4A1/A2, VIM, COL4A2, and VWF	Droplet Microfluidics	Human	[134]
TFAIP3, CXCR4, ZFP36, S100A8, S100A9, CXCL8, FOS, MTND6, HLA-DQA2, MT-ND6, CXCL8, NFKBIA, CD69, CD83, and HLA-DQA2			[127]
CD19 and CCR6		Human, mice	[115]
CD16+, CD162, ABCA1, APOE, PDE3A, IGKC, LGMN, iCD83, FCGR3A, CD16, and FCN1	Droplet Microfluidics	Human	[5]

3.3. Lung Transplantation

Identifying the donor and recipient cells in transplant biology is an important task that is possible with scRNA-seq. One algorithm toward this goal is called scTx, which identifies the donor and recipient genotypes using expressed single nucleotide variants and assigns the cells to a genotype [135]. The authors tested their proposed algorithm on lung transplanted samples and claimed that it could detect two genotypes from post-transplant bronchoalveolar lavage and lungs with chronic lung allograft dysfunction samples.

Because of chronic rejection, bronchiolitis obliterans syndrome is a big challenge and the key reason for weak lung transplantation outcomes. In a recent work [136], researchers used single-cell RNA sequencing to provide an atlas of bronchiolitis obliterans syndrome after lung transplantation outcomes. This atlas can be used to identify the changes in the cell compositions and their individual gene expression profiles during lung rejection. They found that the Mzb1-expressing plasma cell population in the lungs with bronchiolitis obliterans syndrome increased more than the others. Also, CD14-expressing monocytes and PDGFRA-expressing fibroblasts were increased. They also performed pseudo-time and trajectory analysis, based on which they found that a Bhlhe41, Cxcr3, ITGB1-triple positive-B cell subset plays as the progenitor pool for Mzb1+ PCs, which results in IgG2c expression and production in the grafts with bronchiolitis obliterans syndrome.

Researchers have investigated the generation, maintenance, and function of human lung tissue-resident memory T cells in transplanted lung samples [137]. They dynamically tracked the donor and recipient T cells. They realized that the donor T cells remain in the transplanted lungs and highly express their markers, including CD69, CD103, and CD49a; however, the lung-infiltrating recipient T cells acquire the phenotypes over months. By using scRNA-seq, they identified two donor T cell subsets with different marker gene expressions; however, recipient T cells were composed of non-tissue-resident memory T cells and tissue-resident memory T cells-like subpopulations, suggesting de novo TRM generation.

Researchers have argued that RNA sequencing analysis indicates that lung disease after severe and prolonged SARS-CoV-2 infection shows pathological and molecular features similar to the ones in pulmonary fibrosis requiring transplantation [138]. This finding suggests that lung transplantation might be necessary for these affected individuals. They also report successful lung transplantation for these patients.

Figure 4 illustrates the single-cell analysis results of a mouse lung graft in which 11 cell populations were found. In Table 5, some key findings in lung transplantation based on scRNA-seq are listed. Also, related biomarkers in lung transplantation are listed in Table 6.

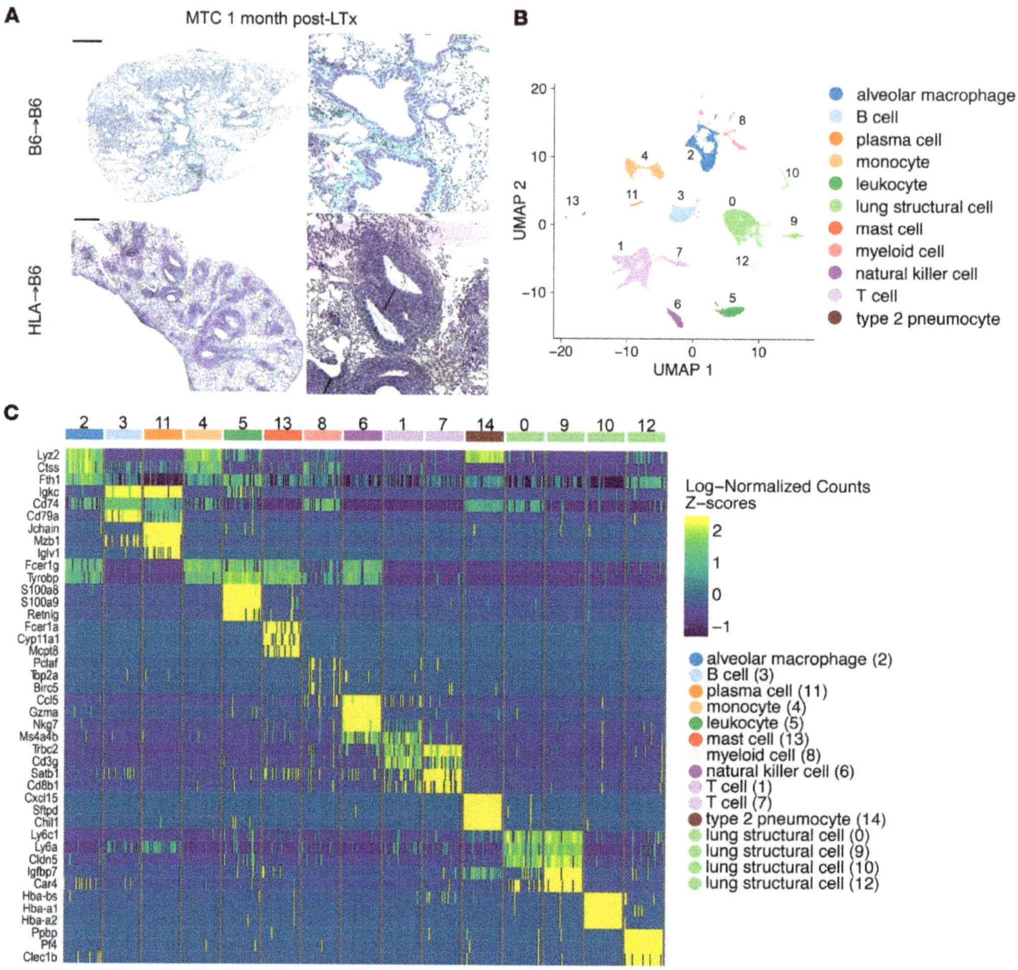

Figure 4. (**A**) Histology sections of control (B6→B6) and bronchiolitis obliterans syndrome (HLA→B6) mouse lung graft one month post-transplantation. Scale bars, 500 μm. Zoomed views are shown on the right. (**B**) UMAP plots of 11 cell populations. (**C**) Heatmap illustrating most upregulated genes in each cell cluster. The figure is taken with permission from [136] under Creative Commons Attribution 4.0 International License (https://creativecommons.org/licenses/by/4.0/), accessed on 1 February 2024).

Table 5. Examples of key findings of applying scRNA-seq to lung transplantation.

Key Findings	Methods/Technologies	Donor/Recipient Species	References
Cellular diversity: The scRNA-seq technique is reported to identify cell populations associated with bronchiolitis obliterans syndrome.	10× Genomics	Mice, human	[136,139–144]
Immune cell populations in rejection: In acute cellular rejection, a clonal population of cytotoxic and effector CD8+ T cells exist in the transplanted lung and remain after treatment.	10× Genomics	Mice, human	[136,139,140,145]

Table 5. *Cont.*

Key Findings	Methods/Technologies	Donor/Recipient Species	References
Potential therapeutic target recognition: A subgroup of innate B-1 cells may contribute to autoimmunity in bronchiolitis obliterans syndrome, which represents a potential therapeutic method.	10× Genomics	Mice, human	[136,139,140]
Biomarker discovery: The scRNA-seq technique has been used to find gene expression signatures that can be considered biomarkers for predicting organ rejection or checking the responses to immunosuppressive therapy.	10× Genomics	Mice, human	[136,143]
Macrophage polarization: Macrophages are found to be heterogeneous cell populations, which upon activation, polarize into various phenotypes. After transplantation, tissue-resident macrophages quickly change their gene expression profile into that of the host organ markers.	10× Genomics	Mice, human	[146,147]
Uncovering novel cell types: For example, additional endothelial and lymphatic cell populations, megakaryocytes, innate lymphoid cells, and mesothelial cells have been identified in mice.	10× Genomics	Mice	[148]

Table 6. Examples of important biomarkers of lung transplantation complications.

References	Methods/Technologies	Cells	Donor/Receptor	Gene Biomarkers
[149]	10× Genomics	Emphysema, cystic fibrosis, sarcoidosis	Human	CD6914, CD103, CD69+, CD137+, CD69+ and/or OX40+
[136]	10× Genomics	COPD, CTD-ILD	Human, mouse	Bhlhe41, Zbtb20, Cxcr3, Itgb1, CD19, CD43, CD5, Xbp1, Sdc1, Mzb1, Irf4, Ighm,
[141]	10× Genomics	Bronchopulmonary dysplasia	Mouse, human	Epcam, Pecam1, Ptprc, Col1a1, Msln,
[142]	10× Genomics	Adenocarcinoma, endobronchial carcinoid, LLL endobronchial typical carcinoid,	Mouse, human	EPCAM, CLDN5, COL1A2, PTPRC, CD31, CD45, KRT5, MKI67, SERPINB3, C20orf85, CLDN5, MYC, ACKR1, ACKR1, GJA5, CCL21, CLDN5 with DAPI, COL1A2, GPC3, Slc7a10, SERPINF1, Pi16, ASPN, COX4I2, COL1A2, APOE, GPR183, Slc7a10,
[144]	RNeasy Plus Mini kit (QIA GEN)	Fibrotic lung disease, idiopathic pulmonary fibrosis, systemic sclerosis-associated interstitial lung disease, interstitial pneumonitis, pneumoconiosis	Mouse, human	CD206, CD169

3.4. Liver Transplantation

Liver transplantation is considered a treatment strategy for patients with hepatocellular carcinoma, and scRNA-seq is a promising technique for detecting associated problems. By using flow cytometry, some cell subpopulations in liver samples can be detected [150,151]; however, unbiased scRNA-seq shows that the heterogeneous cell population in the liver is much wider than what is detected by the flow cytometry and consists of numerous subpopulations [152]. Most importantly, researchers used this technology (10× Genomics) to obtain an unbiased and comprehensive liver transplant cell atlas by collecting liver tissue samples pre-procurement, post-preservation, and two hours post-reperfusion [153]. They identified different cell subgroups, their transcriptome changes, and the interactions between them during liver transplantation. The results of this work can be used to realize the cellular and molecular mechanism of graft ischemia–reperfusion injury during liver transplantation. See Figure 5 for scRNA-seq of endothelial cells in liver grafts.

Another research group used single-cell analysis on transplanted liver samples and detected a subset of CSF3+ Kupffer cells that is related to the injuries associated with graft transplantation injury [154]. They also found higher levels of dendritic cells and CD8+ T cells in the fatty liver donors. In a preprint, authors classified the liver cells into 14 cell types and 29 subpopulations with different cell states [155]. They claimed to have found pathogenic cellular modules associated with early allograft dysfunction, consisting of mucosal-associated invariant T cells, granzyme B, granzyme K, natural killer cells, and S100A12 neutrophils.

Recently, based on scRNA-seq, the complex landscapes of organogenesis containing liver development and decidualization were analyzed [156]. In this study, scRNA-seq and cytometry by time of flight (CyTOF) were used to uncover the cell states and sources involved in liver graft remodeling. They also used transcriptome data to show the interplay among hepatocytes and macrophages. The transcriptomic data they obtained revealed that the complexity of the metabolic remodeling of the transplanted liver is a complex task in which a regulatory network of ligands and receptors among macrophages and hepatocytes is involved.

Subpopulations of various hepatic cell types containing macrophages, epithelial progenitor cells, and myofibroblasts and their behavior were uncovered with the application of scRNA-seq on human and zebrafish livers [157]. They applied scRNA-seq on single cells obtained from the livers of 18-month-old male zebrafish to uncover the transcriptional profiles of the cell types available in the liver and to use them as tools to understand liver function and diseases. They determined the similarities between the transcriptomic data of adult zebrafish liver and the human liver single-cell transcriptome. The next step of this study could be on transplanted livers in zebrafish.

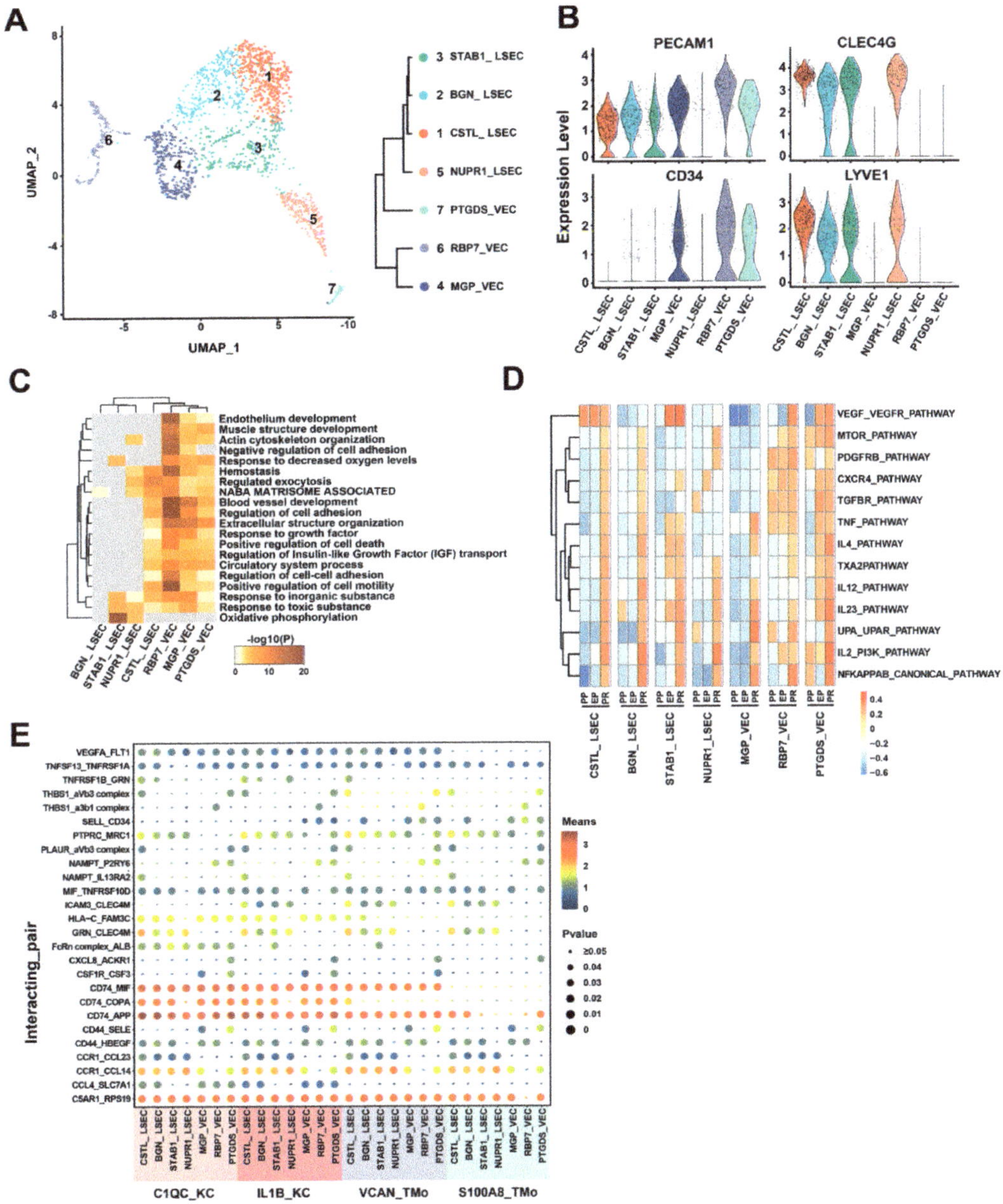

Figure 5. Single-cell RNA-sequencing of endothelial cells after liver transplantation. (**A**) UMAP plot presenting seven endothelial cell clusters (**left**). Dendrogram of the seven clusters (**right**). (**B**) Violin plots demonstrating the normalized expression of PECAM1, CLEC4G, CD34, and LYVE1 genes. (**C**) Gene Ontology enrichment analysis results. (**D**) Gene set variation analysis identifying the pathways. Different colors stand for different activation scores. (**E**) Cell-cell interaction analysis between mononuclear phagocyte clusters and different endothelial cell clusters. The circle size stands for the level of p-value. Colors stand for different mean values. The figure is taken from [153] with permission under a Creative Commons Attribution 4.0 International License (http://creativecommons.org/licenses/by/4.0/, accessed on 1 February 2024).

Key findings and biomarkers of scRNA-seq in liver transplantation are listed in Tables 7 and 8, respectively.

Table 7. Examples of key findings in applying scRNA-seq on liver transplantation.

Key Findings	Methods/Technologies	Donor/Recipient Species	References
Cellular diversity: The scRNA-seq technique has been used to uncover the cellular heterogeneity in the liver, including immune cells, macrophages, IFNg, myeloid, and T cell subclusters. These heterogeneities highlight signatures with different roles in allograft complications.	10× Genomics	Human, rat	[155,158–165]
Immune cell populations in rejection: The scRNA-seq technique has identified various immune cell subsets involved in graft rejection, such as T cells, B cells, neutrophils, myeloid cells, dendritic cells, stromal cells, macrophages, and their transcriptional profiles and functional states during organ rejection. Some cell populations, including IL-7R+CD4+ T cell, and CRTAM+CD8+ T cell, are shown to be reduced in the transplanted liver.	10× Genomics	Human, mouse	
Gene expression profile variations during rejection: Gene expression variation of B cells in bronchiolitis obliterans syndrome is uncovered.	10× Genomics	Mice	[136]
Potential therapeutic target recognition: The scRNA-seq technique has been used to identify novel potential targets for therapeutic purposes in liver transplantation. For example, it helps to understand the heterogeneity of LDLR+MDSC and CTLA4+CD8+ T, especially CD4+CD8+FOXP3 T cells, which may result in finding innovative therapeutic methods.	10× Genomics	Human	[158]
Biomarker discovery: Machine learning and scRNA-seq have helped in identifying novel biomarkers.	10× Genomics	Human	[166]
Macrophage polarization	10× Genomics	Human	[152,167]
Uncovering novel cell types	10× Genomics	Human, rat	[155,158,162,163, 165]

Table 8. Examples of important biomarkers of liver transplantation complications.

References	Sequencing Method	Cells	Donor/Receptor	Gene Biomarkers
[158]	10× Genomics	Liver cirrhosis, hepatocellular carcinoma	human	LDLR, GZMB, GZMA, GZMB, GZMH, NKG7, GZMK, DUSP4, and COTL1
[159]	10× Genomics Chromium	Hepatocellular carcinoma, cirrhosis of the liver	human	TCRs, BCRs, CD3D, KLRF1, CD79A, IGHG1, CD177, CD68, PECAM1, KRT7. CD4+ T cell lineages, CD4+, (Tem, GZMK), CD4+ (CCR7, LEF1), CD4+, (MAIT, SLC4A10), (MKI67)
[155]	10× Genomics	Hepatocellular carcinoma and primary sclerosing cholangitis disease	Human, rat	S100A12, LTF, PRTN3

Table 8. *Cont.*

References	Sequencing Method	Cells	Donor/Receptor	Gene Biomarkers
[152]	10× Genomics	Liver cancer	Human, mouse	KRT19 (CK19), EPCAM, FXDY2, CLDN4, CLDN10, SOX9, MMP7, CXCL1, CFTR, TFF2, KRT7 (CK7), CD24
[160]	10× Genomics	Para-tumor liver tissue, cirrhotic	human	EPCAM, SOX9, AFP, KRT7, S100A6, S100A11, ALB, PCK1, FGG, FGA, TTR, EBPB, APOB, CYP2E1, APOE
[161]	Seven Bridges Genomics	Cholangiosepsis	human	CD15, CD68, CD3, CD8, CD20, FCGR3B, CD68, CD3E, CD4, CD8A, Tregs, FOXP3, NKG7, FLT3, CD24, CD79A, JCHAIN, ALB, FLT1, KRT19, IFITM2, CSF3R, FPR1, FCGR3B, VNN2, G0S2, CXCR2, SOD2. CXCR2, CXCR4, CD83, CCRL2, CCL3, CCL4, ICAM1, VEGFA, CST3, CTSB, MS4A7, MARCH1, CD68, MAFB, CD163, VCAN, CSF1R, LYZ, VCAN, S100A8, S100A9, S100A12, MNDA
[162]	10× Genomics	Chronic hepatitis B (CHB), HBV-associated liver cirrhosis (LC) patients	human	CD3D, KLRF1, CD19, SDC1, CD14, FCGR3A
[155]	10× Genomics	Chronic liver disease	Mice	S100A6, Ccl2, Cxcl1, Cxcl12, Col1a2, Col3a1, Col5a2
[163]	10× Genomics	Human liver cirrhosis	Human, mice	MNDA, CD9, TIMD4
[164]	CEL-seq2	Colorectal cancer metastasis or cholangiocarci-noma	Human, mouse	AKR1B10, MKI67, PCNA, ALB, HP, HNF4A, ASGR1, PROX1, KRT19, CFTR, ASGR1 plus ALB, CXCL8 plus MMP7, PECAM1, CLEC4G, CD34, CLEC4M and FLT1
[165]	10× Chromium Smart- seq2	Hepatocellular carcinoma (HCC)	Human	CD14, CD2, CD3D, CD4, CD68, LYZ, MS4A1
[168]	10× Chromium	Nonalcoholic steatohepatitis (NASH), HCV	Human	CD45, CD31, CD68, CD146, SSC-A, PDPN, CCL21, LYVE1, FLT4, PROX1
[169]	10× Chromium Smart- seq2	Solitary colorectal metastasis	Mouse	Mki67, Col1a1+, NGFR, Adamtsl2
[170]	Droplet-based sequencing and data analysis, 1× Genomics	Cholangiocarcinoma	Mouse	CD68, CK-19, MHCII, MHCI, CD45, CD11b, Ly6G, Ly6C, CD19, CD115, B220, TER-119, Tim4, NK1.1, MERTK, CD8a, CD3e, TCRb, CD206, Lgals3, CD11c, CX3CR1, CCR2, F4/80, CD14, CD64
[171]	DNBSEQ-G400RS (MGI Tech)	Cholangiocarcinoma	Mouse	Alb, Apoa1, Ass1, Spp1, Sox9

3.5. Other Transplants

The scRNA-seq technique is also used in organ transplant recipients with squamous cell carcinoma. This method is combined with T-cell receptor sequencing to define the T-cell prospect in cutaneous squamous cell carcinoma [172]. This method is used to find tumor-infiltrating lymphocyte phenotype in squamous cell carcinoma from immune-competent and immune-suppressed patients. CD8+ T cells from the samples were sequenced to distinguish various T cell populations. Also, the T cell immune response was characterized by sequencing the α and β CDR3 regions.

4. Stem Cell Transplantation

Stem cell transplantation is considered a promising therapeutic method for various diseases, including blood diseases, immune system diseases, neurodegeneration, and cancer [173–175]. However, to be considered safe and reliable, a clear understanding of the cell behavior after transplantation is needed. Stem cells have shown promise in treating lung diseases. Pulmonary fibrosis (PF) is an example of a chronic lung disease for which treatment adipose-derived mesenchymal stem cells (ADSCs) are considered candidates. To understand the underlying mechanisms, the interaction between ADSCs and lung cells was studied at the single-cell level [176]. Using scRNA-seq, the authors realized that ADSC treatment changed both the transcriptomic profile and the composition of the lung cells, especially macrophages. They also identified potential signaling pathways, such as NGR, ANNEXIN, HGF, and PERIOSTIN. They found that the ADSCs increased the Trem2+ anti-inflammatory lung macrophages. They also decreased inflammation and fibrosis in the lung.

The umbilical cord blood (UCB) transplant is considered a promising therapeutic option for multiple diseases (e.g., blood cancers, myeloproliferative disorders, and genetic diseases). Single-cell analysis has revealed the cellular heterogeneity in the nucleated cells in UCB [177]. The authors reported 12 major cell types with multiple subpopulations.

Researchers extracted the single-cell full-length transcriptome data to construct an isoform-based transcriptional atlas of the murine endothelial-to-hematopoietic stem cell transition [178]. They used the obtained atlas to identify the hemogenic signature isoforms and the alternative splicing events. The results are crucially important because the transcribed mRNAs typically undergo alternative splicing, which affects the transcript isoforms and results in different proteins.

Allogeneic hematopoietic stem cell transplantation is considered a treatment method for malignant hematological diseases. Tracking of T-cells in transplantation is important and may uncover information about the graft-versus-leukemia effect. Researchers used single-cell RNA sequencing to extract the transcriptomic data of ~35,000 single T cells in the blood of 14 samples before and after transplantation [179]. They reported a huge drop in unique T-cell clones post-transplantation compared with the donor samples.

To answer the need for cell replacement in diabetes, human pluripotent stem cells differentiated into insulin-secreting β cells in islet organoids can be used. However, the behavior of these cells in vitro compared with native adult β cells is different. Single-cell transcriptomic profiling can detect the transcriptomic changes in these cells. By using this method, researchers have realized that transplanted insulin-secreting β cells show a behavior closer to the adult β cells [180]. They showed an increase in the insulin and IAPP protein secretions after transplantation. The obtained results of this study provide a wealth of information about the human islet cell maturation, as well as the maturation of the insulin-secreting β cells.

Stem cells have been transplanted for brain disorder treatment as well. The cells to be transplanted are usually obtained from the fetal brain tissue or the stem cells. After transplantation, the dopamine neurons are rare, determining the identity of other cell types. Researchers used single-cell RNA sequencing on a rat model to characterize the grafts from the human embryonic stem cells and fetal tissue [181]. They found a high level of neurons and astrocytes in both cases, while they found an additional perivascular-like cell type in the stem cell-derived grafts. Figure 6 illustrates the performed scRNA-seq analysis, as well as the histological validation of the transplanted cells into the midbrain.

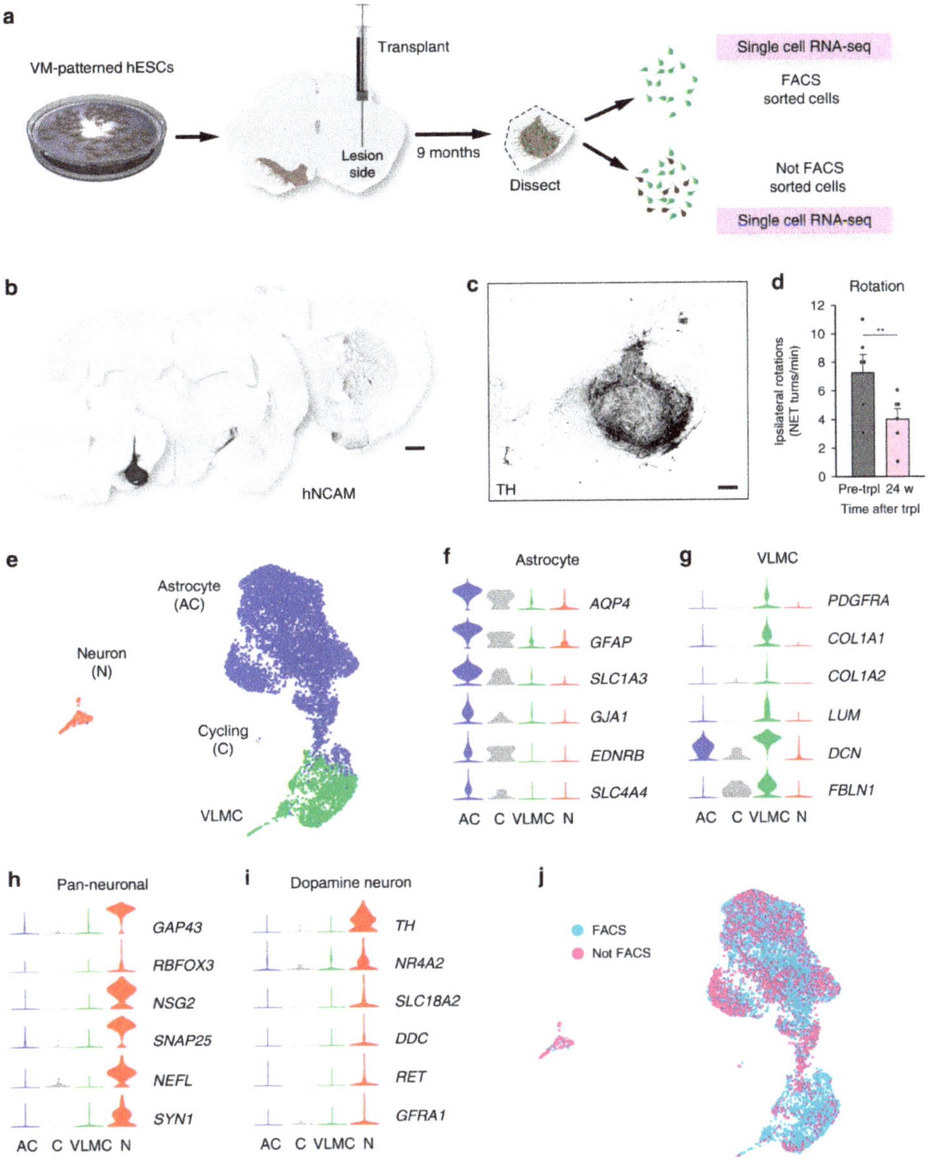

Figure 6. The scRNA-seq technique and histological validation of transplanted cells into the midbrain. (**a**) Schematic of experiments. VM-patterned hESCs: ventral midbrain-patterned human embryonic stem cells. (**b**) Growth of hNCAM fiber from hESC-derived graft. (**c**) Immunohistochemistry. (**d**) Drug-induced rotation test that shows functional recovery in rats after transplantation ($n = 6$ rats; mean ± SEM; ** $p < 0.01$; compared to post-lesion; two-tailed paired t-test). (**e**) UMAP plots of 7875 analyzed cells after transplantation. (**f–i**) Expression level for each cluster of biomarkers. (**j**) UMAP of transplanted cells for cells isolated by FACS (blue circles, $n = 5958$) and not by FACS (magenta circles, $n = 1917$). Scale bars, 1 mm (**b**); 200 μM (**c**). The figure is taken with permission from [181] under a Creative Commons Attribution 4.0 International License (http://creativecommons.org/licenses/by/4.0/, accessed on 1 February 2024).

Muscle stem cells maintain their regenerative capabilities after transplantation into recipient hosts [182,183]. However, primary myoblasts do not show the ability to engraft and proliferate after transplantation. The scRNA-seq technique is used to uncover the transcriptional state and developmental dynamic trajectories of injured muscle stem cells and primary myoblasts [132]. Researchers aligned the obtained transcriptomes of muscle stem cells derived from homeostatic and injured muscles, as well as primary myoblasts, along a pseudo-temporal single trajectory to order them (unsupervised). This method allowed them to describe the progression of the differentiation process at the single-cell level. They found two clusters (i.e., close-to-quiescence and early-activation clusters) with partially overlapping transcriptomes. They used bioinformatic techniques to recognize the difference between the two and place them in distinct space-time pathways.

Researchers have used scRNA-seq to provide the transcriptional landscape of human hematopoietic progenitors at the single-cell level [133]. They then showed that the CD38 antigen, which is usually considered the biomarker to negatively enrich the primitive progenitors for transplantation, is not a good choice. By showing the biological relevance of the CD164 gene in early hematopoiesis, they suggested using this marker instead in clinical transplantation and gene therapy. In another work, where scRNA-seq was used to classify transplanted hematopoietic stem cell identities at various differentiation stages, researchers found that the branching of hematopoietic lineage differentiation in adult marrow occurs at the transcriptional hematopoietic stem cell and transcriptional multipotent progenitor stages [184]. However, they found that the majority of transplanted hematopoietic stem cells committed to transcriptional multipotent progenitors. The proliferation of the donor-derived hematopoietic stem cells surviving after transplantation is accompanied by gradually decreasing the hematopoietic stem cell population. However, a balance between proliferation, differentiation, and stem cell maintenance maintains the cell functions at the bulk level.

To identify the role of tissue-resident memory T cells in the host defense system, samples from allogeneic hematopoietic stem cell transplanted patients were used, and the interindividual variation in host skin tissue T cell maintenance was studied [185]. Long-term persistence of host skin T cells that is reported not to be consistent with the chronic graft-versus-host disease development was seen in a group of patients.

The scRNA-seq technique was used to show that both fetal ventral midbrain and human embryonic stem cells-derived dopamine progenitors increase neurons and astrocytes after grafting [181]. Oligodendrocytes were present in fetal cell grafts; however, a cell type not known as a part of neural grafts was seen in grafts of human embryonic stem cells-derived ventral midbrain-patterned progenitors. The results of scRNA-seq of the cells before transplantation and at the time of transplantation indicated that genes associated with vascular leptomeningeal cells and progenitors were expressed. This finding suggests the potential of human embryonic stem cell-derived progenitors for generating both neural and perivascular cell types at this time. The scRNA-seq results suggest that future studies should investigate the contribution of various cell types to graft function. These studies can open the window to understanding the role of vascular leptomeningeal cells, astrocytes, and oligodendrocytes in the behavior of the graft.

The scRNA-seq technique was used to study the dynamic gene expression profile during limbal stem cell differentiation [186]. Expression heterogeneities among subgroups of the differentiated cells were detected. Epithelial–mesenchymal transition during the differentiation process, which may result in the generation of untargeted cells, was reported. Pseudo-time trajectory showed changes in transcriptions and signs of commitment for limbal stem cells and their progeny. The new markers found for limbal stem cells in this study need further work to identify their origin and accuracies.

Tables 9 and 10 show some key findings and biomarkers of single-cell analysis in stem cell transplantation, respectively.

Table 9. Examples of key findings in applying scRNA-Seq on stem cell transplantation.

Key Findings	Methods/Technologies	Donor/Recipient Species	References
Cellular diversity	Illumina Hiseq platform (Novogene), 10× Genomics	mouse	[184,187,188]
Cell populations: A neutrophil progenitor population that highly expresses S100A gene family members is detected in transplanted hematopoietic stem cells. Combined with FACS, scRNA-seq can ensure cellular purity in samples. Evaluating the ability of the bone marrow-mesenchymal to differentiate into subpopulations is possible.	10× Genomics Chromium	mouse, human	[189,190]
Gene expression profile variations	Illumina Hiseq platform (Novogene), 10× Genomics Chromium	mouse	[180,184,187]
Potential therapeutic target recognition: The scRNA-seq technique identifies therapeutic targets for osteosarcoma.	10× Genomics	mouse, human	[191]
Uncover novel cell types: The scRNA-seq technique assists in uncovering novel cell types.	10× Genomics	mouse, human	[188]

Table 10. Examples of important biomarkers of stem cell transplantation complications.

References	Methods/Technologies	Notes	Donor/Receptor	Gene Biomarkers
[187]	10× Genomics	two iliac cristae, two tibiae and two femora	mice	CD41, CD150
[184]	Illumina Hiseq platform (Novogene)	hematopoietic system	mouse, rat	Lin-Sca1+Kit+CD34-Flk2- Lin-Sca1+Kit+CD34-CD150+CD41- Lin-Sca1+Kit+CD34-CD150-CD41- CD201+CD150+CD48-CD45+ CD201+CD150+CD48- CD45+Sca1+Kit+
[180]	10× Genomics Chromium	Diabetes	mouse	MAFA, FAM159B, NAA20
[188]	10× Genomics	acute myeloid leukemia (AML) patients	mouse, human	CCR10, TNFRSF18, GZMK, CD8A, TNFRSF18, SIGLEC7, GNLY, LGALS3, CCR10, CD4, CLEC4C, PF4, PTCRA, CD8B, ID3, CD79A
[192]	10× Genomics	hematological malignancies,	human	CD3D, CD4, IL7R, CCR7, CCR6, CCL5, TBX21, FOXP3, CD8A, CD8B, CXCR6, RORC, CD69, IFIT3, GZMH, TRGC1, XCL1, XCL2, IL1R1, KIT, IFNG, FCGR3A
[193]	10× Genomics	Wolfram syndrome (WS)	mouse, human	SPINK1, ID3, NKX2-2, MAFB, NKX6-1, NKX2-2, GCK, ISL1, PDX1

5. Tumor Transplantation

Another application area in the field of transplantation is where it is used to study cancer therapeutic methods in vivo. A tumor is transplanted in an animal model, and then its behavior and response to cancer drugs are studied. However, single-cell analysis shows different immune landscapes in transplant and primary tumors and distinct responses to immunotherapy. Kirsch, from Duke University, and his coworkers showed that PD-1 blockade and radiotherapy can be used to cure transplant sarcomas; however, this protocol does not work in autochthonous sarcomas [194]. They found differences in immune landscapes of tumor-infiltrating immune cells from transplanted and primary tumors.

In most preclinical studies in cancer biology, researchers transplant tumors to study them in vivo. With the help of single-cell analysis, researchers have shown that although transplant sarcomas can be treated by programmed cell death-1 (PD-1) blockade and radiotherapy, this treatment cannot cure autochthonous tumors. This group used scRNA-seq and mass cytometry to study tumor-infiltrating immune cells from transplanted and primary tumors before and after radiation therapy and anti-PD-1 immunotherapy and found different immune profiles. They found that transplanted tumors are enriched for activated CD8+ T cells and PD-L1+ macrophages and concluded that PD-1 blockade and radiotherapy may be good treatments for patients with a sarcoma immune phenotype similar to those transplanted tumors [194].

The scRNA-seq technique applied to transplanted breast cancer tumors in mice shows that the aggressive tumor niche is determined by a basal-like population and mixed-lineage cancer cells [195]. The analysis showed two luminal-like populations (i.e., major Luminal 1 and minor Luminal 2). Figure 7 shows the scRNA-seq experiments and the obtained results.

See Tables 11 and 12 for examples of key findings in scRNA-seq applied to tumor transplantation and the related biomarkers, respectively.

Table 11. Examples of key findings in applying scRNA-seq on tumor transplantation.

Key Findings	Methods/Technologies	Donor/Recipient Species	References
Cellular diversity: The intra-individual, interindividual, spatial, functional, and genomic heterogeneity in melanoma cells, as well as tumor factors affecting the microenvironment (e.g., tumor-infiltrating immune cells, tumor-associated fibroblasts, and endothelial cells), are identified.	10× Genomics	human	[196–198]
Key factors: TNF receptor-related factor 3 (Traf3) is found to be significantly mutated in murine intrahepatic cholangiocarcinoma. In human intrahepatic cholangiocarcinoma, an inverse correlation between Traf3 and NF-κB-inducing kinase expression is reported. NF-κB-inducing kinase inhibition damps the growth of intrahepatic cholangiocarcinoma.	DNBSEQ-G400RS (MGI Tech), 10× Genomics	mouse	[171,196,197]
Gene expression profile variations: The scRNA-seq on the liver had identified mostly convergent gene expression alterations when primary biliary cholangitis and primary sclerosing cholangitis were compared to normal controls. Genes expressed by one cell type (e.g., CAFs) may affect the proportion of other cell types (e.g., T cells).	10× Genomics, DNBSEQ-G400RS (MGI Tech)	mouse	[171,196,199]
Potential therapeutic target recognition: The E2 subunit of mitochondrial pyruvate dehydrogenase complex (PDC-E2) is potentially considered for validating potential immunotherapeutic candidate strategies against cholangiocarcinoma.	10× Genomics, DNBSEQ-G400RS (MGI Tech)	mouse	[171,196,199]

Table 11. *Cont.*

Key Findings	Methods/Technologies	Donor/Recipient Species	References
Uncovering novel cell types: The scRNA-seq technique reveals novel cell types and states without biased results. It identifies novel cell subtypes that undergo immune rejection. Primary biliary cholangitis liver and underlying developed cholangiocarcinoma contain several clonotypes, often shared between two tissues.	10× Genomics	mouse, human	[113,200]
Tracking T-cell polarization: scRNA-seq detects genes associated with Th1 and Tc1 lymphocyte subsets in primary biliary cholangitis compared with primary sclerosing cholangitis livers. These T cells were detected within cholangiocarcinoma tumors and draining lymph nodes of mice with primary biliary cholangitis but not primary sclerosing cholangitis. Th1- and Tc1-polarized subsets play a key role in rejecting cholangiocarcinoma tumors.	10× Genomics	mouse	[199]

Table 12. Examples of important biomarkers of tumor transplantation complications.

References	Methods/Technologies	Tumor Types	Donor/Receptor	Gene Biomarkers
[199]	10× Genomics	cholangiocarcinoma	mouse	FoxP3, IFNγ, IL4, IL17a, Cd3g, Cd4, Cd8a, Id2, Tcf7, Eomes Il7r, Prdm1, Il2, Tbx21, Gata3, Il4, Rorc, Bcl6, Foxp3, Gzma, Gzmb, Gzmk, Ifng, Icos, Cd28, Cd27, Tnfrsf4, Tnfrsf9, Tnfrsf18, Cd40lg, Pdcd1, Ctla4, Lag3, Havcr2, Tigit, Btla, Lta, Adora2a, Klrg1, Cd38, Nt5e,
[171]	DNBSEQ-G400RS (MGI Tech)	cholangiocarcinoma	mouse	Alb, Apoa1, Ass1, Spp1, Sox9
[196]	10× Genomics	melanoma tumors	human	CD2, CD3D, CD3E, CD3G, CD19, CD79A, CD79B, BLK, CD163, CD14, CSF1R, PECAM1, VWF, CDH5, FAP, THY1, DCN, COL1A1, COL1A2, COL6A1, COL6A2, COL6A3
[198]	10× Genomics	Non-small cell lung carcinoma (NSCLC), neuroblastoma (NB), MBC, glioblastoma; high-grade glioma, CLL, ovarian, melanoma, sarcoma	human	KRT8, MRC1, TRAC, JCHAIN, TPSAB1, PTPRC, APOE, MAG, THY1, MITF, CA8, CFH, PAX3, CD99, KRT5, SFTPB, FOXJ1, MUC1, CGRP, SFTPC, AGER, FSP1 PECAM1, TH, MYCN, SOX2, STMN2, FDX1, PROM1, PDGFRA, UCHL1, LGALS3, HOPX, VIM
[197]	10× Genomics	Non-small-cell lung cancer (NSCLC), lung squamous carcinoma (LUSC), lung adenocarcinoma (LUAD)	human	TPSAB1, TPSB2, CPA3, HPGDS, CLU, AREG, MS4A2, RGS13, VWA5A, LAPTM4A, C1orf186, SLC18A2, LTC4S, KIT, HDC, MAOB, RGS1, RP11-354E11.2, SAMSN1, RGS2, SLC26A2, PTGS1, NSMCE1

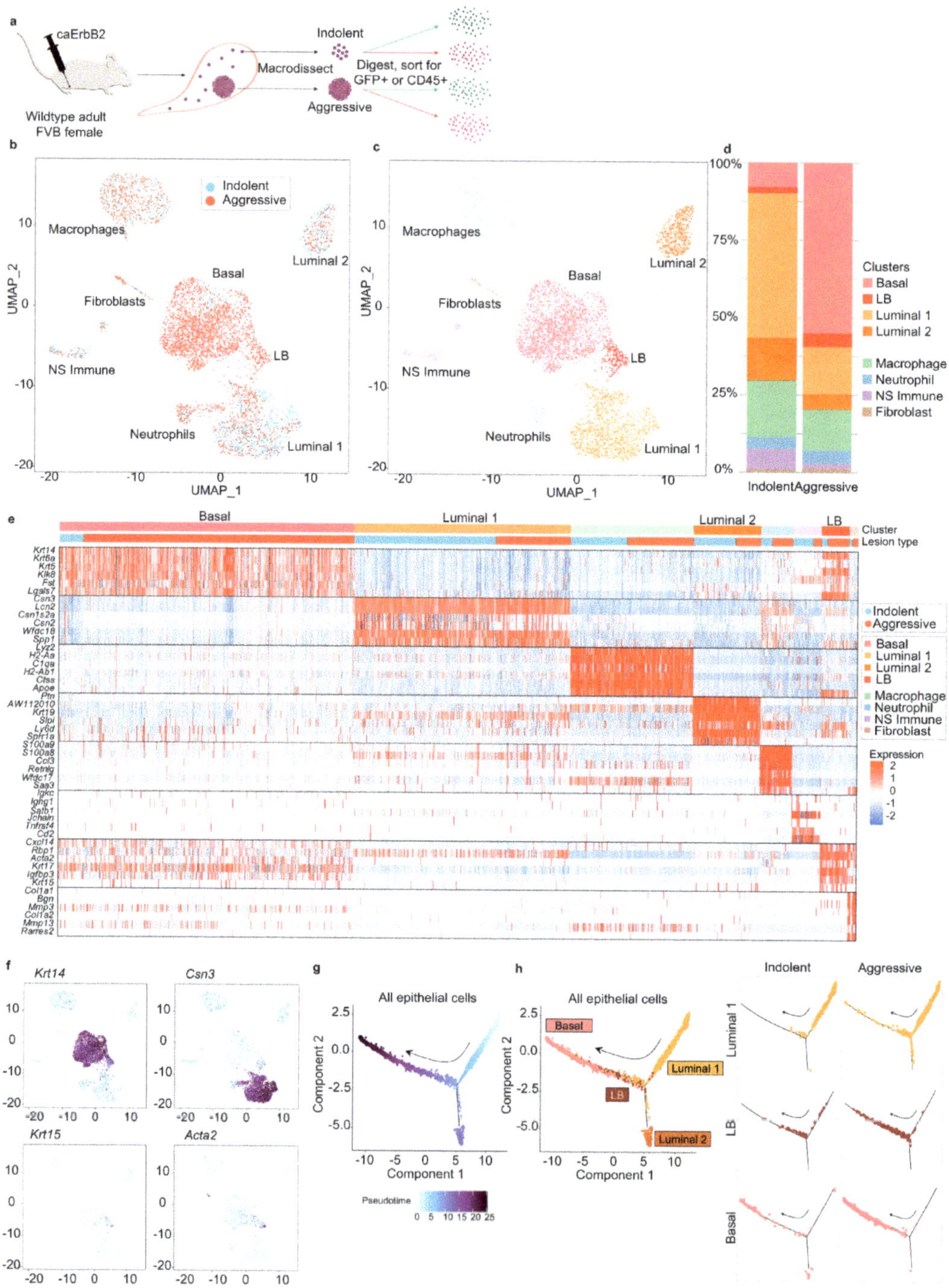

Figure 7. The scRNA-seq experiments on breast tumor transplanted mice. (**a**) Schematic of scRNA-seq experiments. Indolent tumor cells, indolent immune cells, aggressive tumor cells, and aggressive

immune cells were collected from 13 pooled mice. (**b–d**) UMAP plots of the single-cell data. Clusters proportions are depicted in (**d**). (**e**) Heatmap of highly differentially expressed genes for each cluster. (**f**) UMAP plots of all tumor cells, with colors indicating expression of KRT14, Csn3, KRT15, and Acta2 for basal-like cells, Luminal 1 cells, and Luminal-Basal cells, respectively. Trajectory plot of all tumor cells, with colors indicating pseudo-time value (**g**) and cluster ((**h**), **left panel**). ((**h**), **right panel**) Trajectory plots of Luminal 1, LB, and basal cells are shown. The arrows depict increasing pseudo-time value. The figure is taken with permission from [195] under a Creative Commons Attribution 4.0 International License (http://creativecommons.org/licenses/by/4.0/, accessed on 1 February 2024).

6. Conclusions

Single-cell RNA sequencing is an exciting new technology with the capability of driving significant progress in various bio-disciplines. It allows researchers to examine individual cells and uncover the molecular dynamics within them, revolutionizing the way biological processes can be studied. One of the important fields in which scRNA-seq has made fundamental advancements is transplantation.

Transplantation is the care gold standard for end-stage organ diseases; however, not all transplantations are successful. The most frequent complication of transplantation is allograft rejection. Currently, the diagnosis of these complications in the clinical setting needs biopsies obtained from the patients. However, traditionally found biomarkers are not fully reliable for detecting rejection.

The scRNA-seq technique uncovers cell heterogeneity, cell states, and graft complications in solid transplantation and stem cell transplantation. It is also used in studying transplanted cancer cells. The scRNA-seq technique can provide a comprehensive single-cell atlas of gene expression profiles in acute rejection and transplant complications. It uncovers the contribution of T cells and natural killer cells, as well as the association of various subsets of macrophages, including infiltrating (m3 and m4) and resident macrophages (m1 and m2), in graft rejection. Major findings include cell diversity in grafts, gene expression profile variation during graft rejection, identifying novel biomarkers, uncovering macrophage polarization, and immune profile landscapes.

The obtained large and high-dimensional data from single-cell analysis needs computational data processing and analysis. Machine learning methods are employed to develop analysis pipelines and predictive models toward this goal. In the future, more advanced machine learning methods will further contribute to the development of the field.

Funding: This research received no external funding.

Data Availability Statement: Data sharing is not applicable to this article as no new data were created or analyzed in this study.

Conflicts of Interest: The authors declare no conflict of interest.

References

1. Jia, Q.; Chu, H.; Jin, Z.; Long, H.; Zhu, B. High-throughput single-cell sequencing in cancer research. *Signal Transduct. Target. Ther.* **2022**, *7*, 145. [CrossRef] [PubMed]
2. Eum, H.H.; Jeong, D.; Kim, N.; Jo, A.; Na, M.; Kang, H.; Hong, Y.; Kong, J.S.; Jeong, G.H.; Yoo, S.A.; et al. Single-cell RNA sequencing reveals myeloid and T cell co-stimulation mediated by IL-7 anti-cancer immunotherapy. *Br. J. Cancer* **2024**. [CrossRef] [PubMed]
3. Eze, U.C.; Bhaduri, A.; Haeussler, M.; Nowakowski, T.J.; Kriegstein, A.R. Single-cell atlas of early human brain development highlights heterogeneity of human neuroepithelial cells and early radial glia. *Nat. Neurosci.* **2021**, *24*, 584–594. [CrossRef] [PubMed]
4. Dopp, J.; Ortega, A.; Davie, K.; Poovathingal, S.; Baz, E.-S.; Liu, S. Single-cell transcriptomics reveals that glial cells integrate homeostatic and circadian processes to drive sleep–wake cycles. *Nat. Neurosci.* **2024**, *27*, 359–372. [CrossRef] [PubMed]
5. Wu, H.; Malone, A.F.; Donnelly, E.L.; Kirita, Y.; Uchimura, K.; Ramakrishnan, S.M.; Gaut, J.P.; Humphreys, B.D. Single-Cell Transcriptomics of a Human Kidney Allograft Biopsy Specimen Defines a Diverse Inflammatory Response. *J. Am. Soc. Nephrol.* **2018**, *29*, 2069–2080. [CrossRef] [PubMed]

6. Papalexi, E.; Satija, R. Single-cell RNA sequencing to explore immune cell heterogeneity. *Nat. Rev. Immunol.* **2018**, *18*, 35–45. [CrossRef] [PubMed]
7. Tang, Y.; Wang, J.; Zhang, Y.; Li, J.; Chen, M.; Gao, Y.; Dai, M.; Lin, S.; He, X.; Wu, C.; et al. Single-Cell RNA Sequencing Identifies Intra-Graft Population Heterogeneity in Acute Heart Allograft Rejection in Mouse. *Front. Immunol.* **2022**, *13*, 832573. [CrossRef] [PubMed]
8. Thareja, G.; Muthukumar, T. Partners in Crime: Inferring Cell-to-cell Interactions in Kidney Allograft Rejection from Single-cell RNA Sequencing. *Transplantation* **2024**, *108*, 325–326. [CrossRef]
9. Elmentaite, R.; Ross, A.D.B.; Roberts, K.; James, K.R.; Ortmann, D.; Gomes, T.; Nayak, K.; Tuck, L.; Pritchard, S.; Bayraktar, O.A.; et al. Single-Cell Sequencing of Developing Human Gut Reveals Transcriptional Links to Childhood Crohn's Disease. *Dev. Cell* **2020**, *55*, 771–783.e5. [CrossRef]
10. Bye, C.R.; Penna, V.; de Luzy, I.R.; Gantner, C.W.; Hunt, C.P.J.; Thompson, L.H.; Parish, C.L. Transcriptional Profiling of Xenogeneic Transplants: Examining Human Pluripotent Stem Cell-Derived Grafts in the Rodent Brain. *Stem Cell Rep.* **2019**, *13*, 877–890. [CrossRef]
11. Abedini-Nassab, R.; Pouryosef Miandoab, M.; Sasmaz, M. Microfluidic Synthesis, Control, and Sensing of Magnetic Nanoparticles: A Review. *Micromachines* **2021**, *12*, 768. [CrossRef] [PubMed]
12. Mantri, M.; Scuderi, G.J.; Abedini-Nassab, R.; Wang, M.F.Z.; McKellar, D.; Shi, H.; Grodner, B.; Butcher, J.T.; De Vlaminck, I. Spatiotemporal single-cell RNA sequencing of developing chicken hearts identifies interplay between cellular differentiation and morphogenesis. *Nat. Commun.* **2021**, *12*, 1771. [CrossRef] [PubMed]
13. Buchwald, J.E.; Martins, P.N. Designer organs: The future of personalized transplantation. *Artif. Organs* **2022**, *46*, 180–190. [CrossRef] [PubMed]
14. Shi, T.; Roskin, K.; Baker, B.M.; Woodle, E.S.; Hildeman, D. Advanced Genomics-Based Approaches for Defining Allograft Rejection with Single Cell Resolution. *Front. Immunol.* **2021**, *12*, 750754. [CrossRef]
15. Raza, S.S.; Wagner, A.P.; Hussain, Y.S.; Khan, M.A. Mechanisms underlying dental-derived stem cell-mediated neurorestoration in neurodegenerative disorders. *Stem Cell Res. Ther.* **2018**, *9*, 245. [CrossRef] [PubMed]
16. McCauley, H.A.; Guasch, G. Serial orthotopic transplantation of epithelial tumors in single-cell suspension. *Methods Mol. Biol.* **2013**, *1035*, 231–245. [CrossRef] [PubMed]
17. Kim, H.K.; Ha, T.W.; Lee, M.R. Single-Cell Transcriptome Analysis as a Promising Tool to Study Pluripotent Stem Cell Reprogramming. *Int. J. Mol. Sci.* **2021**, *22*, 5988. [CrossRef] [PubMed]
18. Herzenberg, L.A.; Parks, D.; Sahaf, B.; Perez, O.; Roederer, M.; Herzenberg, L.A. The history and future of the fluorescence activated cell sorter and flow cytometry: A view from Stanford. *Clin. Chem.* **2002**, *48*, 1819–1827. [CrossRef] [PubMed]
19. Bonner, W.A.; Hulett, H.R.; Sweet, R.G.; Herzenberg, L.A. Fluorescence activated cell sorting. *Rev. Sci. Instrum.* **1972**, *43*, 404–409. [CrossRef]
20. McKinnon, K.M. Flow Cytometry: An Overview. *Curr. Protoc. Immunol.* **2018**, *120*, 5.1.1–5.1.11. [CrossRef]
21. Koike, Y.; Kodera, S.; Yokoyama, Y.; Hayakawa, T. Real-time irradiation system using patterned light to actuate light-driven on-chip gel actuators. *Robomech J.* **2022**, *9*, 5. [CrossRef]
22. Adam, G.; Chidambaram, S.; Reddy, S.S.; Ramani, K.; Cappelleri, D.J. Towards a Comprehensive and Robust Micromanipulation System with Force-Sensing and VR Capabilities. *Micromachines* **2021**, *12*, 784. [CrossRef] [PubMed]
23. Kato, Y.; Matsumoto, T.; Kino-Oka, M. Effect of liquid flow by pipetting during medium change on deformation of hiPSC aggregates. *Regen. Ther.* **2019**, *12*, 20–26. [CrossRef] [PubMed]
24. Di Carlo, D.; Wu, L.Y.; Lee, L.P. Dynamic single cell culture array. *Lab. Chip* **2006**, *6*, 1445–1449. [CrossRef] [PubMed]
25. Narayanamurthy, V.; Nagarajan, S.; Firus Khan, A.Y.; Samsuri, F.; Sridhar, T.M. Microfluidic hydrodynamic trapping for single cell analysis: Mechanisms, methods and applications. *Anal. Methods* **2017**, *9*, 3751–3772. [CrossRef]
26. Luan, Q.; Macaraniag, C.; Zhou, J.; Papautsky, I. Microfluidic systems for hydrodynamic trapping of cells and clusters. *Biomicrofluidics* **2020**, *14*, 031502. [CrossRef] [PubMed]
27. Abedini-Nassab, R. Magnetophoretic Circuit Biocompatibility. *J. Mech. Med. Biol.* **2020**, *20*, 2050050. [CrossRef]
28. Ahmadi, F.; Tran, H.; Letourneau, N.; Little, S.R.; Fortin, A.; Moraitis, A.N.; Shih, S.C.C. An Automated Single-Cell Droplet-Digital Microfluidic Platform for Monoclonal Antibody Discovery. *Small* **2024**. [CrossRef] [PubMed]
29. Abedini-Nassab, R.; Mahdaviyan, N. A Microfluidic Platform Equipped with Magnetic Nano Films for Organizing Bio-Particle Arrays and Long-Term Studies. *IEEE Sens. J.* **2020**, *20*, 9668–9676. [CrossRef]
30. Abedini-Nassab, R.; Shourabi, R. High-throughput precise particle transport at single-particle resolution in a three-dimensional magnetic field for highly sensitive bio-detection. *Sci. Rep.* **2022**, *12*, 6380. [CrossRef]
31. Zhang, Y.; Zhou, A.; Chen, S.; Lum, G.Z.; Zhang, X. A perspective on magnetic microfluidics: Towards an intelligent future. *Biomicrofluidics* **2022**, *16*, 011301. [CrossRef] [PubMed]
32. Xu, H.; Liao, C.; Zuo, P.; Liu, Z.; Ye, B.C. Magnetic-Based Microfluidic Device for On-Chip Isolation and Detection of Tumor-Derived Exosomes. *Anal. Chem.* **2018**, *90*, 13451–13458. [CrossRef]
33. Abedini-Nassab, R.; Ding, X.; Xie, H. A novel magnetophoretic-based device for magnetometry and separation of single magnetic particles and magnetized cells. *Lab. Chip* **2022**, *22*, 738–746. [CrossRef] [PubMed]
34. Sadeghidelouei, N.; Abedini-Nassab, R. Unidirectional particle transport in microfluidic chips operating in a tri-axial magnetic field for particle concentration and bio-analyte detection. *Microfluid. Nanofluidics* **2023**, *28*, 6. [CrossRef]

35. Yu, E.S.; Lee, H.; Lee, S.M.; Kim, J.; Kim, T.; Lee, J.; Kim, C.; Seo, M.; Kim, J.H.; Byun, Y.T.; et al. Precise capture and dynamic relocation of nanoparticulate biomolecules through dielectrophoretic enhancement by vertical nanogap architectures. *Nat. Commun.* **2020**, *11*, 2804. [CrossRef]
36. Punjiya, M.; Nejad, H.R.; Mathews, J.; Levin, M.; Sonkusale, S. A flow through device for simultaneous dielectrophoretic cell trapping and AC electroporation. *Sci. Rep.* **2019**, *9*, 11988. [CrossRef]
37. Mugele, F.; Baret, J.-C. Electrowetting: From basics to applications. *J. Physics Condens. Matter* **2005**, *17*, R705. [CrossRef]
38. Abedini-Nassab, R.; Wirfel, J.; Talebjedi, B.; Tasnim, N.; Hoorfar, M. Quantifying the dielectrophoretic force on colloidal particles in microfluidic devices. *Microfluid. Nanofluidics* **2022**, *26*, 38. [CrossRef]
39. Yang, S.; Rufo, J.; Zhong, R.; Rich, J.; Wang, Z.; Lee, L.P.; Huang, T.J. Acoustic tweezers for high-throughput single-cell analysis. *Nat. Protoc.* **2023**, *18*, 2441–2458. [CrossRef]
40. Abedini-Nassab, R.; Emami, S.M.; Nowghabi, A.N. Nanotechnology and Acoustics in Medicine and Biology. *Recent Pat. Nanotechnol.* **2022**, *16*, 198–206. [CrossRef]
41. Rufo, J.; Cai, F.; Friend, J.; Wiklund, M.; Huang, T.J. Acoustofluidics for biomedical applications. *Nat. Rev. Methods Primers* **2022**, *2*, 30. [CrossRef]
42. Zhang, Y.; Chen, X. Particle separation in microfluidics using different modal ultrasonic standing waves. *Ultrason. Sonochem.* **2021**, *75*, 105603. [CrossRef]
43. Ohiri, K.A.; Kelly, S.T.; Motschman, J.D.; Lin, K.H.; Wood, K.C.; Yellen, B.B. An acoustofluidic trap and transfer approach for organizing a high density single cell array. *Lab. Chip* **2018**, *18*, 2124–2133. [CrossRef]
44. Volpe, G.; Maragò, O.M.; Rubinsztein-Dunlop, H.; Pesce, G.; Stilgoe, A.B.; Volpe, G.; Tkachenko, G.; Truong, V.G.; Chormaic, S.N.; Kalantarifard, F.; et al. Roadmap for optical tweezers. *J. Phys. Photonics* **2023**, *5*, 022501. [CrossRef]
45. Wang, M.M.; Tu, E.; Raymond, D.E.; Yang, J.M.; Zhang, H.; Hagen, N.; Dees, B.; Mercer, E.M.; Forster, A.H.; Kariv, I.; et al. Microfluidic sorting of mammalian cells by optical force switching. *Nat. Biotechnol.* **2005**, *23*, 83–87. [CrossRef] [PubMed]
46. Yang, Y.; Mao, Y.; Shin, K.S.; Chui, C.O.; Chiou, P.Y. Self-Locking Optoelectronic Tweezers for Single-Cell and Microparticle Manipulation across a Large Area in High Conductivity Media. *Sci. Rep.* **2016**, *6*, 22630. [CrossRef] [PubMed]
47. Schraivogel, D.; Kuhn, T.M.; Rauscher, B.; Rodriguez-Martinez, M.; Paulsen, M.; Owsley, K.; Middlebrook, A.; Tischer, C.; Ramasz, B.; Ordonez-Rueda, D.; et al. High-speed fluorescence image-enabled cell sorting. *Science* **2022**, *375*, 315–320. [CrossRef] [PubMed]
48. Valle, M.; O'Brien, B.; Green, T.D.; Reiner, J.E.; Seashols-Williams, S. Droplet-based optical trapping for cell separation in mock forensic samples. *J. Forensic Sci.* **2024**, *69*, 273–281. [CrossRef] [PubMed]
49. Schneckenburger, H.; Hendinger, A.; Sailer, R.; Gschwend, M.H.; Strauss, W.S.; Bauer, M.; Schütze, K. Cell viability in optical tweezers: High power red laser diode versus Nd:YAG laser. *J. Biomed. Opt.* **2000**, *5*, 40–44. [CrossRef]
50. Konishi, S.; Ohya, C.; Yamada, T. Selective control of the contact and transport between droplet pairs by electrowetting-on-dielectric for droplet-array sandwiching technology. *Sci. Rep.* **2021**, *11*, 12355. [CrossRef]
51. Vallet, M.; Berge, B.; Vovelle, L. Electrowetting of water and aqueous solutions on poly(ethylene terephthalate) insulating films. *Polymer* **1996**, *37*, 2465–2470. [CrossRef]
52. Abedini-Nassab, R.; Sadeghidelouei, N.; Shields Iv, C.W. Magnetophoretic circuits: A review of device designs and implementation for precise single-cell manipulation. *Anal. Chim. Acta* **2023**, *1272*, 341425. [CrossRef] [PubMed]
53. Abedini-Nassab, R. *Magnetomicrofluidic Circuits for Single-Bioparticle Transport*; Springer Nature: Singapore, 2023.
54. Abedini-Nassab, R. Magnetomicrofluidic Platforms for Organizing Arrays of Single-Particles and Particle-Pairs. *J. Microelectromech. Syst.* **2019**, *28*, 732–738. [CrossRef]
55. Dashti, R.; Abedini-Nassab, R. A High-Throughput Hybrid Electromicrofluidic Platform for Organizing Single-Cell Protein Secretion Profiling Assays. *IEEE Sens. J.* **2024**, *24*, 7448–7455. [CrossRef]
56. Au, A.K.; Lai, H.; Utela, B.R.; Folch, A. Microvalves and Micropumps for BioMEMS. *Micromachines* **2011**, *2*, 179–220. [CrossRef]
57. Studer, V.; Hang, G.; Pandolfi, A.; Ortiz, M.; French Anderson, W.; Quake, S.R. Scaling properties of a low-actuation pressure microfluidic valve. *J. Appl. Phys.* **2003**, *95*, 393–398. [CrossRef]
58. Thorsen, T.; Maerkl, S.J.; Quake, S.R. Microfluidic large-scale integration. *Science* **2002**, *298*, 580–584. [CrossRef] [PubMed]
59. Marcus, J.S.; Anderson, W.F.; Quake, S.R. Microfluidic single-cell mRNA isolation and analysis. *Anal. Chem.* **2006**, *78*, 3084–3089. [CrossRef]
60. Macosko, E.Z.; Basu, A.; Satija, R.; Nemesh, J.; Shekhar, K.; Goldman, M.; Tirosh, I.; Bialas, A.R.; Kamitaki, N.; Martersteck, E.M.; et al. Highly Parallel Genome-wide Expression Profiling of Individual Cells Using Nanoliter Droplets. *Cell* **2015**, *161*, 1202–1214. [CrossRef]
61. Jain, A.; Stavrakis, S.; deMello, A. Droplet-based microfluidics and enzyme evolution. *Curr. Opin. Biotechnol.* **2024**, *87*, 103097. [CrossRef]
62. Nan, L.; Zhang, H.; Weitz, D.A.; Shum, H.C. Development and future of droplet microfluidics. *Lab. Chip* **2024**, *24*, 1135–1153. [CrossRef] [PubMed]
63. Shi, J.; Zhang, Y.; Fan, Y.; Liu, Y.; Yang, M. Recent advances in droplet-based microfluidics in liquid biopsy for cancer diagnosis. *Droplet* **2024**, *3*, e92. [CrossRef]
64. Zhang, J.; Xu, W.; Xu, F.; Lu, W.; Hu, L.; Zhou, J.; Zhang, C.; Jiang, Z. Microfluidic droplet formation in co-flow devices fabricated by micro 3D printing. *J. Food Eng.* **2021**, *290*, 110212. [CrossRef]

65. Cramer, C.; Fischer, P.; Windhab, E.J. Drop formation in a co-flowing ambient fluid. *Chem. Eng. Sci.* **2004**, *59*, 3045–3058. [CrossRef]
66. Yao, J.; Lin, F.; Kim, H.S.; Park, J. The Effect of Oil Viscosity on Droplet Generation Rate and Droplet Size in a T-Junction Microfluidic Droplet Generator. *Micromachines* **2019**, *10*, 808. [CrossRef] [PubMed]
67. Ushikubo, F.Y.; Birribilli, F.S.; Oliveira, D.R.B.; Cunha, R.L. Y- and T-junction microfluidic devices: Effect of fluids and interface properties and operating conditions. *Microfluid. Nanofluidics* **2014**, *17*, 711–720. [CrossRef]
68. Garstecki, P.; Fuerstman, M.J.; Stone, H.A.; Whitesides, G.M. Formation of droplets and bubbles in a microfluidic T-junction—Scaling and mechanism of break-up. *Lab. Chip* **2006**, *6*, 437–446. [CrossRef] [PubMed]
69. Yin, Z.; Huang, Z.; Lin, X.; Gao, X.; Bao, F. Droplet Generation in a Flow-Focusing Microfluidic Device with External Mechanical Vibration. *Micromachines* **2020**, *11*, 743. [CrossRef]
70. Dewandre, A.; Rivero-Rodriguez, J.; Vitry, Y.; Sobac, B.; Scheid, B. Microfluidic droplet generation based on non-embedded co-flow-focusing using 3D printed nozzle. *Sci. Rep.* **2020**, *10*, 21616. [CrossRef]
71. Bageritz, J.; Raddi, G. Single-Cell RNA Sequencing with Drop-Seq. *Methods Mol. Biol.* **2019**, *1979*, 73–85. [CrossRef]
72. Klein, A.M.; Mazutis, L.; Akartuna, I.; Tallapragada, N.; Veres, A.; Li, V.; Peshkin, L.; Weitz, D.A.; Kirschner, M.W. Droplet barcoding for single-cell transcriptomics applied to embryonic stem cells. *Cell* **2015**, *161*, 1187–1201. [CrossRef] [PubMed]
73. Lu, Y.; Shiau, F.; Yi, W.; Lu, S.; Wu, Q.; Pearson, J.D.; Kallman, A.; Zhong, S.; Hoang, T.; Zuo, Z.; et al. Single-Cell Analysis of Human Retina Identifies Evolutionarily Conserved and Species-Specific Mechanisms Controlling Development. *Dev. Cell* **2020**, *53*, 473–491.e9. [CrossRef] [PubMed]
74. Gierahn, T.M.; Wadsworth, M.H.; Hughes, T.K.; Bryson, B.D.; Butler, A.; Satija, R.; Fortune, S.; Love, J.C.; Shalek, A.K. Seq-Well: Portable, low-cost RNA sequencing of single cells at high throughput. *Nat. Methods* **2017**, *14*, 395–398. [CrossRef] [PubMed]
75. Ramskold, D.; Luo, S.; Wang, Y.C.; Li, R.; Deng, Q.; Faridani, O.R.; Daniels, G.A.; Khrebtukova, I.; Loring, J.F.; Laurent, L.C.; et al. Author Correction: Full-length mRNA-Seq from single-cell levels of RNA and individual circulating tumor cells. *Nat. Biotechnol.* **2020**, *38*, 374. [CrossRef] [PubMed]
76. Isakova, A.; Neff, N.; Quake, S.R. Single-cell quantification of a broad RNA spectrum reveals unique noncoding patterns associated with cell types and states. *Proc. Natl. Acad. Sci. USA* **2021**, *118*, e2113568118. [CrossRef] [PubMed]
77. Picelli, S.; Faridani, O.R.; Björklund, Å.K.; Winberg, G.; Sagasser, S.; Sandberg, R. Full-length RNA-seq from single cells using Smart-seq2. *Nat. Protoc.* **2014**, *9*, 171–181. [CrossRef] [PubMed]
78. Picelli, S.; Björklund, Å.K.; Faridani, O.R.; Sagasser, S.; Winberg, G.; Sandberg, R. Smart-seq2 for sensitive full-length transcriptome profiling in single cells. *Nat. Methods* **2013**, *10*, 1096–1098. [CrossRef] [PubMed]
79. Hagemann-Jensen, M.; Ziegenhain, C.; Chen, P.; Ramsköld, D.; Hendriks, G.-J.; Larsson, A.J.M.; Faridani, O.R.; Sandberg, R. Single-cell RNA counting at allele and isoform resolution using Smart-seq3. *Nat. Biotechnol.* **2020**, *38*, 708–714. [CrossRef] [PubMed]
80. Cao, J.; Packer, J.S.; Ramani, V.; Cusanovich, D.A.; Huynh, C.; Daza, R.; Qiu, X.; Lee, C.; Furlan, S.N.; Steemers, F.J.; et al. Comprehensive single-cell transcriptional profiling of a multicellular organism. *Science* **2017**, *357*, 661–667. [CrossRef]
81. Sheng, K.; Cao, W.; Niu, Y.; Deng, Q.; Zong, C. Effective detection of variation in single-cell transcriptomes using MATQ-seq. *Nat. Methods* **2017**, *14*, 267–270. [CrossRef]
82. Hashimshony, T.; Wagner, F.; Sher, N.; Yanai, I. CEL-Seq: Single-cell RNA-Seq by multiplexed linear amplification. *Cell Rep.* **2012**, *2*, 666–673. [CrossRef] [PubMed]
83. Hashimshony, T.; Senderovich, N.; Avital, G.; Klochendler, A.; de Leeuw, Y.; Anavy, L.; Gennert, D.; Li, S.; Livak, K.J.; Rozenblatt-Rosen, O.; et al. CEL-Seq2: Sensitive highly-multiplexed single-cell RNA-Seq. *Genome Biol.* **2016**, *17*, 77. [CrossRef] [PubMed]
84. Branton, D.; Deamer, D.W.; Marziali, A.; Bayley, H.; Benner, S.A.; Butler, T.; Di Ventra, M.; Garaj, S.; Hibbs, A.; Huang, X.; et al. The potential and challenges of nanopore sequencing. *Nat. Biotechnol.* **2008**, *26*, 1146–1153. [CrossRef] [PubMed]
85. Philpott, M.; Watson, J.; Thakurta, A.; Brown, T., Jr.; Brown, T., Sr.; Oppermann, U.; Cribbs, A.P. Nanopore sequencing of single-cell transcriptomes with scCOLOR-seq. *Nat. Biotechnol.* **2021**, *39*, 1517–1520. [CrossRef]
86. Shiau, C.K.; Lu, L.; Kieser, R.; Fukumura, K.; Pan, T.; Lin, H.Y.; Yang, J.; Tong, E.L.; Lee, G.; Yan, Y.; et al. High throughput single cell long-read sequencing analyses of same-cell genotypes and phenotypes in human tumors. *Nat. Commun.* **2023**, *14*, 4124. [CrossRef]
87. Abedini-Nassab, R. Nanotechnology and Nanopore Sequencing. *Recent. Pat. Nanotechnol.* **2017**, *11*, 34–41. [CrossRef] [PubMed]
88. Wang, Y.; Zhao, Y.; Bollas, A.; Wang, Y.; Au, K.F. Nanopore sequencing technology, bioinformatics and applications. *Nat. Biotechnol.* **2021**, *39*, 1348–1365. [CrossRef] [PubMed]
89. Louie, S.M.; Moye, A.L.; Wong, I.G.; Lu, E.; Shehaj, A.; Garcia-de-Alba, C.; Ararat, E.; Raby, B.A.; Lu, B.; Paschini, M.; et al. Progenitor potential of lung epithelial organoid cells in a transplantation model. *Cell Rep.* **2022**, *39*, 110662. [CrossRef]
90. Kono, N.; Arakawa, K. Nanopore sequencing: Review of potential applications in functional genomics. *Dev. Growth Differ.* **2019**, *61*, 316–326. [CrossRef]
91. Koenig, A.L.; Shchukina, I.; Amrute, J.; Andhey, P.S.; Zaitsev, K.; Lai, L.; Bajpai, G.; Bredemeyer, A.; Smith, G.; Jones, C.; et al. Single-cell transcriptomics reveals cell-type-specific diversification in human heart failure. *Nat. Cardiovasc. Res.* **2022**, *1*, 263–280. [CrossRef]

92. Paik, D.T.; Tian, L.; Williams, I.M.; Rhee, S.; Zhang, H.; Liu, C.; Mishra, R.; Wu, S.M.; Red-Horse, K.; Wu, J.C. Single-Cell RNA Sequencing Unveils Unique Transcriptomic Signatures of Organ-Specific Endothelial Cells. *Circulation* **2020**, *142*, 1848–1862. [CrossRef] [PubMed]
93. Cai, J.; Deng, J.; Gu, W.; Ni, Z.; Liu, Y.; Kamra, Y.; Saxena, A.; Hu, Y.; Yuan, H.; Xiao, Q.; et al. Impact of Local Alloimmunity and Recipient Cells in Transplant Arteriosclerosis. *Circ. Res.* **2020**, *127*, 974–993. [CrossRef] [PubMed]
94. Kopecky, B.J.; Dun, H.; Amrute, J.M.; Lin, C.Y.; Bredemeyer, A.L.; Terada, Y.; Bayguinov, P.O.; Koenig, A.L.; Frye, C.C.; Fitzpatrick, J.A.J.; et al. Donor Macrophages Modulate Rejection after Heart Transplantation. *Circulation* **2022**, *146*, 623–638. [CrossRef]
95. Hu, Z.; Liu, W.; Hua, X.; Chen, X.; Chang, Y.; Hu, Y.; Xu, Z.; Song, J. Single-Cell Transcriptomic Atlas of Different Human Cardiac Arteries Identifies Cell Types Associated with Vascular Physiology. *Arterioscler. Thromb. Vasc. Biol.* **2021**, *41*, 1408–1427. [CrossRef] [PubMed]
96. Anto Michel, N.; Ljubojevic-Holzer, S.; Bugger, H.; Zirlik, A. Cellular Heterogeneity of the Heart. *Front. Cardiovasc. Med.* **2022**, *9*, 868466. [CrossRef] [PubMed]
97. Wan, J.; Zhang, Z.; Tian, S.; Huang, S.; Jin, H.; Liu, X.; Zhang, W. Single cell study of cellular diversity and mutual communication in chronic heart failure and drug repositioning. *Genomics* **2022**, *114*, 110322. [CrossRef]
98. Chen, Z.; Xu, H.; Li, Y.; Zhang, X.; Cui, J.; Zou, Y.; Yu, J.; Wu, J.; Xia, J. Single-Cell RNA sequencing reveals immune cell dynamics and local intercellular communication in acute murine cardiac allograft rejection. *Theranostics* **2022**, *12*, 6242–6257. [CrossRef] [PubMed]
99. Martini, E.; Kunderfranco, P.; Peano, C.; Carullo, P.; Cremonesi, M.; Schorn, T.; Carriero, R.; Termanini, A.; Colombo, F.S.; Jachetti, E.; et al. Single-Cell Sequencing of Mouse Heart Immune Infiltrate in Pressure Overload–Driven Heart Failure Reveals Extent of Immune Activation. *Circulation* **2019**, *140*, 2089–2107. [CrossRef]
100. Kong, D.; Huang, S.; Miao, X.; Li, J.; Wu, Z.; Shi, Y.; Liu, H.; Jiang, Y.; Yu, X.; Xie, M.; et al. The dynamic cellular landscape of grafts with acute rejection after heart transplantation. *J. Heart Lung Transplant.* **2023**, *42*, 160–172. [CrossRef]
101. Jung, Y.; Kim, J.; Jang, H.; Kim, G.; Kwon, Y.-W. Strategy of Patient-Specific Therapeutics in Cardiovascular Disease through Single-Cell RNA Sequencing. *Korean Circ. J.* **2023**, *53*, 1–16. [CrossRef]
102. Loupy, A.; Duong Van Huyen, J.P.; Hidalgo, L.; Reeve, J.; Racapé, M.; Aubert, O.; Venner, J.M.; Falmuski, K.; Bories, M.C.; Beuscart, T.; et al. Gene Expression Profiling for the Identification and Classification of Antibody-Mediated Heart Rejection. *Circulation* **2017**, *135*, 917–935. [CrossRef] [PubMed]
103. Hua, X.; Hu, G.; Hu, Q.; Chang, Y.; Hu, Y.; Gao, L.; Chen, X.; Yang, P.-C.; Zhang, Y.; Li, M.; et al. Single-Cell RNA Sequencing to Dissect the Immunological Network of Autoimmune Myocarditis. *Circulation* **2020**, *142*, 384–400. [CrossRef]
104. Yang, L.; Zhu, Y.; Tian, D.; Wang, S.; Guo, J.; Sun, G.; Jin, H.; Zhang, C.; Shi, W.; Gershwin, M.E.; et al. Transcriptome landscape of double negative T cells by single-cell RNA sequencing. *J. Autoimmun.* **2021**, *121*, 102653. [CrossRef]
105. Schumacher, D.; Kramann, R. Multiomic Spatial Mapping of Myocardial Infarction and Implications for Personalized Therapy. *Arterioscler. Thromb. Vasc. Biol.* **2023**, *43*, 192–202. [CrossRef]
106. Clark, A.R.; Greka, A. The power of one: Advances in single-cell genomics in the kidney. *Nat. Rev. Nephrol.* **2020**, *16*, 73–74. [CrossRef]
107. Wilson, P.C.; Wu, H.; Kirita, Y.; Uchimura, K.; Ledru, N.; Rennke, H.G.; Welling, P.A.; Waikar, S.S.; Humphreys, B.D. The single-cell transcriptomic landscape of early human diabetic nephropathy. *Proc. Natl. Acad. Sci. USA* **2019**, *116*, 19619–19625. [CrossRef] [PubMed]
108. Young, M.D.; Mitchell, T.J.; Vieira Braga, F.A.; Tran, M.G.B.; Stewart, B.J.; Ferdinand, J.R.; Collord, G.; Botting, R.A.; Popescu, D.M.; Loudon, K.W.; et al. Single-cell transcriptomes from human kidneys reveal the cellular identity of renal tumors. *Science* **2018**, *361*, 594–599. [CrossRef]
109. Muto, Y.; Wilson, P.C.; Ledru, N.; Wu, H.; Dimke, H.; Waikar, S.S.; Humphreys, B.D. Single cell transcriptional and chromatin accessibility profiling redefine cellular heterogeneity in the adult human kidney. *Nat. Commun.* **2021**, *12*, 2190. [CrossRef] [PubMed]
110. Shen, Q.; Wang, Y.; Chen, J.; Ma, L.; Huang, X.; Tang, S.C.W.; Lan, H.; Jiang, H. Single-Cell RNA Sequencing Reveals the Immunological Profiles of Renal Allograft Rejection in Mice. *Front. Immunol.* **2021**, *12*, 693608. [CrossRef]
111. Shi, T.; Burg, A.R.; Caldwell, J.T.; Roskin, K.M.; Castro-Rojas, C.M.; Chukwuma, P.C.; Gray, G.I.; Foote, S.G.; Alonso, J.A.; Cuda, C.M.; et al. Single-cell transcriptomic analysis of renal allograft rejection reveals insights into intragraft TCR clonality. *J. Clin. Investig.* **2023**, *133*, e170191. [CrossRef]
112. Malone, A.F.; Wu, H.; Fronick, C.; Fulton, R.; Gaut, J.P.; Humphreys, B.D. Harnessing Expressed Single Nucleotide Variation and Single Cell RNA Sequencing To Define Immune Cell Chimerism in the Rejecting Kidney Transplant. *J. Am. Soc. Nephrol.* **2020**, *31*, 1977–1986. [CrossRef] [PubMed]
113. Liu, Y.; Hu, J.; Liu, D.; Zhou, S.; Liao, J.; Liao, G.; Yang, S.; Guo, Z.; Li, Y.; Li, S.; et al. Single-cell analysis reveals immune landscape in kidneys of patients with chronic transplant rejection. *Theranostics* **2020**, *10*, 8851–8862. [CrossRef] [PubMed]
114. Zhuang, Q.; Li, H.; Peng, B.; Liu, Y.; Zhang, Y.; Cai, H.; Liu, S.; Ming, Y. Single-Cell Transcriptomic Analysis of Peripheral Blood Reveals a Novel B-Cell Subset in Renal Allograft Recipients with Accommodation. *Front. Pharmacol.* **2021**, *12*, 706580. [CrossRef] [PubMed]

115. Asano, Y.; Daccache, J.; Jain, D.; Ko, K.; Kinloch, A.; Veselits, M.; Wolfgeher, D.; Chang, A.; Josephson, M.; Cunningham, P.; et al. Innate-like self-reactive B cells infiltrate human renal allografts during transplant rejection. *Nat. Commun.* **2021**, *12*, 4372. [CrossRef]
116. Lamarthee, B.; Callemeyn, J.; Van Herck, Y.; Antoranz, A.; Anglicheau, D.; Boada, P.; Becker, J.U.; Debyser, T.; De Smet, F.; De Vusser, K.; et al. Transcriptional and spatial profiling of the kidney allograft unravels a central role for FcyRIII+ innate immune cells in rejection. *Nat. Commun.* **2023**, *14*, 4359. [CrossRef]
117. van der List, A.C.J.; Litjens, N.H.R.; Brouwer, R.W.W.; Klepper, M.; den Dekker, A.T.; van Ijcken, W.F.J.; Betjes, M.G.H. Single-Cell RNA Sequencing of Donor-Reactive T Cells Reveals Role of Apoptosis in Donor-Specific Hyporesponsiveness of Kidney Transplant Recipients. *Int. J. Mol. Sci.* **2023**, *24*, 14463. [CrossRef] [PubMed]
118. Dangi, A.; Natesh, N.R.; Husain, I.; Ji, Z.; Barisoni, L.; Kwun, J.; Shen, X.; Thorp, E.B.; Luo, X. Single cell transcriptomics of mouse kidney transplants reveals a myeloid cell pathway for transplant rejection. *JCI Insight* **2020**, *5*, e141321. [CrossRef]
119. Wang, J.; Luo, P.; Zhao, J.; Tan, J.; Huang, F.; Ma, R.; Huang, P.; Huang, M.; Huang, Y.; Wei, Q.; et al. Profiling the Resident and Infiltrating Monocyte/Macrophages during Rejection following Kidney Transplantation. *J. Immunol. Res.* **2020**, *2020*, 5746832. [CrossRef] [PubMed]
120. Subramanian, A.; Sidhom, E.H.; Emani, M.; Vernon, K.; Sahakian, N.; Zhou, Y.; Kost-Alimova, M.; Slyper, M.; Waldman, J.; Dionne, D.; et al. Single cell census of human kidney organoids shows reproducibility and diminished off-target cells after transplantation. *Nat. Commun.* **2019**, *10*, 5462. [CrossRef]
121. Garreta, E.; Nauryzgaliyeva, Z.; Montserrat, N. Human induced pluripotent stem cell-derived kidney organoids toward clinical implementations. *Curr. Opin. Biomed. Eng.* **2021**, *20*, 100346. [CrossRef]
122. Melo Ferreira, R.; Sabo, A.R.; Winfree, S.; Collins, K.S.; Janosevic, D.; Gulbronson, C.J.; Cheng, Y.H.; Casbon, L.; Barwinska, D.; Ferkowicz, M.J.; et al. Integration of spatial and single-cell transcriptomics localizes epithelial cell-immune cross-talk in kidney injury. *JCI Insight* **2021**, *6*, e147703. [CrossRef]
123. Zheng, Y.; Lu, P.; Deng, Y.; Wen, L.; Wang, Y.; Ma, X.; Wang, Z.; Wu, L.; Hong, Q.; Duan, S.; et al. Single-Cell Transcriptomics Reveal Immune Mechanisms of the Onset and Progression of IgA Nephropathy. *Cell Rep.* **2020**, *33*, 108525. [CrossRef] [PubMed]
124. Lubetzky, M.L.; Salinas, T.; Schwartz, J.E.; Suthanthiran, M. Urinary Cell mRNA Profiles Predictive of Human Kidney Allograft Status. *Clin. J. Am. Soc. Nephrol.* **2021**, *16*, 1565–1577. [CrossRef] [PubMed]
125. Azim, S.; Zubair, H.; Rousselle, T.; McDaniels, J.M.; Shetty, A.C.; Kuscu, C.; Kuscu, C.; Talwar, M.; Eason, J.D.; Maluf, D.G.; et al. Single-cell RNA sequencing reveals peripheral blood mononuclear immune cell landscape associated with operational tolerance in a kidney transplant recipient. *Am. J. Transplant.* **2023**, *23*, 1434–1445. [CrossRef]
126. Muthukumar, T.; Yang, H.; Belkadi, A.; Thareja, G.; Li, C.; Snopkowski, C.; Chen, K.; Salinas, T.; Lubetzky, M.; Lee, J.; et al. Single Cell Rna-Sequencing of Urinary Cells and Defining the Immune Landscape of Rejection in Human Kidney Allografts. *Am. J. Transplant* **2021**, *21*, 305.
127. Kong, F.; Ye, S.; Zhong, Z.; Zhou, X.; Zhou, W.; Liu, Z.; Lan, J.; Xiong, Y.; Ye, Q. Single-Cell Transcriptome Analysis of Chronic Antibody-Mediated Rejection after Renal Transplantation. *Front. Immunol.* **2021**, *12*, 767618. [CrossRef]
128. Wen, N.; Wu, J.; Li, H.; Liao, J.; Lan, L.; Yang, X.; Zhu, G.; Lei, Z.; Dong, J.; Sun, X. Immune landscape in rejection of renal transplantation revealed by high-throughput single-cell RNA sequencing. *Front. Cell Dev. Biol.* **2023**, *11*, 1208566. [CrossRef] [PubMed]
129. Suryawanshi, H.; Yang, H.; Lubetzky, M.; Morozov, P.; Lagman, M.; Thareja, G.; Alonso, A.; Li, C.; Snopkowski, C.; Belkadi, A. Detection of infiltrating fibroblasts by single-cell transcriptomics in human kidney allografts. *PLoS ONE* **2022**, *17*, e0267704. [CrossRef]
130. Pang, Q.; Chen, L.; An, C.; Zhou, J.; Xiao, H. Single-cell and bulk RNA sequencing highlights the role of M1-like infiltrating macrophages in antibody-mediated rejection after kidney transplantation. *Heliyon* **2024**, *10*, e27865. [CrossRef]
131. Park, J.; Shrestha, R.; Qiu, C.; Kondo, A.; Huang, S.; Werth, M.; Li, M.; Barasch, J.; Suszták, K. Single-cell transcriptomics of the mouse kidney reveals potential cellular targets of kidney disease. *Science* **2018**, *360*, 758–763. [CrossRef]
132. Dell'Orso, S.; Juan, A.H.; Ko, K.D.; Naz, F.; Perovanovic, J.; Gutierrez-Cruz, G.; Feng, X.; Sartorelli, V. Correction: Single cell analysis of adult mouse skeletal muscle stem cells in homeostatic and regenerative conditions. *Development* **2019**, *146*, dev181743. [CrossRef]
133. Pellin, D.; Loperfido, M.; Baricordi, C.; Wolock, S.L.; Montepeloso, A.; Weinberg, O.K.; Biffi, A.; Klein, A.M.; Biasco, L. A comprehensive single cell transcriptional landscape of human hematopoietic progenitors. *Nat. Commun.* **2019**, *10*, 2395. [CrossRef]
134. Rashmi, P.; Sur, S.; Sigdel, T.K.; Boada, P.; Schroeder, A.W.; Damm, I.; Kretzler, M.; Hodgin, J.; Hartoularos, G.; Ye, C.J. Multiplexed droplet single-cell sequencing (Mux-Seq) of normal and transplant kidney. *Am. J. Transplant.* **2022**, *22*, 876–885. [CrossRef] [PubMed]
135. Wilson, G.W.; Moshkelgosha, S.; Duong, A.; Keshavjee, S.; Martinu, T.; Juvet, S.; Yeung, J.C. Deconvolution of Donor and Recipient Cells from Lung Transplant Single Cell RNA-seq Data. *J. Heart Lung Transplant.* **2021**, *40*, S149. [CrossRef]
136. Smirnova, N.F.; Riemondy, K.; Bueno, M.; Collins, S.; Suresh, P.; Wang, X.; Patel, K.N.; Cool, C.; Königshoff, M.; Sharma, N.S. Single-cell transcriptome mapping identifies a local, innate B cell population driving chronic rejection after lung transplantation. *JCI Insight* **2022**, *7*, e156648. [CrossRef]

137. Snyder, M.E.; Finlayson, M.O.; Connors, T.J.; Dogra, P.; Senda, T.; Bush, E.; Carpenter, D.; Marboe, C.; Benvenuto, L.; Shah, L.; et al. Generation and persistence of human tissue-resident memory T cells in lung transplantation. *Sci. Immunol.* **2019**, *4*, eaav5581. [CrossRef]
138. Bharat, A.; Querrey, M.; Markov, N.S.; Kim, S.; Kurihara, C.; Garza-Castillon, R.; Manerikar, A.; Shilatifard, A.; Tomic, R.; Politanska, Y.; et al. Lung transplantation for patients with severe COVID-19. *Sci. Transl. Med.* **2020**, *12*, eabe4282. [CrossRef] [PubMed]
139. Wanczyk, H.; Jensen, T.; Weiss, D.J.; Finck, C. Advanced single-cell technologies to guide the development of bioengineered lungs. *Am. J. Physiol. Lung Cell. Mol. Physiol.* **2021**, *320*, L1101–L1117. [CrossRef] [PubMed]
140. Mahata, B.; Zhang, X.; Kolodziejczyk, A.A.; Proserpio, V.; Haim-Vilmovsky, L.; Taylor, A.E.; Hebenstreit, D.; Dingler, F.A.; Moignard, V.; Göttgens, B. Single-cell RNA sequencing reveals T helper cells synthesizing steroids de novo to contribute to immune homeostasis. *Cell Rep.* **2014**, *7*, 1130–1142. [CrossRef]
141. Hurskainen, M.; Mižíková, I.; Cook, D.P.; Andersson, N.; Cyr-Depauw, C.; Lesage, F.; Helle, E.; Renesme, L.; Jankov, R.P.; Heikinheimo, M. Single cell transcriptomic analysis of murine lung development on hyperoxia-induced damage. *Nat. Commun.* **2021**, *12*, 1565. [CrossRef]
142. Travaglini, K.J.; Nabhan, A.N.; Penland, L.; Sinha, R.; Gillich, A.; Sit, R.V.; Chang, S.; Conley, S.D.; Mori, Y.; Seita, J. A molecular cell atlas of the human lung from single-cell RNA sequencing. *Nature* **2020**, *587*, 619–625. [CrossRef] [PubMed]
143. Silva, T.d.; Voisey, J.; Hopkins, P.; Apte, S.; Chambers, D.; O'Sullivan, B. Markers of rejection of a lung allograft: State of the art. *Biomark. Med.* **2022**, *16*, 483–498. [CrossRef] [PubMed]
144. Misharin, A.V.; Morales-Nebreda, L.; Reyfman, P.A.; Cuda, C.M.; Walter, J.M.; McQuattie-Pimentel, A.C.; Chen, C.-I.; Anekalla, K.R.; Joshi, N.; Williams, K.J. Monocyte-derived alveolar macrophages drive lung fibrosis and persist in the lung over the life span. *J. Exp. Med.* **2017**, *214*, 2387–2404. [CrossRef] [PubMed]
145. Lunardi, F.; Abbrescia, D.I.; Vedovelli, L.; Pezzuto, F.; Fortarezza, F.; Comacchio, G.M.; Guzzardo, V.; Ferrigno, P.; Loy, M.; Giraudo, C. Molecular Profiling of Tissue Samples with Chronic Rejection from Patients with Chronic Lung Allograft Dysfunction: A Pilot Study in Cystic Fibrosis Patients. *Biomolecules* **2023**, *13*, 97. [CrossRef] [PubMed]
146. Malone, A.F. Monocytes and Macrophages in Kidney Transplantation and Insights from Single Cell RNA-Seq Studies. *Kidney360* **2021**, *2*, 1654–1659. [CrossRef] [PubMed]
147. Lavin, Y.; Winter, D.; Blecher-Gonen, R.; David, E.; Keren-Shaul, H.; Merad, M.; Jung, S.; Amit, I. Tissue-resident macrophage enhancer landscapes are shaped by the local microenvironment. *Cell* **2014**, *159*, 1312–1326. [CrossRef]
148. Reyfman, P.A.; Walter, J.M.; Joshi, N.; Anekalla, K.R.; McQuattie-Pimentel, A.C.; Chiu, S.; Fernandez, R.; Akbarpour, M.; Chen, C.-I.; Ren, Z.; et al. Single-Cell Transcriptomic Analysis of Human Lung Provides Insights into the Pathobiology of Pulmonary Fibrosis. *Am. J. Respir. Crit. Care Med.* **2018**, *199*, 1517–1536. [CrossRef] [PubMed]
149. Snyder, M.E.; Moghbeli, K.; Bondonese, A.; Craig, A.; Popescu, I.; Fan, L.; Tabib, T.; Lafyatis, R.; Chen, K.; Trejo Bittar, H.E. Human lung tissue resident memory T cells are re-programmed but not eradicated with systemic glucocorticoids after acute cellular rejection. *medRxiv* **2021**. [CrossRef]
150. Lee, S.M.L.; Bertinetti-Lapatki, C.; Schiergens, T.S.; Jauch, K.W.; Roth, A.B.; Thasler, W.E. Concurrent isolation of hepatic stem cells and hepatocytes from the human liver. *In Vitro Cell Dev. Biol. Anim.* **2020**, *56*, 253–260. [CrossRef]
151. Shi, W.; Wang, Y.; Zhang, C.; Jin, H.; Zeng, Z.; Wei, L.; Tian, Y.; Zhang, D.; Sun, G. Isolation and purification of immune cells from the liver. *Int. Immunopharmacol.* **2020**, *85*, 106632. [CrossRef]
152. MacParland, S.A.; Liu, J.C.; Ma, X.Z.; Innes, B.T.; Bartczak, A.M.; Gage, B.K.; Manuel, J.; Khuu, N.; Echeverri, J.; Linares, I.; et al. Single cell RNA sequencing of human liver reveals distinct intrahepatic macrophage populations. *Nat. Commun.* **2018**, *9*, 4383. [CrossRef] [PubMed]
153. Wang, L.; Li, J.; He, S.; Liu, Y.; Chen, H.; Yin, M.; Zou, D.; Chen, S.; Luo, T.; Yu, X.; et al. Resolving the graft ischemia-reperfusion injury during liver transplantation at the single cell resolution. *Cell Death Dis.* **2021**, *12*, 589. [CrossRef] [PubMed]
154. Yang, X.; Lu, D.; Wang, R.; Lian, Z.; Lin, Z.; Zhuo, J.; Chen, H.; Yang, M.; Tan, W.; Wei, X.; et al. Single-cell profiling reveals distinct immune phenotypes that contribute to ischaemia-reperfusion injury after steatotic liver transplantation. *Cell Prolif.* **2021**, *54*, e13116. [CrossRef] [PubMed]
155. Wang, Z.; Shao, X.; Wang, K.; Lu, X.; Zhuang, L.; Yang, X.; Zhang, P.; Yang, P.; Zheng, S.; Xu, X.; et al. Single-cell analysis reveals a pathogenic cellular module associated with early allograft dysfunction after liver transplantation. *bioRxiv* **2022**. [CrossRef]
156. Huang, H.; Chen, R.; Lin, Y.; Jiang, J.; Feng, S.; Zhang, X.; Zhang, C.; Ji, Q.; Chen, H.; Xie, H.; et al. Decoding Single-cell Landscape and Intercellular Crosstalk in the Transplanted Liver. *Transplantation* **2023**, *107*, 890–902. [CrossRef]
157. Morrison, J.K.; DeRossi, C.; Alter, I.L.; Nayar, S.; Giri, M.; Zhang, C.; Cho, J.H.; Chu, J. Single-cell transcriptomics reveals conserved cell identities and fibrogenic phenotypes in zebrafish and human liver. *Hepatol. Commun.* **2022**, *6*, 1711–1724. [CrossRef] [PubMed]
158. Li, X.; Li, S.; Wu, B.; Xu, Q.; Teng, D.; Yang, T.; Sun, Y.; Zhao, Y.; Li, T.; Liu, D. Landscape of immune cells heterogeneity in liver transplantation by single-cell RNA sequencing analysis. *Front. Immunol.* **2022**, *13*, 890019. [CrossRef] [PubMed]
159. Shan, Y.; Qi, D.; Zhang, L.; Wu, L.; Li, W.; Liu, H.; Li, T.; Fu, Z.; Bao, H.; Song, S. Single-cell RNA-seq revealing the immune features of donor liver during liver transplantation. *Front. Immunol.* **2023**, *14*, 1096733. [CrossRef] [PubMed]
160. Tang, H.; Yuan, J.; Gong, Y.-F.; Zhang, C.-Y.; Liu, M.; Luo, S.-X. Single-cell transcriptome sequencing reveals potential novel combination of biomarkers for antibody-based cancer therapeutics in hepatocellular carcinoma. *Front. Genet.* **2022**, *13*, 928256. [CrossRef]

161. Hautz, T.; Salcher, S.; Fodor, M.; Sturm, G.; Ebner, S.; Mair, A.; Trebo, M.; Untergasser, G.; Sopper, S.; Cardini, B. Immune cell dynamics deconvoluted by single-cell RNA sequencing in normothermic machine perfusion of the liver. *Nat. Commun.* **2023**, *14*, 2285. [CrossRef]
162. Zhao, J.; Zhang, S.; Liu, Y.; He, X.; Qu, M.; Xu, G.; Wang, H.; Huang, M.; Pan, J.; Liu, Z. Single-cell RNA sequencing reveals the heterogeneity of liver-resident immune cells in human. *Cell Discov.* **2020**, *6*, 22. [CrossRef]
163. Ramachandran, P.; Dobie, R.; Wilson-Kanamori, J.; Dora, E.; Henderson, B.; Luu, N.; Portman, J.; Matchett, K.; Brice, M.; Marwick, J. Resolving the fibrotic niche of human liver cirrhosis at single-cell level. *Nature* **2019**, *575*, 512–518. [CrossRef]
164. Aizarani, N.; Saviano, A.; Mailly, L.; Durand, S.; Herman, J.S.; Pessaux, P.; Baumert, T.F.; Grün, D. A human liver cell atlas reveals heterogeneity and epithelial progenitors. *Nature* **2019**, *572*, 199–204. [CrossRef]
165. Zhang, Q.; He, Y.; Luo, N.; Patel, S.J.; Han, Y.; Gao, R.; Modak, M.; Carotta, S.; Haslinger, C.; Kind, D. Landscape and dynamics of single immune cells in hepatocellular carcinoma. *Cell* **2019**, *179*, 829–845.e20. [CrossRef]
166. Barbetta, A.; Rocque, B.; Sarode, D.; Bartlett, J.A.; Emamaullee, J. Revisiting transplant immunology through the lens of single-cell technologies. *Semin. Immunopathol.* **2023**, *45*, 91–109. [CrossRef]
167. Roushansarai, N.S.; Pascher, A.; Becker, F. Innate Immune Cells during Machine Perfusion of Liver Grafts—The Janus Face of Hepatic Macrophages. *J. Clin. Med.* **2022**, *11*, 6669. [CrossRef]
168. Tamburini, B.A.J.; Finlon, J.M.; Gillen, A.E.; Kriss, M.S.; Riemondy, K.A.; Fu, R.; Schuyler, R.P.; Hesselberth, J.R.; Rosen, H.R.; Burchill, M.A. Chronic liver disease in humans causes expansion and differentiation of liver lymphatic endothelial cells. *Front. Immunol.* **2019**, *10*, 1036. [CrossRef]
169. Dobie, R.; Wilson-Kanamori, J.R.; Henderson, B.E.; Smith, J.R.; Matchett, K.P.; Portman, J.R.; Wallenborg, K.; Picelli, S.; Zagorska, A.; Pendem, S.V. Single-cell transcriptomics uncovers zonation of function in the mesenchyme during liver fibrosis. *Cell Rep.* **2019**, *29*, 1832–1847.e8. [CrossRef]
170. Loeuillard, E.; Yang, J.; Buckarma, E.; Wang, J.; Liu, Y.; Conboy, C.; Pavelko, K.D.; Li, Y.; O'Brien, D.; Wang, C. Targeting tumor-associated macrophages and granulocytic myeloid-derived suppressor cells augments PD-1 blockade in cholangiocarcinoma. *J. Clin. Investig.* **2020**, *130*, 5380–5396. [CrossRef] [PubMed]
171. Shiode, Y.; Kodama, T.; Shigeno, S.; Murai, K.; Tanaka, S.; Newberg, J.Y.; Kondo, J.; Kobayashi, S.; Yamada, R.; Hikita, H. TNF receptor–related factor 3 inactivation promotes the development of intrahepatic cholangiocarcinoma through NF-κB-inducing kinase–mediated hepatocyte transdifferentiation. *Hepatology* **2022**, *77*, 395–410. [CrossRef] [PubMed]
172. Frazzette, N.; Khodadadi-Jamayran, A.; Doudican, N.; Santana, A.; Felsen, D.; Pavlick, A.C.; Tsirigos, A.; Carucci, J.A. Decreased cytotoxic T cells and TCR clonality in organ transplant recipients with squamous cell carcinoma. *NPJ Precis. Oncol.* **2020**, *4*, 13. [CrossRef]
173. Blau, H.M.; Daley, G.Q. Stem Cells in the Treatment of Disease. *N. Engl. J. Med.* **2019**, *380*, 1748–1760. [CrossRef]
174. Karam, D.; Gertz, M.; Lacy, M.; Dispenzieri, A.; Hayman, S.; Dingli, D.; Buadi, F.; Kapoor, P.; Kourelis, T.; Warsame, R.; et al. Impact of maintenance therapy post autologous stem cell transplantation for multiple myeloma in early and delayed transplant. *Bone Marrow Transplant.* **2022**, *57*, 803–809. [CrossRef]
175. Fast, E.M.; Sporrij, A.; Manning, M.; Rocha, E.L.; Yang, S.; Zhou, Y.; Guo, J.; Baryawno, N.; Barkas, N.; Scadden, D.; et al. External signals regulate continuous transcriptional states in hematopoietic stem cells. *Elife* **2021**, *10*, e66512. [CrossRef] [PubMed]
176. Rahman, M.; Wang, Z.Y.; Li, J.X.; Xu, H.W.; Wang, R.; Wu, Q. Single-Cell RNA Sequencing Reveals the Interaction of Injected ADSCs with Lung-Originated Cells in Mouse Pulmonary Fibrosis. *Stem Cells Int.* **2022**, *2022*, 9483166. [CrossRef]
177. Zhao, Y.; Li, X.; Zhao, W.; Wang, J.; Yu, J.; Wan, Z.; Gao, K.; Yi, G.; Wang, X.; Fan, B.; et al. Single-cell transcriptomic landscape of nucleated cells in umbilical cord blood. *Gigascience* **2019**, *8*, giz047. [CrossRef]
178. Wang, F.; Tan, P.; Zhang, P.; Ren, Y.; Zhou, J.; Li, Y.; Hou, S.; Li, S.; Zhang, L.; Ma, Y.; et al. Single-cell architecture and functional requirement of alternative splicing during hematopoietic stem cell formation. *Sci. Adv.* **2022**, *8*, eabg5369. [CrossRef]
179. Wittenbecher, F.; Keilholz, L.; Obermayer, B.; Conrad, T.; Frentsch, M.; Blau, I.W.; Vuong, L.G.; Borchert, F.; Lesch, S.; Movasshagi, K.; et al. Single-Cell Clonal Tracking in Allogeneic Hematopoietic Stem Cell Transplantation Reveals Time Dependent and Distinct Functional Patterns in Traceable Donor T Cell Clones. *Blood* **2021**, *138*, 335. [CrossRef]
180. Augsornworawat, P.; Maxwell, K.G.; Velazco-Cruz, L.; Millman, J.R. Single-cell transcriptome profiling reveals β cell maturation in stem cell-derived islets after transplantation. *Cell Rep.* **2020**, *32*, 108067. [CrossRef]
181. Tiklová, K.; Nolbrant, S.; Fiorenzano, A.; Björklund, Å.; Sharma, Y.; Heuer, A.; Gillberg, L.; Hoban, D.B.; Cardoso, T.; Adler, A.F.; et al. Single cell transcriptomics identifies stem cell-derived graft composition in a model of Parkinson's disease. *Nat. Commun.* **2020**, *11*, 2434. [CrossRef]
182. Arjona, M.; Goshayeshi, A.; Rodriguez-Mateo, C.; Brett, J.O.; Both, P.; Ishak, H.; Rando, T.A. Tubastatin A maintains adult skeletal muscle stem cells in a quiescent state ex vivo and improves their engraftment ability in vivo. *Stem Cell Rep.* **2022**, *17*, 82–95. [CrossRef] [PubMed]
183. Montarras, D.; Morgan, J.; Collins, C.; Relaix, F.; Zaffran, S.; Cumano, A.; Partridge, T.; Buckingham, M. Direct isolation of satellite cells for skeletal muscle regeneration. *Science* **2005**, *309*, 2064–2067. [CrossRef] [PubMed]
184. Dong, F.; Hao, S.; Zhang, S.; Zhu, C.; Cheng, H.; Yang, Z.; Hamey, F.K.; Wang, X.; Gao, A.; Wang, F.; et al. Differentiation of transplanted haematopoietic stem cells tracked by single-cell transcriptomic analysis. *Nat. Cell Biol.* **2020**, *22*, 630–639. [CrossRef] [PubMed]

185. de Almeida, G.P.; Lichtner, P.; Eckstein, G.; Brinkschmidt, T.; Chu, C.-F.; Sun, S.; Reinhard, J.; Mädler, S.C.; Kloeppel, M.; Verbeek, M.; et al. Human skin-resident host T cells can persist long term after allogeneic stem cell transplantation and maintain recirculation potential. *Sci. Immunol.* **2022**, *7*, eabe2634. [CrossRef] [PubMed]
186. Sun, C.; Wang, H.; Ma, Q.; Chen, C.; Yue, J.; Li, B.; Zhang, X. Time-course single-cell RNA sequencing reveals transcriptional dynamics and heterogeneity of limbal stem cells derived from human pluripotent stem cells. *Cell Biosci.* **2021**, *11*, 24. [CrossRef] [PubMed]
187. Grover, A.; Sanjuan-Pla, A.; Thongjuea, S.; Carrelha, J.; Giustacchini, A.; Gambardella, A.; Macaulay, I.; Mancini, E.; Luis, T.C.; Mead, A. Single-cell RNA sequencing reveals molecular and functional platelet bias of aged haematopoietic stem cells. *Nat. Commun.* **2016**, *7*, 11075. [CrossRef] [PubMed]
188. Zheng, G.X.; Terry, J.M.; Belgrader, P.; Ryvkin, P.; Bent, Z.W.; Wilson, R.; Ziraldo, S.B.; Wheeler, T.D.; McDermott, G.P.; Zhu, J. Massively parallel digital transcriptional profiling of single cells. *Nat. Commun.* **2017**, *8*, 14049. [CrossRef] [PubMed]
189. Maxwell, K.G.; Augsornworawat, P.; Velazco-Cruz, L.; Kim, M.H.; Asada, R.; Hogrebe, N.J.; Morikawa, S.; Urano, F.; Millman, J.R. Gene-edited human stem cell–derived β cells from a patient with monogenic diabetes reverse preexisting diabetes in mice. *Sci. Transl. Med.* **2020**, *12*, eaax9106. [CrossRef] [PubMed]
190. Oguma, Y.; Kuroda, Y.; Wakao, S.; Kushida, Y.; Dezawa, M. Single-cell RNA sequencing reveals different signatures of mesenchymal stromal cell pluripotent-like and multipotent populations. *iScience* **2022**, *25*, 105395. [CrossRef]
191. Cesaro, A.; Defrene, J.; Lachhab, A.; Page, N.; Tardif, M.R.; Al-Shami, A.; Oravecz, T.; Fortin, P.R.; Daudelin, J.F.; Labrecque, N.; et al. Enhanced myelopoiesis and aggravated arthritis in *S100a8*-deficient mice. *PLoS ONE* **2019**, *14*, e0221528. [CrossRef]
192. Bode, D.; Cull, A.H.; Rubio-Lara, J.A.; Kent, D.G. Exploiting Single-Cell Tools in Gene and Cell Therapy. *Front. Immunol.* **2021**, *12*, 702636. [CrossRef]
193. You, G.; Zhang, M.; Bian, Z.; Guo, H.; Xu, Z.; Ni, Y.; Lan, Y.; Yue, W.; Gong, Y.; Chang, Y. Decoding lymphomyeloid divergence and immune hyporesponsiveness in G-CSF-primed human bone marrow by single-cell RNA-seq. *Cell Discov.* **2022**, *8*, 59. [CrossRef]
194. Wisdom, A.J.; Mowery, Y.M.; Hong, C.S.; Himes, J.E.; Nabet, B.Y.; Qin, X.; Zhang, D.; Chen, L.; Fradin, H.; Patel, R.; et al. Single cell analysis reveals distinct immune landscapes in transplant and primary sarcomas that determine response or resistance to immunotherapy. *Nat. Commun.* **2020**, *11*, 6410. [CrossRef]
195. Sinha, V.C.; Rinkenbaugh, A.L.; Xu, M.; Zhou, X.; Zhang, X.; Jeter-Jones, S.; Shao, J.; Qi, Y.; Zebala, J.A.; Maeda, D.Y.; et al. Single-cell evaluation reveals shifts in the tumor-immune niches that shape and maintain aggressive lesions in the breast. *Nat. Commun.* **2021**, *12*, 5024. [CrossRef]
196. Tirosh, I.; Izar, B.; Prakadan, S.M.; Wadsworth, M.H.; Treacy, D.; Trombetta, J.J.; Rotem, A.; Rodman, C.; Lian, C.; Murphy, G. Dissecting the multicellular ecosystem of metastatic melanoma by single-cell RNA-seq. *Science* **2016**, *352*, 189–196. [CrossRef]
197. Lambrechts, D.; Wauters, E.; Boeckx, B.; Aibar, S.; Nittner, D.; Burton, O.; Bassez, A.; Decaluwé, H.; Pircher, A.; Van den Eynde, K. Phenotype molding of stromal cells in the lung tumor microenvironment. *Nat. Med.* **2018**, *24*, 1277–1289. [CrossRef]
198. Slyper, M.; Porter, C.B.; Ashenberg, O.; Waldman, J.; Drokhlyansky, E.; Wakiro, I.; Smillie, C.; Smith-Rosario, G.; Wu, J.; Dionne, D. A single-cell and single-nucleus RNA-Seq toolbox for fresh and frozen human tumors. *Nat. Med.* **2020**, *26*, 792–802. [CrossRef]
199. Paillet, J.; Plantureux, C.; Lévesque, S.; Le Naour, J.; Stoll, G.; Sauvat, A.; Caudana, P.; Tosello Boari, J.; Bloy, N.; Lachkar, S. Autoimmunity affecting the biliary tract fuels the immunosurveillance of cholangiocarcinoma. *J. Exp. Med.* **2021**, *218*, e20200853. [CrossRef] [PubMed]
200. Noé, A.; Cargill, T.N.; Nielsen, C.M.; Russell, A.J.C.; Barnes, E. The Application of Single-Cell RNA Sequencing in Vaccinology. *J. Immunol. Res.* **2020**, *2020*, 8624963. [CrossRef] [PubMed]

biosensors

MDPI

Review

Photonic Microfluidic Technologies for Phytoplankton Research

José Francisco Algorri [1,2,3,*], Pablo Roldán-Varona [1,2,33], José Miguel López-Higuera [1,2,3], Luis Rodriguez-Cobo [1,2,3] and Adolfo Cobo-García [1,2,3]

1 Photonics Engineering Group, Universidad de Cantabria, 39005 Santander, Spain
2 CIBER de Bioingeniera, Biomateriales y Nanomedicina, Instituto de Salud Carlos III, 28029 Madrid, Spain
3 Instituto de Investigación Sanitaria Valdecilla (IDIVAL), 39011 Santander, Spain
* Correspondence: algorrijf@unican.es

Abstract: Phytoplankton is a crucial component for the correct functioning of different ecosystems, climate regulation and carbon reduction. Being at least a quarter of the biomass of the world's vegetation, they produce approximately 50% of atmospheric O_2 and remove nearly a third of the anthropogenic carbon released into the atmosphere through photosynthesis. In addition, they support directly or indirectly all the animals of the ocean and freshwater ecosystems, being the base of the food web. The importance of their measurement and identification has increased in the last years, becoming an essential consideration for marine management. The gold standard process used to identify and quantify phytoplankton is manual sample collection and microscopy-based identification, which is a tedious and time-consuming task and requires highly trained professionals. Microfluidic Lab-on-a-Chip technology represents a potential technical solution for environmental monitoring, for example, in situ quantifying toxic phytoplankton. Its main advantages are miniaturisation, portability, reduced reagent/sample consumption and cost reduction. In particular, photonic microfluidic chips that rely on optical sensing have emerged as powerful tools that can be used to identify and analyse phytoplankton with high specificity, sensitivity and throughput. In this review, we focus on recent advances in photonic microfluidic technologies for phytoplankton research. Different optical properties of phytoplankton, fabrication and sensing technologies will be reviewed. To conclude, current challenges and possible future directions will be discussed.

Keywords: phytoplankton; microfluidics; photonics

1. Introduction

Phytoplankton are all planktonic autotrophic aquatic organisms with photosynthetic capacity that live dispersed in the water. This name comes from the Greek terms, *φυτον* (phyton, "plant") and *πλαγκτος* ("plánktos", "wanderer" or "the one that wanders about"). They are also called microalgae, but despite all phytoplankton being microalgae, not all microalgae occur in plankton. Through photosynthesis, they produce energy-rich organic material that captures CO_2 (almost a third of the anthropogenic carbon released into the atmosphere) and release oxygen (approximately 50% of atmospheric O_2 [1]), thus helping to ameliorate greenhouse gases [2]. Phytoplankton is a taxonomically diverse group, consisting of more than ten thousand species and taxa that contribute to at least a quarter of the biomass of the world's vegetation and constitute the base of the food web that supports either directly or indirectly all the animal populations of the open sea [3]. In addition, it has been introduced by some chefs into gourmet cooking recently [4]. As can be observed, phytoplankton is a key component for the correct functioning of different ecosystems, climate regulation and carbon reduction. The importance of its measurement and identification has increased in the last years, becoming an essential consideration for marine management [5]. Furthermore, due to climate change and human contamination [6,7] phytoplankton populations are being affected. Events in which phytoplankton undergo rapid population increase are known as algal blooms. A harmful algal bloom (HAB) occurs when it causes negative impacts by producing toxins, which can cause illness to mammals, fish, corals and other marine organisms [8–10]. These events are colloquially known as "red tides" since these organisms sometimes stain the water that colour. Therefore, HABs constitute a serious threat to public health, causing huge losses in fisheries and other productive sectors worldwide. The HAB problem is widespread on all the world's seas and its trend is increasing. Red tides cause far-reaching economic damage to extractive and aquaculture activities. The group of phytoplankton that has caused the most toxic episodes is the group of dinoflagellates, specifically, Gymnodinus Gymnodinium catenatum and Gonyaulax tamarensis Lebour (both producers of paralytic toxin) as well as Dinophysis acuta Ehremberg and Dinophysis acuminata (responsible for gastrointestinal disorders due to shellfish ingestion). In many cases, the traditional classification of dinoflagellates based on morphological characteristics is insufficient. Therefore, it is necessary to develop other techniques to identify those harmful organisms. Another important application in which phytoplankton detection is gaining attention is in ships' ballast water. Ballast water is used in ships to maintain safe operating conditions during transit. Although most phytoplankton die due to the environmental conditions in the ballast tank, some can survive. Due to these species that may survive and establish a reproductive population (with a huge number of phytoplankton and zooplankton), serious ecological, economic and health problems can be caused. This phenomenon was first studied between 1903 and 1907 in the North Sea [11]; in this article, Ostenfeld recognised the invasion of the Asian phytoplankton algae Odontella (Biddulphia Sinensis). Despite this, it was not until the 1970s that the scientific community reviewed the problem in detail. According to the International Maritime Organization (IMO), in the late 1980s, Canada and Australia were among the countries experiencing particular problems with invasive species. They brought their concerns to the attention of IMO's Marine Environment Protection Committee (MEPC). Direct and indirect health effects are becoming increasingly serious and environmental damage is often irreversible. Recognising the possible severity of the consequences, action has been taken by the IMO, adopting the "International Convention on the Management of Ships Ballast Water and Sediments" [12]. Therefore, detecting microalgae and bacteria in the ship's ballast water not only involves analysing the related quality of ballast water, but also is aimed at balancing the ecological environment and economic interests of each country [13].

As can be observed, the identification and measurement of specific characteristics of phytoplankton are essential for controlling pollution of the marine environment, as well as for aquaculture and the shellfish industry. The gold standard process used to identify and quantify phytoplankton is manual sample collection and microscopy-based identification, which is a tedious and time-consuming task and requires highly trained professionals. Flow cytometry is used to automatise the measuring process [14]. Flow cytometry gives the classification and identification of phytoplankton species, quantitative analysis and the extraction of parameters at the individual level. Generally, a typical flow cytometer can process thousands of cells per minute, much faster than manual observation using light microscopy. However, the expensive and bulky instruments make in situ measurement very difficult. For these reasons, the routine quantitative monitoring of phytoplankton in water is costly and challenging, requiring sophisticated equipment, a lab and almost unique expertise. Much recent activity has focused on developing microfabricated flow cytometers, which integrate inexpensive optical components to solve these problems. In fact, it can rapidly count cells and probe cellular populations at the single cell level [15]. Through microfabricated devices and microfluidic Lab-On-a-Chip (LOC) systems, it is possible to create a completely autonomous and portable integrated system. Microfluidic channels can handle tiny fluid volumes down to picoliters in a controlled microenvironment. For this reason, they allow the precise control and manipulation of fluids, typically in a passive way. Microfluidic LOC technology represents a potential technical solution for environmental monitoring, for example, identification and classification of particles in water [16,17] like phytoplankton. Its main advantages are miniaturisation, portability, reduced reagent/sample consumption and automation and cost reduction. Among all the microfluidic technologies [18–21], photonic microfluidic chips that rely on optical sensing have emerged as powerful tools that can be used to identify and analyse phytoplankton with high specificity, sensitivity and throughput.

Few literature reviews can be found related to microfluidics for marine research. For example, in [22], microfluidic systems for microalgal biotechnology are reviewed. Specifically, different applications are reported, e.g., microalgal biofuel applications, cultivation, downstream processing, microalgae-based microbial fuel cells and microalgae-based biosensors. A more general approach can be found in [23], focused on how microfluidic platforms address some challenges of plankton research. For example, analysis of a low density of organisms in environmental samples, difficulties in cultivating plankton, pre-concentrating, detecting and sorting them and how analytical microfluidic platforms are dedicated to the interactions between plankton and their environment. Finally, in a more recent work, the authors summarised some optofluidic systems and techniques for microalgal detection and characterisation [24]. For this reason, this review will focus only on recent developments in photonic microfluidics dedicated to phytoplankton research from a technological point of view. Optofluidic systems are the most widely used techniques for phytoplankton detection and characterisation. They are mainly based on three optical properties: fluorescence, scattering and imaging. Each of them can achieve cellular, lipid content, metabolic heterogeneity and count. This article is organised as follows: Section 2 will focus on the optical characteristics of phytoplankton (absorption, scattering and fluorescence) and Section 3 on the fabrication of microfluidic systems. Then, Section 4 will present the last advances in phytoplankton microfluidic technologies organised by phytoplankton optical properties. Finally, a discussion section will summarise and discuss the presented works. To conclude, we will present some current challenges and possible future directions of this promising technology.

2. Optical Characteristics of Phytoplankton

As mentioned before, the principal optical properties of phytoplankton are related to absorption, scattering and fluorescence. These optical processes are affected by the phytoplankton's composition, specifically, many different pigments, with chlorophyll being the most important. Furthermore, depending on the group or taxon under consideration, the concentration of other pigments in their cells varies [25]. For this reason, phytoplankton has an important effect on the colour of the ocean and the measurement of these properties allows the study of their ecology and evolution over time.

2.1. Absorption

Absorption is the process by which light is absorbed and converted into energy. This radiation, when absorbed, can be re-emitted or transformed into another type of energy, such as heat. Phytoplankton absorb sunlight and use this energy to produce chemical energy (photosynthesis). Two dominant peaks in the absorption spectrum exist in all phytoplankton cells (determined by chlorophylls). The primary one is in the blue (440 nm) and the second is in the red part of the spectrum (675 nm). Spectral light absorption curves in phytoplankton populations have been extensively studied (Babin et al. 2003), showing that the absorption coefficient ($a_{\phi}(\lambda)$) varies according to the presence of other pigments (depending on species and taxa) that will cause the blue peak to broaden and the appearance of additional absorption maxima. In addition, the packet effect due to phytoplankton size and changes in the physiological state of cells also affect these spectra [26,27]. These taxon-specific absorption peaks have been used as a tool for in situ optical detection [28], as well as for the development of remote sensing algorithms [29,30]. It has to be noted that absorption by phytoplankton is not a simple sum of the absorption coefficients of individual pigments. The absorption spectrum of phytoplankton varies in magnitude and shape due to the composition of different cell pigments [27]. In addition, specific proteins in the cells produce changes in the absorption spectrum, with cell pigment concentration and size also being influential [31]. However, as a simple approximation the so-called "package effect index", $Q_a^*(\lambda)$ (Equation (1)) is used.

$$Q_a^*(\lambda) = \frac{a_\phi(\lambda)}{a_{sol}(\lambda)}, \quad (1)$$

where $a_{sol}(\lambda)$ is the absorption coefficient of the same material, which would be dispersed into the solution. The absorption coefficients $a_{sol}(\lambda)$ (in m^{-1}) can be obtained by summing the contributions of all individual pigments, using the relationship:

$$a_{sol}(\lambda) = \sum C_i a_{sol,i}^*(\lambda), \quad (2)$$

where the $a_{sol,i}^*$ coefficients are the weight-specific absorption spectra of individual pigments (in $m^2 \cdot mg^{-1}$) and C_i are their concentrations in the medium (in $mg \cdot m^{-3}$). As stated in ref. [27], the $a_{sol,i}^*$ spectra of Figure 1 were estimated by scaling the absorption spectra of individual pigments in solvent, measured in relative values by high-pressure liquid chromatography, to the weight-specific absorption coefficients proposed in [32] and then shifting the positions of maxima to their in vivo positions, as in [33].

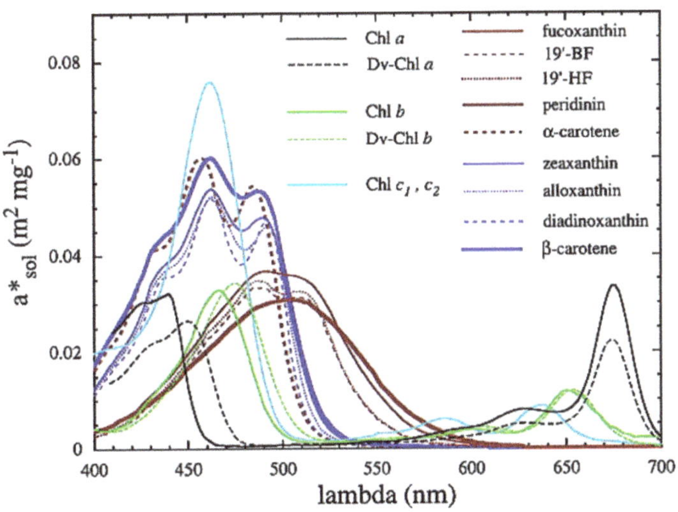

Figure 1. Weight-specific absorption spectra of the main pigments in vivo, $a_{sol,i}^*(\lambda)$ (in m^2 mg^{-1}), which are derived from absorption spectra of individual pigments in solvent. The red and blue lines represent the absorption spectra of photosynthetic and nonphotosynthetic carotenoids, respectively. Reprinted with permission from Ref. [27]. Copyright 2004, American Geophysical Union.

2.2. Light Scattering

When electromagnetic radiation strikes a molecule or particle, it can be scattered, i.e., the incident radiation is re-emitted in a direction different from that of the original radiation. In the case of elastic scattering (involving negligible energy transfer, same frequency), depending on the particle size with respect to the light wavelength (the colour), the re-emission intensity depends on the colour (Rayleigh scattering) or is independent of colour but has a certain direction (Mie scattering). Due to their morphology, phytoplankton cells are dominated by scattering or forward scattering rather than backscattering. Despite this, phytoplankton backscattering properties are also extremely important, as they are directly related to reflectance calculations, which is an essential measurement in oceanography [34]. In fact, the parameter "Remote Sensing Reflectance" (R_{rs}) is grossly proportional to backscattering divide by absorption (b_b/a) [35]. In Figure 2, the resulting contribution of phytoplankton $R_{rs}(\Phi)$ to the total R_{rs} is shown. As can be observed, the contributions of the phytoplankton to the total, as a percentage of total b_b/a, vary with biomass (mg·m^{-3}). Compared to the rest of the oceanic particles, the values of the scattering coefficients of phytoplankton are relatively low since they contain a large number of water molecules and exhibit strong absorption properties [36]. The exception to the rule is the coccolithophorid phytoplankton that produce small calcium carbonate flakes, which make them very effective dispersers and allow the blooms to be seen from space [37]. The phytoplankton scattering and backscattering coefficients and the scattering volume function can be obtained both from theoretical models (Mie theory) or measurements [38–40]. The properties depend on the size, shape and refractive index of all components of the phytoplankton cell [41]. Knowledge about the scattering angular distribution for phytoplankton is scarce due to the small number of experimental data sets [42,43]. Several commercial light scattering instruments for in situ measurements are reviewed by Moore et al. [44].

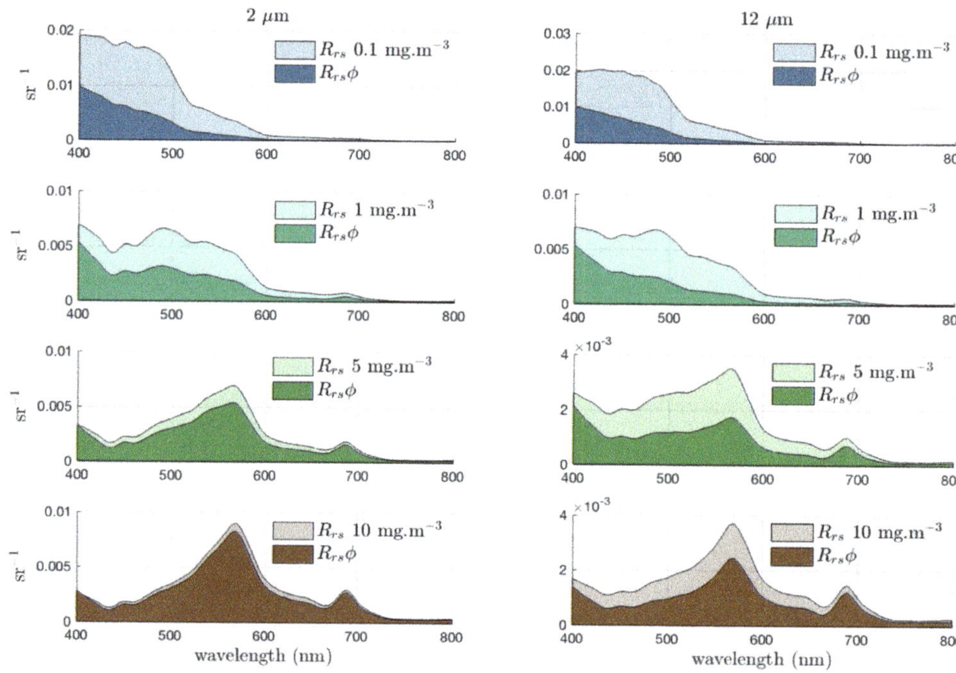

Figure 2. Relative contribution of phytoplankton to total R_{rs} (with $a_{gd}(400) = 0.07 \cdot [Clh_a]^{0.75}$ and $b_{bnap}(550) = 0.005$ m^{-1}) for increasing biomass with effective diameter, D_{eff} = 2 and 12 μm. The absorption caused by coloured dissolved organic matter (a_{gd}) covaries with Chl_a, whereas non-algal backscatter b_{bnap} is constant. These populations are idealised examples and not intended to represent any observed relationship between Chl_a concentration and D_{eff}. Reprinted with permission from [34]. Copyright 2018, MDPI.

In recent years, Raman scattering has also been investigated as a possible technique in phytoplankton research. The Raman effect is the inelastic scattering of light (with different frequency) by a substance. The Raman shift is independent of the excitation wavelength, whereas the scattered light intensity is inversely proportional. The interest in the phenomenon lies in the fact that the colour (wavelength) change is characteristic of the substance and gives information about its composition, that is, the type of chemical bonds present and its atomic structure. For this reason, Raman spectroscopy was proposed as a fast and sensitive method to measure phytoplankton and its composition in marine environments [45]. The use of visible Raman spectroscopy to identify the nutrient composition or depletion, the detection of invasive, problematic phytoplankton species and cell viability have been reported in [45–47], respectively. However, there are some disadvantages when it is used to measure pigment-rich biological material. The most important is that Raman intensity is too low (only 1 in 10^6–10^8 photons undergo Raman scattering) so the presence of fluorescence in phytoplankton (see next section) can mask the signals [48]. This effect can be reduced by using different excitation wavelengths and data processing (see Section 4.1). Another problem is that some pigment has a large range of bands while other biomolecule signals, such as fats and proteins can be masked [49]. Finally, the Raman spectra's complexity limits chemometric methods' application. Some recent works have proposed some solutions as using longer wavelengths (NIR) [50] and Fourier spectroscopy with multivariate data analysis [51] (see Section 4.2).

2.3. Fluorescence

Fluorescence is the property of some atoms and molecules to absorb light at a specific wavelength followed by light emission at a longer wavelength. The difference between scattering and fluorescence is their origin; whereas fluorescence occurs from a relatively long-lived excited electronic state, scattering occurs via the emission of a photon from a short-lived excited "virtual" state. In addition, the fluorescence emission wavelength is generally independent of the exciting wavelength, whereas light scattering increases with increases in the exciting wavelength. In the case of phytoplankton, several pigments (chlorophylls, pheopigments and phycobilins) fluoresce, with chlorophyll being the most important. One of the first proposals to use fluorescence as sensing parameter was presented in 1966 [52]. This work proposed the monitoring of phytoplankton biomass-chlorophyll by fluorescence measurements. Nowadays, this technique is commonly used in commercial devices and sensors (fluorometers, radiance and irradiance meters, flow cytometers). One of the main advantages is the possibility of measuring from the sea surface by aeroplanes and satellites [53]. Although fluorescence spectroscopy can detect small concentrations of chlorophyll and other pigments, it is not very specific. The main problem is that phytoplankton cells' fluorescence is highly variable due to their physiological conditions. Phytoplankton fluorescence depends on several parameters, e.g., the taxonomic position, the pigment content and ratio, the nutrient conditions, stage of growth, photoadaptation, physiological state, etc. [54–56]. Thus, deriving the biomass of spectral groups using the spectral fluorescence of multicomponent natural samples is not trivial [57]. For this reason, it is generally used to either approximate total plankton concentrations or detect the presence of harmful species; a recent comprehensive review can be found in ref. [56]. A simple equation can express fluorescence from phytoplankton chlorophyll, $F = I_i \cdot chl_a \cdot a^*_{phyto} \cdot \phi_f$ [58], with I_i as the impinging light intensity, chl_a as chlorophyll concentration, a^*_{phyto} as chlorophyll-specific phytoplankton absorption coefficient and ϕ_f as the fluorescence quantum yield.

As an example, in Figure 3, the spectral signature of some algal groups that mainly differ in their pigments can be observed. As mentioned in [57], all these measurements can be used as starting point to identify the phytoplankton.

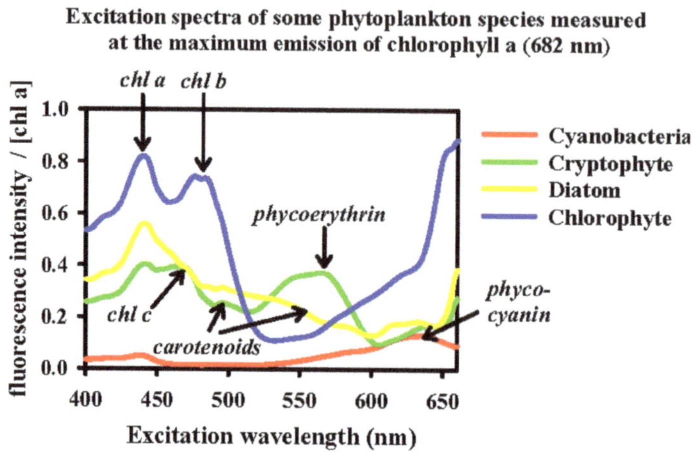

Figure 3. Fluorescence excitation spectra for different chemotaxonomic phytoplankton pigment groups. Reprinted with permission from [57]. Copyright 2003, International Council for the Exploration of the Sea (ICES).

3. Fabrication of Photonic Microfluidic Systems

A microfluidic system is based on a device in which one or more fluids flow through micrometer-sized channels (10^{-4}–10^{-6} m). They can handle tiny fluid volumes down to picoliters in a controlled microenvironment; the flow is laminar and characterised by low thermal and chemical diffusion times. For this reason, they allow the precise control and manipulation of fluids, typically in a passive way, with some components being necessary such as micro-pumps or micro-valves for active devices [20]. Microfluidic chips or systems have applications in various fields, such as chemistry, environmental sciences or medical research. For example, microfluidic chips can be used to mimic the complex structure, microenvironment and physiological function of human organs [59], drug screening [60], toxicity testing [61] or stem cell testing [62], among other applications. In addition, microfludidic systems can be integrated with other photonic components to produce a photonic biosensor. Photonic biosensors have several advantages, such as immunity to electromagnetic interference, high-speed operation, low power consumption, use potential in harsh environments, miniaturisation capacity, multiplexing possibility, mechanical stability, low manufacturing and integration cost and real-time detection, directly and without labels (label-free) [63,64]. Photonic biosensors based on microfluidic channels are optical transducers. They modify a specific light parameter (intensity, phase or frequency) in the presence of an analyte. They typically consist of photodetectors (commonly used with light emission detection techniques, such as fluorescence), interferometers (with an analyte-sensitive arm) and/or resonant structures, such as microresonators (which also detect refractive index changes). All these sensors can be integrated on a micrometer scale, especially with the development of microfluidic systems in recent years.

The combination of microfluidics and photonics was proposed at the beginning of this century, when these two fields were maturing [65,66] . As mentioned below, despite the main fabrication techniques being based on lithographic methods, fs-lasers have been proven to fabricate microfluidic lasers [67], microfluidic channels and waveguides [68] in a 3D arrangement [69,70]. In addition, it is also possible to fabricate waveguides in commercial microfluidic chips by fs-laser, facilitating the fabrication process [71]. These advantages enable its use for LOC and bio-photonics applications [72]. Sensor characteristics are quantified in terms of selectivity (detecting a particular element) and sensitivity. Usually, to improve these parameters, receptors specific to the component to be measured are included. It should be noted that the comparison of the parameters reported in the literature is not simple, sometimes data are given in mass concentration (mass of a solute in relation to the volume of the solution) which is difficult to compare with other analytes or in the case of giving more comparable data, such as the refractive index units (RIU), it is not normalised to the size of the sensor. The most commonly used transduction signals in optical sensing are fluorescence, scattering and refractive index. There are two types of detectors, those specific to measuring the optical signals (fluorescence and scattering) or spectroscopic (label-free, to measure refractive index changes). In the first case, fluorescence has the highest sensitivity, down to the single-cell limit. Scattering, however, can be masked from the previous signal (see more details in the following sections). These systems can be supported by engineered microfluidic channels that stretch and sort the phytoplankton cells [23,73]. In the case of LOC interferometric sensors, a miniaturised system is used, with two separate arms in which one is the reference and the other is the measurement arm (in which the receiver and the analyte are included). When the analyte is on the reference arm, the light passing through it will have a phase difference with respect to the light from the other arm. When the light from the two arms recombines, it can be measured to measure the phase delay, which will be greater the higher the rate of the analyte. Although it is difficult to achieve the same selectivity and sensitivity as the previous sensors, they are widely used in environmental measurement applications such as air quality, greenhouse gases, detection of chemical, biological, radiological, nuclear, explosive agents, etc. [74]. In the case of phytoplankton, no sensors have been reported yet.

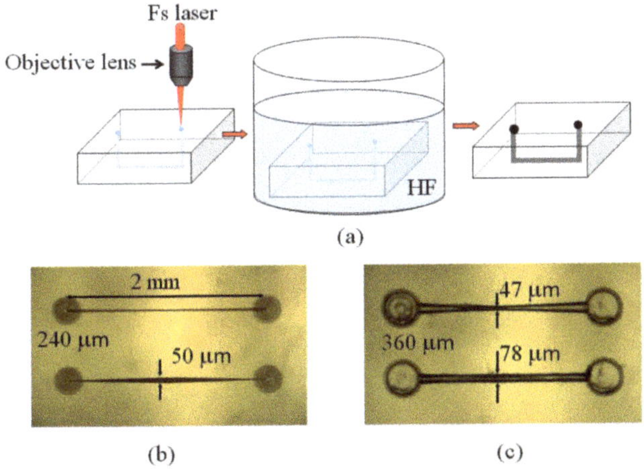

Figure 4. Femtosecond laser microfabrication of opto-microfluidic devices: (**a**) fabrication procedures, (**b**) U-shape-microchannels in fused silica before etching, and (**c**) after etching for 5 h in 20% HF acid within a shaker. Reprinted with permission from [75]. Copyright 2011, MDPI.

The fabrication processes in glass or silicon depend on the type of fabricated structure; extended details can be found in [20,76]. For example, soft lithography techniques have been mainly used to make microfluidic channels [77]. For this, polydimethylsiloxane (PDMS) elastomer can be used to have a fast and cheap fabrication process. However, the main problem is fabricating 3D structures, which requires additional stacking and bonding processes [20]. The same problem can be found in other typical techniques such as planar microfabrication (e.g., injection moulding or semiconductor processes based on photolithography). To solve this problem, femtosecond (fs) laser fabrication is the best option, as has been demonstrated in the 3D fabrication of transparent structures [78,79]. Some of the most important characteristics of fs-laser are: it is a single-step and maskless process, it can be applied to several materials (e.g., glasses, crystals, polymers) by changing the irradiation parameters; it can be used in 3D as depth irradiation can be easily modified; the fabrication of several components in different steps is possible. As an example, a schematic depiction of the fabrication procedure can be found in Figure 4a. Figure 4b,c show real examples of fabricated devices and Figure 5 illustrates how optical waveguides are integrated into a commercial LOC.

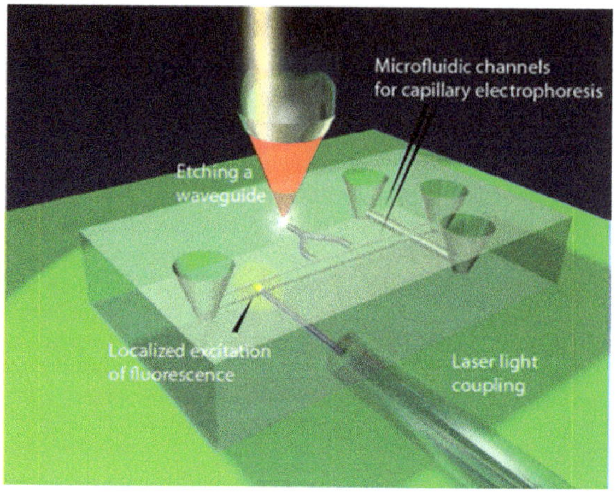

Figure 5. A schematic representation of an LOC device designed for capillary electrophoresis fabricated with a femtosecond-laser. The fluorescence is excited in a highly localised region in the microchannel by the optical waveguides and optical fibres. Moreover, the waveguides guide emissions to a detector. Reprinted with permission from [76]. Copyright 2014, SPIE.

In the case of microfluidic channels and other optical components (e.g., microlenses, hollow optical waveguides, optical micro-resonators [80–82]), the fabrication process involves the use of a wet chemical etching (HF or KOH) after the application of fs-laser [83]. KOH can produce defects such as relatively high surface roughness (of a few tens of nanometers). Despite this, this can be solved by polishing the accessible regions or using post-processing heat techniques for internal walls, e.g., oxygen/hydrogen flame polishing [80], annealing in an oven [82] and CO_2 laser reflow [84]. Nowadays, the main accepted hypothesis for the etching rate variation in fused-silica regions where the fs-laser is applied is that the laser beam reduces the Si-O-Si bond angle [85]. Other studies have tested the effect of the polarisation in the etching selectivity, as the absorption of the laser energy is spatially modulated, producing the so-called nanogratings perpendicular to the polarisation direction that enhances the etching selectivity [86]. Another option to avoid etching is fs-laser drilling of glass immersed in distilled water (liquid-assisted fs-laser drilling) [87]. In this case, the immersion is intended to remove some of the ablation debris that can contaminate the channel and restrict the size of the microstructures [88]. Some works have also proposed using porous glass (10 nm pores uniformly distributed) [89]. The process avoids wet etching and the porous element can be sealed by annealing the glass. Compared to fused silica, the channels can have arbitrary geometries, unlimited lengths and features sized beyond the optical diffraction limit [20]. Finally, it has to be noted that fs-laser can also produce typical waveguides based on refractive index modulation, which is caused by a localised nonlinear absorption at the laser focus region. Low index contrast waveguides have both a low transmission loss and excellent mode overlap with optical fibre, meaning that the overall extraction efficiency of a circuit is high [90]. When low intensities are used, the temperature rises and decreases with each pulse, resulting in a refractive index change that depends on the glass density and cooling rate. For higher intensities, the mechanism is based on the plasma formed inside the glass, which creates a Coulomb explosion and a shockwave; this induces an inhomogeneous material distribution and a refractive index change. Several works have demonstrated the feasibility of an fs-laser in terms of creating waveguides in different glasses [90–94]. For example, in [91] waveguides in fused silica glass are demonstrated with an RI change of $\sim 4 \times 10^{-3}$) and low propagation loss of

0.12 dB/cm. However, some issues such as spherical aberration, self-focusing and nonlinear absorption are common and produce an asymmetrical mode field pattern [94–96]. To control this parameter, the use of spatio temporally focused beams [97], spatial light modulators [98,99], thermal annealing [100] or polarisation control [101] have been proposed. As can be observed, the field has been very active this last decade.

4. Phytoplankton Microfluidic Technologies

As mentioned before, the most traditional method for identifying and quantifying phytoplankton is manual sample collection and microscopy-based identification, which is a tedious and time-consuming task and requires highly trained professionals. Standard flow cytometry has recently been used to automatise the phytoplankton measuring process. However, the expensive and bulky instruments make the in situ measurement very difficult. Microfluidic flow cytometry provides a solution by producing portable devices that enable on-site phytoplankton analysis and classification. However, conventional microfluidic flow cytometry generally relies on light scattering (Section 4.1) or fluorescence properties (Section 4.2), which cannot offer spatially resolved characterisation and distinguish through differences in surface morphology. For these reasons, imaging-based flow cytometry has been recently proposed to enhance phytoplankton measurement (Section 4.3). Impedance measurements sometimes support these techniques, so a brief review is also included at the end of this section (Section 4.4).

4.1. Technologies Based on Scattering

Absorption, scattering and side scattering of phytoplankton have been traditionally used in their characterisation (see Section 2). The cell properties are also measured in commercial flow cytometers using scattered light. Being an elastic scattering process, the measured wavelength will be the same as the source. The forward and side light scattering is collected from narrow angles and light diffracted around the cell, respectively. These measurements are demonstrated to give information about cell size and shape. Despite this, the measurement of these properties in LOC devices is not so usual and they are used in combination with other techniques.

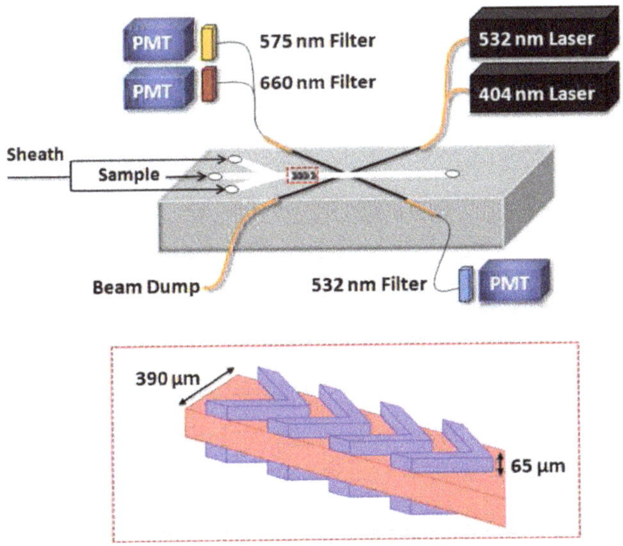

Figure 6. A schematic depiction of the proposed cytometer. (Top) The optical and microfluidics setup. (Bottom) A zoomed-in view showing the chevron grooves extending into the PDMS substrate. Reprinted with permission from [102]. Copyright 2011, AIP Publishing.

For example, in [103], Hashemi et al. developed a lab on chip based on autofluorescence and light scattering, which consisted of a special microfluidic chip with chevron grooves that can focus the flow in two dimensions. By using two grooves (on the top and bottom of the microchannel), they created two symmetrical sheath flows that wrap around a central flow (see Figure 6 bottom). A 488 nm argon laser and three photomultiplier tubes were used to record the fluorescence and light scattering signals. The proposed system was capable of detecting the picoplankton *Synechococcus* sp.with diameters lower than 1 μm and phytoplankton species as long as 80 μm (Nitzschia d.). In a later work, the authors replaced the large 488 nm argon laser with two solid-state laser sources (404 nm and one 532 nm), which are both small for in situ measurement (see Figure 6 top). In addition, the provided wavelength is closer to the maximum absorbance of chlorophyll and phycoerythrin [102].

In ref. [104], side light scattering is measured in combination with chlorophyll fluorescence and resistance pulse sensing (RPS) (see following sections). As shown in Figure 7, the chlorophyll fluorescence with 680 nm wavelength is detected from the positive Z-axis and the side light scattering with 480 nm wavelength is detected from the negative Z-axis. At the same time, the signal of resistance pulse sensing of the phytoplankton is acquired from the difference between the two sense arms of RPS+ and RPS− [104]. The light scattering measurements provide information about the intracellular contents and the size and surface roughness of the cell. On the other hand, the chlorophyll fluorescence is related to the activity and the RPS to the size of the cell.

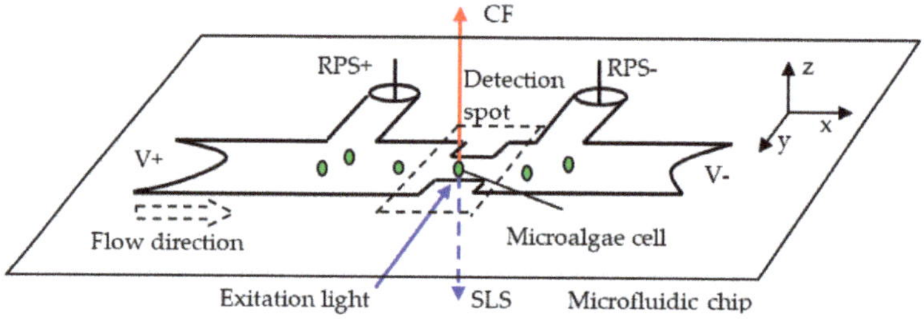

Figure 7. Schematic representation of the operation principle of single microalgae cell classification. The system is based on the simultaneous detection of chlorophyll fluorescence (CF), side light scattering (SLS) and resistance pulse sensing (RPS) signals. Reprinted with permission from [104]. Copyright 2016, MDPI.

As mentioned in Section 2.2, in recent years, Raman scattering has also been proposed as a possible technique for phytoplankton research [45,105,106]. In the case of microfluidic devices, Raman has been proposed in several works to act as a sorting method (Raman-activated cell sorting, RACS) [107]. Despite this, Raman signals are too low, requiring long interrogation times as compared with fluorescence detection (in the order of seconds to minutes in comparison with microseconds in fluorescence). One solution can be the isolation [108] or immobilisation [109] of single cells. In the first work, a highly motile species (Euglena Gracilis) is isolated by semiclosed microchannels with liquid flow only, whereas in the second one, optical tweezers are used (Raman tweezers, see [110] for more information). Another innovative solution was a "trap-free" RACS in a flow that allows continuous and automated sorting of individual cells [111]. In this case, the authors provide a stable flow field in the detection region by using two pressure dividers that eliminate local pressure fluctuations (see Figure 8). They achieved a 96.3% purity of the selected cells at a speed of 0.5 Hz.

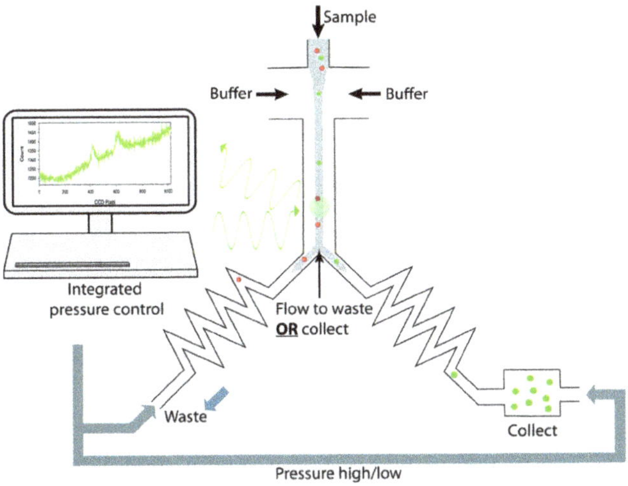

Figure 8. Schematic representation of the proposed device. A stream of cells were hydrodynamically focused in the detection channel for continuous Raman acquisition; on-the-fly classification was carried out to identify target cells and was immediately followed by alternating the pressures applied to the waste and collection channels, to direct the target cells to the collection chamber. Integrated software was developed to synchronise and automate all the operations. Reprinted with permission from [111]. Copyright 2016, Royal Society of Chemistry.

Another proposed technique to improve the throughput has been Raman-activated droplet sorting (RADS). For example, Kim et al. recently reported a method that uses Raman spectroscopy with PDMS-based microfluidic devices to perform on-chip, droplet-based in vivo phytoplankton lipid analysis with single-cell resolution [112]. The time-course tracking and study of lipid accumulation in C. reinhardtii cells under eight different culture conditions was successfully conducted, demonstrating the potential for Raman-microfluidics-based lipidomics. Another RADS microfluidic system for functional screening of live cells in a label-free and high-throughput manner was presented in [113]. In this case, the sorting is achieved by dielectrophoresis. The system is based on the level of astaxanthin content within phytoplankton cells, which has a high detection and sorting efficiency of approximately 260 cells/min and high accuracy of 98%. Furthermore, 92.73% of the selected cells remained alive and could proliferate. Haematococcus pluvialis cells in the microchannel were hydrodynamically focused in a single line after squeezing two buffer streams. When the cell passed through the detection region, the astaxanthin content in the cell was measured with Raman spectroscopy; then, the detected cell was encapsulated in the droplet for sorting the next step. Positive dielectrophoresis was used to manipulate the cell in the droplet with efficient trap and release, thus forcing cells with different astaxanthin contents into the pre-designed collection channel or waste channel based on their Raman spectroscopic responses.

4.2. Technologies Based on Fluorescence

Optical fluorescence-based detection methods are among the most popular and widely used in the detection and characterisation of biological and biochemical samples in microfluidic chips, including phytoplankton analysis. The fluorescence of phytoplankton can be the result of endogenous pigments (see Section 2.2) and different species of phytoplankton have their unique fluorescence spectra due to the different pigment ratios. The most common technique that uses fluorescence properties is flow cytometry. A comprehensive review can be found in [14]. However, the introduction of microfluidics reduces the device size through LOC devices. Based on this principle, Benazzi et al. developed a high-speed microfluidic platform to measure fluorescence from single cells at three different wavelength ranges (using a luminescence spectrometer with a 532 and 633 nm laser), achieving cell discrimination at a flow rate of 3 cm·s^{-1} [114].

Some years later, Schaap et al. [115] developed an LOC that could distinguish phytoplankton species with a more straightforward setup. Only one focalised laser source and a single quadrant-cell photodetector were utilised (see Figure 9). A curved waveguide guided the laser light to the microchannel and the different cells produced distinctive wavelets dependent on the phytoplankton geometry and size. As mentioned in the previous section, the same year Hashemi et al. developed another autofluorescence-based lab on chip which consisted of an optofluidic cytometer [102,103]. As stated by the authors, the differences in fluorescence signals were used to reveal the different ratios of chlorophyll and phycobilins. In contrast, the differences in light scattering signals were used to assess the size and shape of phytoplankton cells. Even the smallest species, with a size of 1 µm could be identified (*Synechococcus* sp.). A few years later, Wang et al. [116] proposed a simple optofluidic device for fluorescence-based phytoplankton detection. It consisted of one sample channel between two sheath channels. Thanks to this configuration, the two branch channels' laminar flows forced the main channel's phytoplankton cells to line up into one line. A 488 nm laser diode was used as the excitation light to illuminate the sample cells. A photodiode was selected to measure the fluorescence of chlorophyll with an output voltage corresponding to the intensity of the fluorescence. It can be used to identify dead cells and living cells by calibrating the fluorescence intensity. The results confirmed that the developed system based on chlorophyll fluorescence could not only detect the living status of single phytoplankton cells but also can evaluate their viability quantitatively [116]. The same principle was subsequently used in combination with impedance measurements [117]. In recent work, the authors developed a ballast water rapid detection device based on the previous fluorescence microfluidic sensor [118]. The authors concluded that obtained results agree with the laboratory standard test measurements, with the advantage of on-site real-time ballast water detection. Another interesting study was presented in [119]. In that work, a label-free analysis and sorting of phytoplankton in microdroplets by chlorophyll fluorescence was presented.

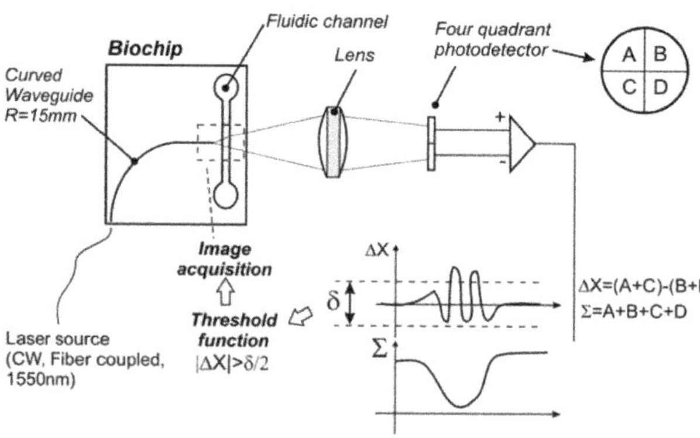

Figure 9. Schematic of biochip working principle: the biochip consists of a fluidic channel and a curved waveguide buried in the glass. A Gaussian beam emitted by a single-mode fibre is coupled into the biochip waveguide and diverges to illuminate a small length of the fluidic channel. Objects that pass through the fluidic channel momentarily distort the beam intensity profile. The light from the biochips is then refocused onto a four-quad detector to monitor small intensity changes. Reprinted with permission from [115]. Copyright 2011, Optica Publishing Group.

As can be observed in Figure 10, the electrical signal of the detector was used to trigger a deflecting system based on voltage. As the authors state, this technique can be applied as a screening tool for microalgal libraries. Moreover, as the method allows the measurement of intrinsic chlorophyll per cell and total chlorophyll per droplet, the cell number and biomass evolution over time can be measured.

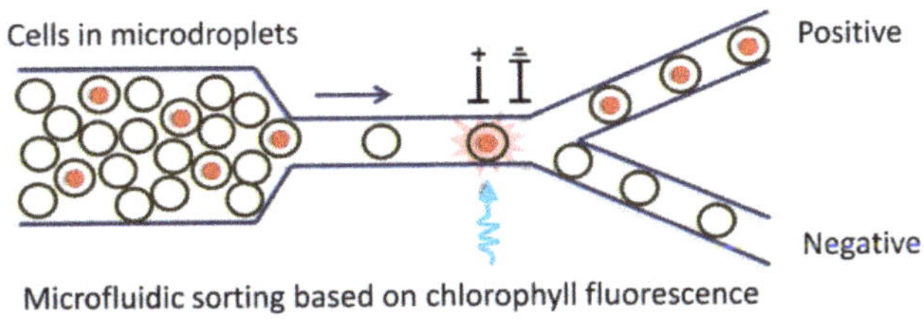

Figure 10. Schematic depiction of the fluorescence-based sorting method. Reprinted with permission from [119]. Copyright 2016, ACS.

4.3. Imaging Flow Cytometry Technologies

As mentioned before, a conventional cytometer generally relies on phytoplankton's light scattering or fluorescence properties, which cannot offer spatial size characterisation and distinguish through differences in surface morphologies. To overcome this challenge, imaging flow cytometry (IFC) was first proposed in 1979 [120] and further developed in the 1980s [121]. Thanks to microfluidics, image-based microfluidic flow cytometers have been recently proposed to create novel and high-efficiency platforms that combine the high-throughput nature of conventional flow cytometry techniques with the optical resolution of microscopy [15].

This technique combines speed and significant sample size capabilities of flow cytometry and zooming capabilities of microscopy (up to 60×), contributing significantly to the advancement of phytoplankton analysis [122]. Using this technique, the measurement of phytoplankton morphology, cellular processes, cell-to-cell interactions, population dynamics and ecology has been improved [123]. Still, there are some issues to be solved, for example, their low analytical throughput (typically between 2000 and 3000 cells/s at 20× magnification); more than one order of magnitude lower than non-imaging flow cytometers [124]. Moreover, the limited depth of field of the objective causes a limitation in the size of the channel and measured cells, requiring other methods such as acoustic focusing to improve it [125]. Finally, the bulky devices make it difficult to use them in situ. To solve the issues mentioned above, different solutions have been proposed [124]. Despite this, for phytoplankton there are additional issues such as the considerable amount of different species in the same sample. In [122,123], other applications of IFC technology for analysing microalgae cultures and phytoplankton are thoroughly reviewed (until 2017). Ref. [126] systematically reviews articles from 2017 to 2020 using the commercial device FlowCam for phytoplankton research. For this reason, in this review, we will focus only on novel proposals of the last five years.

Holographic and multispectral techniques have been proposed to implement different devices. In 2018, a holographic device capable of detecting phytoplankton flowing through a 0.8 mm thick microfluidic chip was proposed (Figure 11b) [127]. A deep convolutive network is used to reconstruct the acquired holograms automatically. This device allows the real-time imaging of highly dense samples. Specifically, 24 microalgae species were identified in flow-through water samples with a high flow rate of 100 mL/h. In addition, the concentration measurement of the potentially harmful microalgae Pseudo-nitzschia is also reported. In 2021, the same authors proposed new analysis methods to perform an automated and high-throughput phenotypic inspection of microalgae populations in the presence of pollutants within the water sample [128].

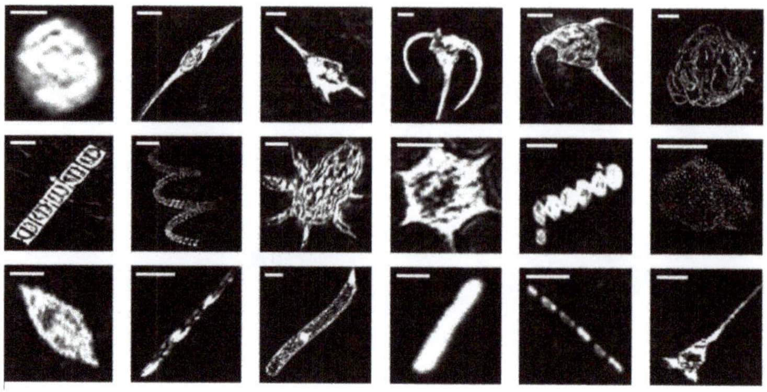

(a)

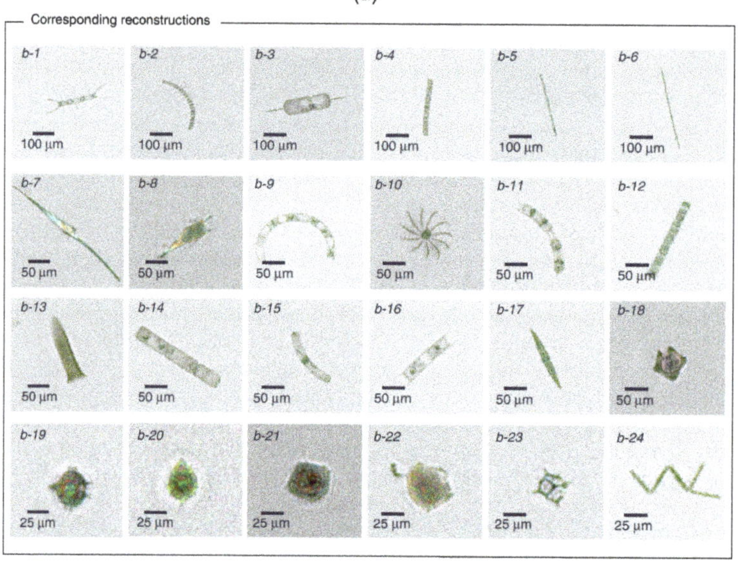

(b)

Figure 11. Examples of various ocean planktons. (**a**) Amplitude images reconstructed and detected from specific focal planes for each plankton class. From top left to lower right: *Alexandrium tamarense, Ceratium fusus, Ceratium lineatum, Ceratium longpipes, Ceratium* sp., *Chaetoceros socialis, Chaetoceros straight, Chaetoceros* sp., *Crustacean, Dictyocha speculum, Melosira octagona, Parvicorbicula socialis, Prorocentrum micans, Pseudo-nitchzia arctica, Rhizosolenia setigera, Rods, Skeletonema costatum, Tintinnid.* All images are segmented to 128 × 128 pixels and scale bars represent 50 µm. (**b**) Planktons detected at the Los Angeles coastline, represented by their phase contrast reconstructions following phase recovery. The organisms were identified as (**b1**) Chaetoceros lorenzianus, (**b2**) Chaetoceros debilis, (**b3**) Ditylum brightwellii, (**b4**) Lauderia, (**b5**) Leptocylindrus, (**b6**) Pseudo-nitzschia, (**b7**) Ceratium fusus, (**b8**) Ceratium furca, (**b9**) Eucampia cornuta, (**b10**) Bacteriastrum, (**b11**) Hemiaulus, (**b12**) Skeletonema, (**b13**) Ciliate, (**b14**) Cerataulina, (**b15**) Guinardia striata, (**b16**) Lithodesmium, (**b17**) Pleurosigma, (**b18**) Protoperidinium claudicans, (**b19**) Protoperidinium steinii, (**b20**) Prorocentrum micans, (**b21**) Lingulodinium polyedra, (**b22**) Dinophysis, (**b23**) Dictyocha fibula (silica skeleton) and (**b24**) Thalassionema. Reprinted from (a) [129] Copyright 2021 (b) [127], Copyright 2018, Springer Nature.

The same year, using holography and deep learning, the capability to produce and reconstruct sharp images of important plankton groups from both culture and environmental samples was demonstrated [129] (Figure 11a). On the topic of imaging flow cytometry combined with deep learning, the following recent papers demonstrated the efficiency of those techniques for identification and classification of protozoa [130,131]. They have the potential of being extended to the identification of classification of phytoplankton.

One problem of previous systems is that current optical imaging technologies still lack the practical speed and sensitivity for measuring thousands to millions of cells down to single-cell precision [132]. A solution to this problem is optofluidic time-stretch microscopy, which can perform high-throughput imaging flow cytometry up to 100,000 cells per second. Since it was first demonstrated in 2009 [133], this technique has been demonstrated in on-chip microfluidics for several applications. In the case of phytoplankton, several works have been reported since 2016 [134–140]. This kind of system is composed of several key components, i.e., a broadband pulsed laser (fs-laser), a temporal disperser (usually a long disperser fibre optic), two spatial dispersers (diffraction gratings), two objective lenses, a microfluidic device, a single-pixel photodetector, an oscilloscope and a digital signal processor. For this reason, LOC systems can be limited by the size of some of the components mentioned above. Readers can check ref. [141] for more information and detailed instructions to fabricate an optofluidic time-stretch microscopy and measure cells in microfluidic channels.

4.4. Electrochemical Impedance Spectroscopy

Although the measurement of dielectric properties of phytoplankton does not fall within the scope of this review, we have briefly included it because recently proposed devices employ multiparametric sensing, combining dielectric and optical properties. One of the most used techniques to characterise the electric properties of electrochemical systems is electrochemical impedance spectroscopy (EIS). It can characterise the dynamics of bound or mobile charges in the bulk or interfacial regions of any solid or liquid material (ionic, semiconductor and dielectrics). The main parameter EIS gives impedance, a concept proposed by Oliver Heaviside in 1880. Impedance is the complex ratio of the voltage to the current for an alternating current, $Z = V(\omega)/I(\omega)$, translating the concept of resistance to AC systems (possessing both magnitude and phase). In recent years, this technique has grown tremendously and is now widely employed in various scientific fields. The medical field is one of the most demanding sectors for this technique: clinical scales to measure corporal parameters, the efficacy of medicinal products [142], cancer detection [143], etc. Another important sector is material engineering: the study of new materials, batteries, metals (corrosion process) [144], etc. In addition, there is a growing interest in using this technique in bioengineering [145]. This technique gives useful information about tissue or cells. The operation principle of single-cell impedance spectroscopy for high-speed analysis has been reported since the beginning of the century [146–148]. Some comprehensive reviews can be found in [149,150]. In the simplest case, phytoplankton behaves like a spherical shell, describing each cell as a perfect sphere with a conductive outer shell and membrane and a resistive interior (see Figure 12) [114]. For other types of phytoplankton, this model has to be modified, e.g., phytoplankton without cell walls or with other biological configurations.

As the frequency response of phytoplankton is complex, the EIS technique is appropriate for characterising their impedance. For example, in [114], different phytoplanktonic species' impedances were measured (*Isochrysis galbana*, *Synechococcus* sp. and *Rhodosorus m.*). The authors conclude that low-frequency signals could be used to measure the size of the particles, but the system is not sufficiently sensitive to detect the smaller cells (<2 µm). In more recent work, phytoplankton bioimpedance was performed using interdigitated electrodes and an impedance analyser [151]. The results showed no significant difference in the extracted cytoplasm conductivity, whereas the specific membrane capacitance (membrane capacitance per unit area) between Chlamydomonas and Selenastrum cells differed significantly.

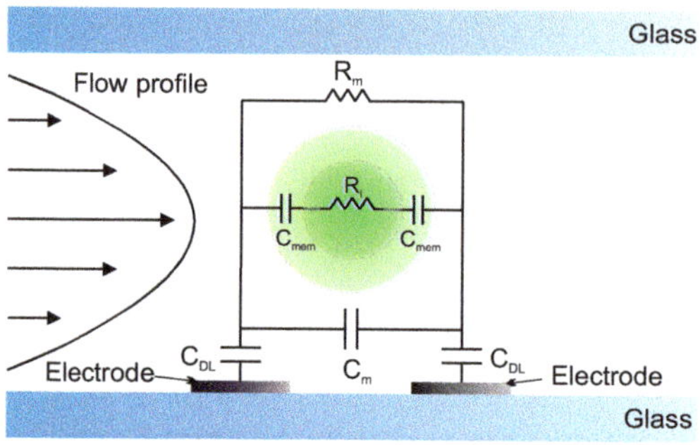

Figure 12. Equivalent electrical circuit of a phytoplankton cell in a microfluidic channel. R_m and C_m are the resistance and capacitance of the medium, respectively, C_{mem} is the capacitance of the cell membrane, R_i is the resistance of the cell cytoplasm and C_{DL} the capacitance of the electrical double layer (DL). The values of the individual electrical components are determined by the dielectric properties of the suspending medium, the geometry of the chip and the dielectric properties of an individual cell [114].

Another technique that simplifies the system (an impedance analyser is unnecessary) is Resistive Pulse Sensing, which is based on the Coulter principle. Particles in low concentrations suspended in the medium can be counted by passing them through a microfluidic channel. A particle passing through the channel displaces a volume of medium equivalent to the submerged volume of the particle from the detection zone. This causes a short-term change in impedance across the aperture. This change can be measured as a voltage or current pulse. The height of the pulse is proportional to the volume of the detected particle. Assuming a constant particle density, the pulse height is also proportional to the particle's mass. For example, this technique was demonstrated by Song et al. [152,153] in an analysis of ships' ballast water, detecting and counting phytoplankton cells and measuring their sizes. In a subsequent work [117], the combination with chlorophyll autofluorescence intensity detection allows for measuring cell viability, excluding interference of other particles and dead cells and producing more accurate results. The system was integrated into an underwater device with advantages such as automation, portability, low cost and easy operation. After that, the system continued being tested to comply with the Standard D-2 performance [13].

One drawback of this kind of system is the lack of close contact between cells and electrodes when passing through the microchannel. This issue could lead to current leakage where electric signals circumvent the cells under measurement by travelling through solutions surrounding the cells [150]. Another issue could be the position dependence of the measurements and the overlapping of two or more cells in the channel. One solution could be the use of narrower channels and sorting methods. In this regard, sorting plankton in microfluidic devices is an effervescent research field where numerous methods have been developed and are continuously improved through the years and applications [23]. The topic is thoroughly reviewed in the latter reference.

5. Discussion

The identification and measurement of specific characteristics of phytoplankton are essential for controlling pollution of the marine environment and aquaculture and the shellfish industry. Although manual sample collection and microscopy-based identification is the traditional method to measure phytoplankton, some flow cytometry techniques have improved the throughput. However, the expensive and bulky instruments make in situ measurement difficult. Precise control, manipulation and measurement are possible through LOC systems incorporating microlfuidic channels and basic optical components. Several techniques are used in this kind of device, along with their advantages and drawbacks. Moreover, each can detect different parameters of the phytoplankton, so their combination is used in several proposed devices (see Table 1).

Table 1. Comparative table of different microfluidic technologies for phytoplankton research.

Modality	Source	Parameters Measured	Main Challenges and Limitations
Scattering	Visible light.	Cell size and shape.	Low intensity in comparison to fluorescence.
Raman scattering	>700 nm (avoid fluorescence).	Composition, nutrient contents.	Low intensity and long interrogation times.
Fluorescence	450 nm (clorophyl abs.), 550 nm (phycoerythrin abs.).	Concentrations of chlorophyll and other pigments, living status.	Highly variable due to their physiological conditions.
Imaging	Visible light.	Intracellular contents, size and surface roughness.	Limited throughput.
Time-stretch	Broadband fs-laser.	Same as imaging.	Complex/bulky setup.
Impedance	AC Voltage (1Hz-1MHz).	Cytoplasm conductivity and membrane capacitance.	Some species are difficult to model.
RPS	DC Voltage.	Size.	Lack of close contact and possible overlapping of cells.

For example, scattering-based technologies can measure the size and surface roughness of the cell; they are used in remote measurements but also in microfluidic channels. The forward and side light scattering is collected from narrow angles and light diffracted around the cell, respectively. These measurements are demonstrated to give information about cell size and shape. Despite this, the low intensity of the signal with respect to fluorescence can be a drawback. For this reason, the measurement of these properties in LOC devices is not so usual and when they are used, they are in combination with other techniques. On the other hand, microfluidic technology combined with Raman spectroscopy precisely identifies the phytoplankton composition and nutrient contents. However, the low intensity of the Raman signal in comparison with normal scattering and fluorescence (only 1 in 10^6–10^8 photons undergoes Raman scattering) leads to errors during data analysis, limiting the measure to species with low fluorescence. This effect can be reduced by using different excitation wavelengths and data processing. Another problem is that some pigments have an extensive range of bands while other biomolecule signals, such as fats and proteins, can be masked. Another consideration is that PDMS is not recommended to fabricate the microfluidic channels as this material can also generate Raman scattering. Finally, the Raman spectra's complexity limits the application of chemometric methods. As mentioned before, recent works have proposed some solutions using longer wavelengths (NIR) and Fourier spectroscopy with multivariate data analysis. To increase the throughput, some solutions have been the isolation or immobilisation of single cells and droplet microfluidic devices (RADS).

In the case of fluorescence, phytoplankton have their unique fluorescence spectra due to the different pigment ratios. For this reason, it can be combined with other techniques to classify species. Fluorescence spectroscopy can detect small concentrations of chlorophyll and other pigments, but it is not very specific. The main problem is that phytoplankton cells' fluorescence is highly variable due to their physiological conditions. Phytoplankton fluorescence depends on several parameters, e.g., the taxonomic position, the pigment content and ratio, the nutrient conditions, the growth stage, photoadaptation and physiological state. Thus, deriving the biomass of spectral groups using the spectral fluorescence of multicomponent natural samples is not trivial. In the case of microfluidics, the size reduction increases the throughput; some works have demonstrated the detection of the living status of single phytoplankton cells and their viability quantitatively. In addition, due to the sort interrogation times, it can be used in sorting devices.

By leveraging algorithms such as deep learning and compression detection, flow cytometry based on microfluidic imaging could perform the powerful function of automatically identifying and counting phytoplankton. Particularly enabled by new multifunctional imaging techniques, the microfluidic flow cytometer can offer information regarding cell morphology, intracellular lipid and pigment. Compared to microfluidic devices based on fluorescence detection and Raman spectroscopy, the imaging-based flow cytometer offers flexibility and feasibility to perform machine automation without human intervention. On the other hand, the automatic identification and classification of the phytoplankton require a database for machine training. Therefore the demand for sample preparatory images based on high experience must be implemented before on-site use. Another issue is the limitation of processed cells per second due to the limited capture speed of current CCD and CMOS systems (limited throughput). A solution to this problem can be optofluidic time-stretch microscopy, which can perform high-throughput imaging flow cytometry up to 100,000 cells per second. However, the complexity of the setup (fs-laser, dispersers, etc.) makes it difficult to make portable devices.

Finally, impedance measurements can be an excellent complementary option, as they can be combined perfectly with previous techniques. The measurements can give information about cytoplasm conductivity and membrane capacitance, producing a characteristic signature for each species. The main problem is that phytoplankton without cell walls or with other biologic configurations are challenging to model. Another inconvenience is the requirement of impedance analysers that can be bulky for portable systems; despite this, custom circuits could be made to solve this issue. On the other hand, to simplify the setup, RPS can provide information about the size with a simple setup. The main drawback of this kind of system is the lack of close contact between cells and electrodes when passing through the microchannel and the possible overlapping of two or more cells in the channel. One solution could be the use of narrower channels and sorting methods.

6. Conclusions

We have presented an overview of the status and challenges of the most relevant microfluidic systems for phytoplankton research. There is a clear need for such innovative microscale technology as several applications would benefit significantly. Lab on chip systems with microfluidic channels possess many advantages, such as fast measurements, high sensitivity, multifunctionality and possible portability. Despite this, these systems are mainly in the research stage, with few examples of commercial products. One reason could be the different modalities that can be used and their difficulty of integration. Each of these modalities offer unique capabilities but also pose some limitations. For this reason, sorting methods, multiparametric systems, multifunctional imaging and deep learning can bring solutions to actual issues. Nevertheless, some emerging proposals are opening new avenues towards enhanced phytoplankton measurement and the next decade seems promising.

Author Contributions: J.F.A. Conceptualisation, Investigation, Methodology, Writing—Original draft, Writing—Review and editing, P.R.-V. Conceptualisation, Investigation, Methodology, Writing—Original draft, Writing—Review and editing, M.G.F.-M. Conceptualisation, Investigation, Methodology, Writing—Original draft, Writing—Review and editing, J.M.L.-H. Writing—Original draft, Writing—Review and editing, Project administration, L.R.-C. Investigation, Methodology, Writing—Original draft, Writing—Review and editing, Project administration, A.C.-G. Writing—Original draft, Writing—Review and editing, Project administration. All authors have read and agreed to the published version of the manuscript.

Funding: This work was supported by Ministerio de Ciencia e Innovación and Agencia Estatal de Investigación (PID2019-107270RB-C21/AIE/10.13039/501100011033. J.F.A. received funding from Ministerio de Ciencia, Innovación y Universidades of Spain under Juan de la Cierva Incorporación grant.

Data Availability Statement: Data underlying the results presented in this paper are not publicly available at this time but may be obtained from the authors upon reasonable request.

Conflicts of Interest: The authors declare no conflict of interest.

References

1. Field, C.B.; Behrenfeld, M.J.; Randerson, J.T.; Falkowski, P. Primary Production of the Biosphere: Integrating Terrestrial and Oceanic Components. *Science* **1998**, *281*, 237–240. [CrossRef] [PubMed]
2. Sabine, C.L.; Feely, R.A.; Gruber, N.; Key, R.M.; Lee, K.; Bullister, J.L.; Wanninkhof, R.; Wong, C.S.; Wallace, D.W.R.; Tilbrook, B.; et al. The Oceanic Sink for Anthropogenic CO_2. *Science* **2004**, *305*, 367–371. [CrossRef] [PubMed]
3. Jeffrey, S.; Wright, S.W.; Zapata, M. Microalgal classes and their signature pigments. In *Phytoplankton Pigments: Characterization, Chemotaxonomy and Applications in Oceanography*; Roy, S., Llewellyn, C.A., Egeland, E.S., Johnsen, G., Eds.; Cambridge Environmental Chemistry Series; Cambridge University Press: London, UK, 2011; pp. 3–77. [CrossRef]
4. Pérez-Lloréns, J.L. Microalgae: From staple foodstuff to avant-garde cuisine. *Int. J. Gastron. Food Sci.* **2020**, *21*, 100221. [CrossRef]
5. Tweddle, J.F.; Gubbins, M.; Scott, B.E. Should phytoplankton be a key consideration for marine management? *Mar. Policy* **2018**, *97*, 1–9. [CrossRef]
6. Nehring, S. Establishment of thermophilic phytoplankton species in the North Sea: Biological indicators of climatic changes? *ICES J. Mar. Sci.* **1998**, *55*, 818–823. [CrossRef]
7. Thyssen, M.; Mathieu, D.; Garcia, N.; Denis, M. Short-term variation of phytoplankton assemblages in Mediterranean coastal waters recorded with an automated submerged flow cytometer. *J. Plankton Res.* **2008**, *30*, 1027–1040. [CrossRef]
8. Harvell, C.D.; Kim, K.; Burkholder, J.M.; Colwell, R.R.; Epstein, P.R.; Grimes, D.J.; Hofmann, E.E.; Lipp, E.K.; Osterhaus, A.D.M.E.; Overstreet, R.M.; et al. Emerging Marine Diseases–Climate Links and Anthropogenic Factors. *Science* **1999**, *285*, 1505–1510. [CrossRef]
9. Knap, A.; Dewailly, E.; Furgal, C.; Galvin, J.; Baden, D.; Bowen, R.E.; Depledge, M.; Duguay, L.; Fleming, L.E.; Ford, T.; et al. Indicators of ocean health and human health: Developing a research and monitoring framework. *Environ. Health Perspect.* **2002**, *110*, 839–845. [CrossRef]
10. Landsberg, J.H. The Effects of Harmful Algal Blooms on Aquatic Organisms. *Rev. Fish. Sci.* **2002**, *10*, 113–390. [CrossRef]
11. Ostenfeld, C. On the immigration of Biddulphia sinensis Grev. and its occurrence in the North Sea during 1903–1907. *Medd. Komm. Havunders. Ser. Plankton* **1908**, *1*, 1–44.
12. Gollasch, S.; David, M.; Voigt, M.; Dragsund, E.; Hewitt, C.; Fukuyo, Y. Critical review of the IMO international convention on the management of ships' ballast water and sediments. *Harmful Algae* **2007**, *6*, 585–600. [CrossRef]
13. Maw, M.M.; Pan, X.; Peng, Z.; Wang, Y.; Zhao, L.; Dai, B.; Wang, J. A Changeable Lab-on-a-Chip Detector for Marine Nonindigenous Microorganisms in Ship's Ballast Water. *Micromachines* **2018**, *9*, 20. [CrossRef] [PubMed]
14. Sosik, H.M.; Olson, R.J.; Armbrust, E.V., Flow Cytometry in Phytoplankton Research. In *Chlorophyll a Fluorescence in Aquatic Sciences: Methods and Applications*; Suggett, D.J., Prášil, O., Borowitzka, M.A., Eds.; Springer: Dordrecht, The Netherlands, 2010; pp. 171–185. [CrossRef]
15. Stavrakis, S.; Holzner, G.; Choo, J.; deMello, A. High-throughput microfluidic imaging flow cytometry. *Curr. Opin. Biotechnol.* **2019**, *55*, 36–43. [CrossRef] [PubMed]
16. Liu, P.; Chin, L.; Ser, W.; Ayi, T.; Yap, P.; Bourouina, T.; Leprince-Wang, Y. An optofluidic imaging system to measure the biophysical signature of single waterborne bacteria. *Lab Chip* **2014**, *14*, 4237–4243. [CrossRef] [PubMed]
17. Elsayed, A.A.; Erfan, M.; Sabry, Y.M.; Dris, R.; Gaspéri, J.; Barbier, J.S.; Marty, F.; Bouanis, F.; Luo, S.; Nguyen, B.T.T.; et al. A microfluidic chip enables fast analysis of water microplastics by optical spectroscopy. *Sci. Rep.* **2021**, *11*, 1–11. [CrossRef]
18. Wu, J.; Gu, M. Microfluidic sensing: State of the art fabrication and detection techniques. *J. Biomed. Opt.* **2011**, *16*, 080901. [CrossRef]
19. Demirci, U.; Khademhosseini, A.; Langer, R.; Blander, J. *Microfluidic Technologies for Human Health*; World Scientific: Singapore, 2013. [CrossRef]
20. He, F.; Liao, Y.; Lin, J.; Song, J.; Qiao, L.; Cheng, Y.; Sugioka, K. Femtosecond Laser Fabrication of Monolithically Integrated Microfluidic Sensors in Glass. *Sensors* **2014**, *14*, 19402–19440. [CrossRef]
21. Yew, M.; Ren, Y.; Koh, K.S.; Sun, C.; Snape, C. A Review of State-of-the-Art Microfluidic Technologies for Environmental Applications: Detection and Remediation. *Glob. Chall.* **2019**, *3*, 1800060. [CrossRef]
22. Kim, H.S.; Devarenne, T.P.; Han, A. Microfluidic systems for microalgal biotechnology: A review. *Algal Res.* **2018**, *30*, 149–161. [CrossRef]
23. Girault, M.; Beneyton, T.; del Amo, Y.; Baret, J.C. Microfluidic technology for plankton research. *Curr. Opin. Biotechnol.* **2019**, *55*, 134–150. [CrossRef]
24. Zheng, X.; Duan, X.; Tu, X.; Jiang, S.; Song, C. The Fusion of Microfluidics and Optics for On-Chip Detection and Characterization of Microalgae. *Micromachines* **2021**, *12*, 1137. [CrossRef] [PubMed]
25. Jeffrey, S.W.; Vesk, M. Introduction to marine phytoplankton and their pigment signatures. *Phytoplankton Pigment. Oceanogr.* 1997, 37–84.
26. Kirk, J.T.O. *Light and Photosynthesis in Aquatic Ecosystems*, 2nd ed.; Cambridge University Press: London, UK, 1994. [CrossRef]
27. Bricaud, A.; Claustre, H.; Ras, J.; Oubelkheir, K. Natural variability of phytoplanktonic absorption in oceanic waters: Influence of the size structure of algal populations. *J. Geophys. Res. Ocean.* **2004**, *109*. [CrossRef]
28. Kirkpatrick, G.J.; Millie, D.F.; Moline, M.A.; Schofield, O. Optical discrimination of a phytoplankton species in natural mixed populations. *Limnol. Oceanogr.* **2000**, *45*, 467–471. [CrossRef]

29. Subramaniam, A.; Brown, C.W.; Hood, R.R.; Carpenter, E.J.; Capone, D.G. Detecting Trichodesmium blooms in SeaWiFS imagery. *Deep. Sea Res. Part Ii: Top. Stud. Oceanogr.* **2001**, *49*, 107–121. [CrossRef]
30. Tomlinson, M.; Wynne, T.; Stumpf, R. An evaluation of remote sensing techniques for enhanced detection of the toxic dinoflagellate, Karenia brevis. *Remote Sens. Environ.* **2009**, *113*, 598–609. [CrossRef]
31. Morel, A.; Bricaud, A. Theoretical results concerning light absorption in a discrete medium, and application to specific absorption of phytoplankton. *Deep. Sea Res. Part A Oceanogr. Res. Pap.* **1981**, *28*, 1375–1393. [CrossRef]
32. Goericke, R.; Repeta, D. Chlorophyll-A and chlorophyll-B and divinyl chlorophyll-A and chlorophyll-B in the open subtropical North Atlantic Ocean. *Mar. Ecol. Prog. Ser.* **1993**, *101*, 307–313. [CrossRef]
33. Bidigare, R.R.; Ondrusek, M.E.; Morrow, J.H.; Kiefer, D.A. *In-Vivo Absorption Properties of Algal Pigments*; Ocean Optics, X., Spinrad, R.W., Eds.; International Society for Optics and Photonics, SPIE: Bellingham, WA, USA, 1990; Volume 1302, pp. 290–302. [CrossRef]
34. Lain, L.R.; Bernard, S. The Fundamental Contribution of Phytoplankton Spectral Scattering to Ocean Colour: Implications for Satellite Detection of Phytoplankton Community Structure. *Appl. Sci.* **2018**, *8*, 2681. [CrossRef]
35. Stramska, M.; Stramski, D.; Mitchell, B.G.; Mobley, C.D. Estimation of the absorption and backscattering coefficients from in-water radiometric measurements. *Limnol. Oceanogr.* **2000**, *45*, 628–641. [CrossRef]
36. Aas, E. Refractive index of phytoplankton derived from its metabolite composition. *J. Plankton Res.* **1996**, *18*, 2223–2249. [CrossRef]
37. Balch, W.M.; Kilpatrick, K.A.; Trees, C.C. The 1991 coccolithophore bloom in the central North Atlantic. 1. Optical properties and factors affecting their distribution. *Limnol. Oceanogr.* **1996**, *41*, 1669–1683. [CrossRef]
38. Wyatt, P.J.; Jackson, C. Discrimination of phytoplankton via light-scattering properties. *Limnol. Oceanogr.* **1989**, *34*, 96–112. [CrossRef]
39. Sullivan, J.M.; Twardowski, M.S. Angular shape of the oceanic particulate volume scattering function in the backward direction. *Appl. Opt.* **2009**, *48*, 6811–6819. [CrossRef]
40. Measurement and modeling of volume scattering functions for phytoplankton from Norwegian coastal waters. *J. Mar. Res.* **2017**, *75*, 579–603. [CrossRef]
41. Volten, H.; de Haan, J.F.; Hovenier, J.W.; Schreurs, R.; Vassen, W.; Dekker, A.G.; Hoogenboom, H.J.; Charlton, F.; Wouts, R. Laboratory measurements of angular distributions of light scattered by phytoplankton and silt. *Limnol. Oceanogr.* **1998**, *43*, 1180–1197. [CrossRef]
42. Lotsberg, J.; Marken, E.; Stamnes, J.; Erga, S.; Aursland, K.; Olseng, C. Laboratory measurements of light scattering from marine particles. *Limnol. Oceanogr. Methods* **2007**, *5*, 34–40. [CrossRef]
43. Zugger, M.E.; Messmer, A.; Kane, T.J.; Prentice, J.; Concannon, B.; Laux, A.; Mullen, L. Optical scattering properties of phytoplankton: Measurements and comparison of various species at scattering angles between 1° and 170°. *Limnol. Oceanogr.* **2008**, *53*, 381–386. [CrossRef]
44. Moore, C.; Barnard, A.; Fietzek, P.; Lewis, M.R.; Sosik, H.M.; White, S.; Zielinski, O. Optical tools for ocean monitoring and research. *Ocean. Sci.* **2009**, *5*, 661–684. [CrossRef]
45. Heraud, P.; Wood, B.R.; Beardall, J.; McNaughton, D. Probing the Influence of the Environment on Microalgae Using Infrared and Raman Spectroscopy. In *New Approaches in Biomedical Spectroscopy*; ACS Publications: Washington, DC, USA, 2007; Chapter 7; pp. 85–106. [CrossRef]
46. Andreasen, M.; Lundgreen, K.; Holbech, H.; Hedegaard, M.A. Raman spectroscopy as a tool for viability assessment of planktonic organisms in UV treated ballast water. *Vib. Spectrosc.* **2020**, *110*, 103142. [CrossRef]
47. He, S.; Xie, W.; Zhang, P.; Fang, S.; Li, Z.; Tang, P.; Gao, X.; Guo, J.; Tlili, C.; Wang, D. Preliminary identification of unicellular algal genus by using combined confocal resonance Raman spectroscopy with PCA and DPLS analysis. *Spectrochim. Acta Part Mol. Biomol. Spectrosc.* **2018**, *190*, 417–422. [CrossRef] [PubMed]
48. Barletta, R.E.; Krause, J.W.; Goodie, T.; El Sabae, H. The direct measurement of intracellular pigments in phytoplankton using resonance Raman spectroscopy. *Mar. Chem.* **2015**, *176*, 164–173. [CrossRef]
49. Hoskins, L.C.; Alexander, V. Determination of carotenoid concentrations in marine phytoplankton by resonance Raman spectrometry. *Anal. Chem.* **1977**, *49*, 695–697. [CrossRef] [PubMed]
50. Novikova, N.I.; Matthews, H.; Williams, I.; Sewell, M.A.; Nieuwoudt, M.K.; Simpson, M.C.; Broderick, N.G.R. Detecting Phytoplankton Cell Viability Using NIR Raman Spectroscopy and PCA. *ACS Omega* **2022**, *7*, 5962–5971. [CrossRef] [PubMed]
51. Ahmmed, F.; Fraser-Miller, S.J.; Garagoda Arachchige, P.S.; Schallenberg, M.; Novis, P.; Gordon, K.C. Lake snow caused by the invasive diatom Lindavia intermedia can be discriminated from different sites and from other algae using vibrational spectroscopy. *J. Raman Spectrosc.* **2021**, *52*, 2597–2608. [CrossRef]
52. Lorenzen, C.J. A method for the continuous measurement of in vivo chlorophyll concentration. *Deep. Sea Res. Oceanogr. Abstr.* **1966**, *13*, 223–227. [CrossRef]
53. Nair, A.; Sathyendranath, S.; Platt, T.; Morales, J.; Stuart, V.; Forget, M.H.; Devred, E.; Bouman, H. Remote sensing of phytoplankton functional types. *Remote Sens. Environ.* **2008**, *112*, 3366–3375. [CrossRef]
54. Prézelin, B.B.; Alberte, R.S. Photosynthetic characteristics and organization of chlorophyll in marine dinoflagellates. *Proc. Natl. Acad. Sci. USA* **1978**, *75*, 1801–1804. [CrossRef]

55. Falkowski, P.G.; Owens, T.G. Light—Shade Adaptation 1: Two strategies in marine phytoplankton. *Plant Physiol.* **1980**, *66*, 592–595. [CrossRef]
56. Babin, M.; Roesler, C.S.; Cullen, J.J. *Real-Time Coastal Observing Systems for Marine Ecosystem Dynamics and Harmful Algal Blooms: Theory, Instrumentation and Modelling*; UNESCO: Paris, France, 2008. [CrossRef]
57. Seppälä, J.V. *Spectral Absorption and Fluorescence Characteristics of the Baltic Sea Phytoplankton*; International Council for the Exploration of the Sea (ICES), København, Denmark, 2003.
58. Falkowski, P.; Kiefer, D.A. Chlorophyll a fluorescence in phytoplankton: Relationship to photosynthesis and biomass*. *J. Plankton Res.* **1985**, *7*, 715–731. [CrossRef]
59. Syama, S.; Mohanan, P. Microfluidic based human-on-a-chip: A revolutionary technology in scientific research. *Trends Food Sci. Technol.* **2021**, *110*, 711–728. [CrossRef]
60. Sun, J.; Warden, A.R.; Ding, X. Recent advances in microfluidics for drug screening. *Biomicrofluidics* **2019**, *13*, 061503. [CrossRef] [PubMed]
61. Dai, J.; Hamon, M.; Jambovane, S. Microfluidics for Antibiotic Susceptibility and Toxicity Testing. *Bioengineering* **2016**, *3*. [CrossRef] [PubMed]
62. van Noort, D.; Ong, S.M.; Zhang, C.; Zhang, S.; Arooz, T.; Yu, H. Stem cells in microfluidics. *Biotechnol. Prog.* **2009**, *25*, 52–60. [CrossRef]
63. Chung, C.J.; Xu, X.; Pan, Z.; Mokhtari-Koushyar, F.; Wang, R.; Yan, H.; Subbaraman, H.; Chen, R.T. Silicon-Based Hybrid Integrated Photonic Chip for K $_u$ band Electromagnetic Wave Sensing. *J. Light. Technol.* **2018**, *36*, 1568–1575. [CrossRef]
64. Liu, Q.; Shin, Y.; Kee, J.S.; Kim, K.W.; Mohamed Rafei, S.R.; Perera, A.P.; Tu, X.; Lo, G.Q.; Ricci, E.; Colombel, M.; et al. Mach–Zehnder interferometer (MZI) point-of-care system for rapid multiplexed detection of microRNAs in human urine specimens. *Biosens. Bioelectron.* **2015**, *71*, 365–372. [CrossRef]
65. Psaltis, D.; Quake, S.R.; Yang, C. Developing optofluidic technology through the fusion of microfluidics and optics. *Nature* **2006**, *442*, 381–386. [CrossRef]
66. Monat, C.; Domachuk, P.; Eggleton, B. Integrated optofluidics: A new river of light. *Nat. Photonics* **2007**, *1*, 106–114. [CrossRef]
67. Cheng, Y.; Sugioka, K.; Midorikawa, K. Microfluidic laser embedded in glass by three-dimensional femtosecond laser microprocessing. *Opt. Lett.* **2004**, *29*, 2007–2009. [CrossRef]
68. Sun, H.; He, F.; Zhou, Z.; Cheng, Y.; Xu, Z.; Sugioka, K.; Midorikawa, K. Fabrication of microfluidic optical waveguides on glass chips with femtosecond laser pulses. *Opt. Lett.* **2007**, *32*, 1536–1538. [CrossRef]
69. Osellame, R.; Hoekstra, H.; Cerullo, G.; Pollnau, M. Femtosecond laser microstructuring: An enabling tool for optofluidic lab-on-chips. *Laser Photonics Rev.* **2011**, *5*, 442–463. [CrossRef]
70. Sugioka, K.; Cheng, Y. Femtosecond laser processing for optofluidic fabrication. *Lab Chip* **2012**, *12*, 3576–3589. [CrossRef] [PubMed]
71. Vazquez, R.M.; Osellame, R.; Nolli, D.; Dongre, C.; van den Vlekkert, H.; Ramponi, R.; Pollnau, M.; Cerullo, G. Integration of femtosecond laser written optical waveguides in a lab-on-chip. *Lab Chip* **2009**, *9*, 91–96. [CrossRef] [PubMed]
72. Schmidt, H.; Hawkins, A.R. The photonic integration of non-solid media using optofluidics. *Nat. Photonics* **2011**, *5*, 598–604. [CrossRef]
73. Schaap, A.; Dumon, J.; den Toonder, J. Sorting algal cells by morphology in spiral microchannels using inertial microfluidics. *Microfluid. Nanofluidics* **2016**, *20*, 1–11. [CrossRef]
74. Chocarro-Ruiz, B.; Fernández-Gavela, A.; Herranz, S.; Lechuga, L.M. Nanophotonic label-free biosensors for environmental monitoring. *Curr. Opin. Biotechnol.* **2017**, *45*, 175–183. [CrossRef]
75. Zhang, D.; Men, L.; Chen, Q. Microfabrication and Applications of Opto-Microfluidic Sensors. *Sensors* **2011**, *11*, 5360–5382. [CrossRef]
76. Righini, G.C.; Chiappini, A. Glass optical waveguides: A review of fabrication techniques. *Opt. Eng.* **2014**, *53*, 1–15. [CrossRef]
77. McDonald, J.C.; Duffy, D.C.; Anderson, J.R.; Chiu, D.T.; Wu, H.; Schueller, O.J.A.; Whitesides, G.M. Fabrication of microfluidic systems in poly(dimethylsiloxane). *Electrophoresis* **2000**, *21*, 27–40. [CrossRef]
78. Gattass, R.R.; Mazur, E. Femtosecond laser micromachining in transparent materials. *Nat. Photonics* **2008**, *2*, 219–225. [CrossRef]
79. Sugioka, K.; Cheng, Y. Ultrafast lasers—Reliable tools for advanced materials processing. *Light. Sci. Appl.* **2014**, *3*, 149–149. [CrossRef]
80. He, F.; Cheng, Y.; Xu, Z.; Liao, Y.; Xu, J.; Sun, H.; Wang, C.; Zhou, Z.; Sugioka, K.; Midorikawa, K.; et al. Direct fabrication of homogeneous microfluidic channels embedded in fused silica using a femtosecond laser. *Opt. Lett.* **2010**, *35*, 282–284. [CrossRef] [PubMed]
81. He, F.; Cheng, Y.; Qiao, L.; Wang, C.; Xu, Z.; Sugioka, K.; Midorikawa, K.; Wu, J. Two-photon fluorescence excitation with a microlens fabricated on the fused silica chip by femtosecond laser micromachining. *Appl. Phys. Lett.* **2010**, *96*, 041108. [CrossRef]
82. He, F.; Lin, J.; Cheng, Y. Fabrication of hollow optical waveguides in fused silica by three-dimensional femtosecond laser micromachining. *Appl. Phys. B* **2011**, *105*, 379–384. [CrossRef]
83. Marcinkevičius, A.; Juodkazis, S.; Watanabe, M.; Miwa, M.; Matsuo, S.; Misawa, H.; Nishii, J. Femtosecond laser-assisted three-dimensional microfabrication in silica. *Opt. Lett.* **2001**, *26*, 277–279. [CrossRef]
84. Lin, J.; Yu, S.; Ma, Y.; Fang, W.; He, F.; Qiao, L.; Tong, L.; Cheng, Y.; Xu, Z. On-chip three-dimensional high-Q microcavities fabricated by femtosecond laser direct writing. *Opt. Express* **2012**, *20*, 10212–10217. [CrossRef]

85. Ponader, C.W.; Schroeder, J.F.; Streltsov, A.M. Origin of the refractive-index increase in laser-written waveguides in glasses. *J. Appl. Phys.* **2008**, *103*, 063516. [CrossRef]
86. Hnatovsky, C.; Taylor, R.S.; Simova, E.; Bhardwaj, V.R.; Rayner, D.M.; Corkum, P.B. Polarization-selective etching in femtosecond laser-assisted microfluidic channel fabrication in fused silica. *Opt. Lett.* **2005**, *30*, 1867–1869. [CrossRef]
87. Li, Y.; Itoh, K.; Watanabe, W.; Yamada, K.; Kuroda, D.; Nishii, J.; Jiang, Y. Three-dimensional hole drilling of silica glass from the rear surface with femtosecond laser pulses. *Opt. Lett.* **2001**, *26*, 1912–1914. [CrossRef]
88. Li, Y.; Qu, S. Femtosecond laser-induced breakdown in distilled water for fabricating the helical microchannels array. *Opt. Lett.* **2011**, *36*, 4236–4238. [CrossRef]
89. Liao, Y.; Shen, Y.; Qiao, L.; Chen, D.; Cheng, Y.; Sugioka, K.; Midorikawa, K. Femtosecond laser nanostructuring in porous glass with sub-50 nm feature sizes. *Opt. Lett.* **2013**, *38*, 187–189. [CrossRef] [PubMed]
90. Meany, T.; Gräfe, M.; Heilmann, R.; Perez-Leija, A.; Gross, S.; Steel, M.J.; Withford, M.J.; Szameit, A. Laser written circuits for quantum photonics. *Laser Photonics Rev.* **2015**, *9*, 363–384. [CrossRef]
91. Nasu, Y.; Kohtoku, M.; Hibino, Y. Low-loss waveguides written with a femtosecond laser for flexible interconnection in a planar light-wave circuit. *Opt. Lett.* **2005**, *30*, 723–725. [CrossRef] [PubMed]
92. Nandi, P.; Jose, G.; Jayakrishnan, C.; Debbarma, S.; Chalapathi, K.; Alti, K.; Dharmadhikari, A.K.; Dharmadhikari, J.A.; Mathur, D. Femtosecond laser written channel waveguides in tellurite glass. *Opt. Express* **2006**, *14*, 12145–12150. [CrossRef] [PubMed]
93. Psaila, N.; Thomson, R.; Bookey, H.; Kar, A.; Chiodo, N.; Osellame, R.; Cerullo, G.; Brown, G.; Jha, A.; Shen, S. Femtosecond laser inscription of optical waveguides in Bismuth ion doped glass. *Opt. Express* **2006**, *14*, 10452–10459. [CrossRef] [PubMed]
94. Tan, D.; Sharafudeen, K.N.; Yue, Y.; Qiu, J. Femtosecond laser induced phenomena in transparent solid materials: Fundamentals and applications. *Prog. Mater. Sci.* **2016**, *76*, 154–228. [CrossRef]
95. Ams, M.; Marshall, G.D.; Withford, M.J. Study of the influence of femtosecond laser polarisation on direct writing of waveguides. *Opt. Express* **2006**, *14*, 13158–13163. [CrossRef]
96. Little, D.J.; Ams, M.; Dekker, P.; Marshall, G.D.; Dawes, J.M.; Withford, M.J. Femtosecond laser modification of fused silica: The effect of writing polarization on Si-O ring structure. *Opt. Express* **2008**, *16*, 20029–20037. [CrossRef]
97. He, F.; Xu, H.; Cheng, Y.; Ni, J.; Xiong, H.; Xu, Z.; Sugioka, K.; Midorikawa, K. Fabrication of microfluidic channels with a circular cross section using spatiotemporally focused femtosecond laser pulses. *Opt. Lett.* **2010**, *35*, 1106–1108. [CrossRef]
98. De la Cruz, A.R.; Ferrer, A.; Gawelda, W.; Puerto, D.; Sosa, M.G.; Siegel, J.; Solis, J. Independent control of beam astigmatism and ellipticity using a SLM for fs-laser waveguide writing. *Opt. Express* **2009**, *17*, 20853–20859. [CrossRef]
99. Salter, P.S.; Jesacher, A.; Spring, J.B.; Metcalf, B.J.; Thomas-Peter, N.; Simmonds, R.D.; Langford, N.K.; Walmsley, I.A.; Booth, M.J. Adaptive slit beam shaping for direct laser written waveguides. *Opt. Lett.* **2012**, *37*, 470–472. [CrossRef] [PubMed]
100. Witcher, J.J.; Reichman, W.J.; Fletcher, L.B.; Troy, N.W.; Krol, D.M. Thermal annealing of femtosecond laser written structures in silica glass. *Opt. Mater. Express* **2013**, *3*, 502–510. [CrossRef]
101. Tan, D.; Sun, X.; Qiu, J. Femtosecond laser writing low-loss waveguides in silica glass: Highly symmetrical mode field and mechanism of refractive index change. *Opt. Mater. Express* **2021**, *11*, 848–857. [CrossRef]
102. Hashemi, N.; Erickson, J.S.; Golden, J.P.; Ligler, F.S. Optofluidic characterization of marine algae using a microflow cytometer. *Biomicrofluidics* **2011**, *5*, 032009. [CrossRef]
103. Hashemi, N.; Erickson, J.S.; Golden, J.P.; Jackson, K.M.; Ligler, F.S. Microflow Cytometer for optical analysis of phytoplankton. *Biosens. Bioelectron.* **2011**, *26*, 4263–4269. [CrossRef]
104. Wang, J.; Zhao, J.; Wang, Y.; Wang, W.; Gao, Y.; Xu, R.; Zhao, W. A New Microfluidic Device for Classification of Microalgae Cells Based on Simultaneous Analysis of Chlorophyll Fluorescence, Side Light Scattering, Resistance Pulse Sensing. *Micromachines* **2016**, *7*. [CrossRef]
105. Samek, O.; Jonáš, A.; Pilát, Z.; Zemánek, P.; Nedbal, L.; Tříska, J.; Kotas, P.; Trtílek, M. Raman Microspectroscopy of Individual Algal Cells: Sensing Unsaturation of Storage Lipids in vivo. *Sensors* **2010**, *10*, 8635–8651. [CrossRef] [PubMed]
106. Chan, J.W. Recent advances in laser tweezers Raman spectroscopy (LTRS) for label-free analysis of single cells. *J. Biophotonics* **2013**, *6*, 36–48. [CrossRef] [PubMed]
107. Huang, W.E.; Li, M.; Jarvis, R.M.; Goodacre, R.; Banwart, S.A. Chapter 5-Shining Light on the Microbial World: The Application of Raman Microspectroscopy. In *Advances in Applied Microbiology*; Academic Press: Cambridge, MA, USA, 2010; Volume 70, pp. 153–186. [CrossRef]
108. Ota, N.; Yonamine, Y.; Asai, T.; Yalikun, Y.; Ito, T.; Ozeki, Y.; Hoshino, Y.; Tanaka, Y. Isolating Single Euglena gracilis Cells by Glass Microfluidics for Raman Analysis of Paramylon Biogenesis. *Anal. Chem.* **2019**, *91*, 9631–9639. [CrossRef]
109. Pilát, Z.; Ježek, J.; Kaňka, J.; Zemánek, P. Raman tweezers in microfluidic systems for analysis and sorting of living cells. In *Imaging, Manipulation and Analysis of Biomolecules, Cells and Tissues XII*; Farkas, D.L., Nicolau, D.V., Leif, R.C., Eds.; International Society for Optics and Photonics, SPIE: Bellingham, WA, USA, 2014; Volume 8947, p. 89471M. [CrossRef]
110. Snook, R.D.; Harvey, T.J.; Correia Faria, E.; Gardner, P. Raman tweezers and their application to the study of singly trapped eukaryotic cells. *Integr. Biol.* **2009**, *1*, 43–52. [CrossRef]
111. McIlvenna, D.; Huang, W.E.; Davison, P.; Glidle, A.; Cooper, J.; Yin, H. Continuous cell sorting in a flow based on single cell resonance Raman spectra. *Lab Chip* **2016**, *16*, 1420–1429. [CrossRef] [PubMed]
112. Kim, H.S.; Waqued, S.C.; Nodurft, D.T.; Devarenne, T.P.; Yakovlev, V.V.; Han, A. Raman spectroscopy compatible PDMS droplet microfluidic culture and analysis platform towards on-chip lipidomics. *Analyst* **2017**, *142*, 1054–1060. [CrossRef] [PubMed]

113. Wang, X.; Ren, L.; Su, Y.; Ji, Y.; Liu, Y.; Li, C.; Li, X.; Zhang, Y.; Wang, W.; Hu, Q.; et al. Raman-Activated Droplet Sorting (RADS) for Label-Free High-Throughput Screening of Microalgal Single-Cells. *Anal. Chem.* **2017**, *89*, 12569–12577. [CrossRef] [PubMed]
114. Morgan, G.B.H.S.M. Discrimination and analysis of phytoplankton using a microfluidic cytometer. *IET Nanobiotechnol.* **2007**, *1*, 94–101.
115. Schaap, A.; Bellouard, Y.; Rohrlack, T. Optofluidic lab-on-a-chip for rapid algae population screening. *Biomed. Opt. Express* **2011**, *2*, 658–664. [CrossRef] [PubMed]
116. Wang, J.; Sun, J.; Song, Y.; Xu, Y.; Pan, X.; Sun, Y.; Li, D. A Label-Free Microfluidic Biosensor for Activity Detection of Single Microalgae Cells Based on Chlorophyll Fluorescence. *Sensors* **2013**, *13*, 16075–16089. [CrossRef] [PubMed]
117. Wang, J.; Song, Y.; Maw, M.M.; Song, Y.; Pan, X.; Sun, Y.; Li, D. Detection of size spectrum of microalgae cells in an integrated underwater microfluidic device. *J. Exp. Mar. Biol. Ecol.* **2015**, *473*, 129–137. [CrossRef]
118. Ding, G.; Wang, J.; Tian, P.; Zhang, J.; Zou, J.; Wu, Y.; Wu, X.; Pan, X. A Novel Handheld High-Throughput Device for Rapid Detection of Phytoplankton in Ship's Ballast Water. *IEEE Trans. Instrum. Meas.* **2021**, *70*, 1–13. [CrossRef]
119. Best, R.J.; Lyczakowski, J.J.; Abalde-Cela, S.; Yu, Z.; Abell, C.; Smith, A.G. Label-Free Analysis and Sorting of Microalgae and Cyanobacteria in Microdroplets by Intrinsic Chlorophyll Fluorescence for the Identification of Fast Growing Strains. *Anal. Chem.* **2016**, *88*, 10445–10451. [CrossRef]
120. Kachel, V.; Benker, G.; Lichtnau, K.; Valet, G.; Glossner, E. Fast imaging in flow: A means of combining flow cytometry and image analysis. *J. Histochem. Cytochem.* **1979**, *27*, 335–341. [CrossRef]
121. Ong, S.; Horne, D.; Yeung, C.; Nickolls, P.; Cole, T. Development of an imaging flow cytometer. *Anal. Quant. Cytol. Histol.* **1987**, *9*, 375—382. [PubMed]
122. Hildebrand, M.; Davis, A.; Abbriano, R.; Pugsley, H.R.; Traller, J.C.; Smith, S.R.; Shrestha, R.P.; Cook, O.; Sánchez-Alvarez, E.L.; Manandhar-Shrestha, K.; et al. Applications of Imaging Flow Cytometry for Microalgae. In *Imaging Flow Cytometry: Methods and Protocols*; Barteneva, N.S., Vorobjev, I.A., Eds.; Springer: New York, NY, USA, 2016; pp. 47–67. [CrossRef]
123. Dashkova, V.; Malashenkov, D.; Poulton, N.; Vorobjev, I.; Barteneva, N.S. Imaging flow cytometry for phytoplankton analysis. *Methods* **2017**, *112*, 188–200. [CrossRef] [PubMed]
124. Holzner, G.; Mateescu, B.; van Leeuwen, D.; Cereghetti, G.; Dechant, R.; Stavrakis, S.; deMello, A. High-throughput multiparametric imaging flow cytometry: Toward diffraction-limited sub-cellular detection and monitoring of sub-cellular processes. *Cell Rep.* **2021**, *34*, 108824. [CrossRef] [PubMed]
125. Olson, R.J.; Shalapyonok, A.; Kalb, D.J.; Graves, S.W.; Sosik, H.M. Imaging FlowCytobot modified for high throughput by in-line acoustic focusing of sample particles. *Limnol. Oceanogr. Methods* **2017**, *15*, 867–874. [CrossRef]
126. Owen, B.M.; Hallett, C.S.; Cosgrove, J.J.; Tweedley, J.R.; Moheimani, N.R. Reporting of methods for automated devices: A systematic review and recommendation for studies using FlowCam for phytoplankton. *Limnol. Oceanogr. Methods* **2022**, *20*, 400–427. [CrossRef]
127. Göröcs, Z.; Tamamitsu, M.; Bianco, V.; Wolf, P.; Roy, S.; Shindo, K.; Yanny, K.; Wu, Y.; Koydemir, H.C.; Rivenson, Y.; et al. A deep learning-enabled portable imaging flow cytometer for cost-effective, high-throughput and label-free analysis of natural water samples. *Light. Sci. Appl.* **2018**, *7*, 1–12. [CrossRef]
128. Işil, C.; de Haan, K.; Göröcs, Z.; Koydemir, H.C.; Peterman, S.; Baum, D.; Song, F.; Skandakumar, T.; Gumustekin, E.; Ozcan, A. Phenotypic Analysis of Microalgae Populations Using Label-Free Imaging Flow Cytometry and Deep Learning. *ACS Photonics* **2021**, *8*, 1232–1242. [CrossRef]
129. MacNeil, L.; Missan, S.; Luo, J.; Trappenberg, T.; LaRoche, J. Plankton classification with high-throughput submersible holographic microscopy and transfer learning. *BMC Ecol. Evol.* **2021**, *21*, 1–11. [CrossRef]
130. Luo, S.; Shi, Y.; Chin, L.K.; Hutchinson, P.E.; Zhang, Y.; Chierchia, G.; Talbot, H.; Jiang, X.; Bourouina, T.; Liu, A.Q. Machine-Learning-Assisted Intelligent Imaging Flow Cytometry: A Review. *Adv. Intell. Syst.* **2021**, *3*, 2100073. [CrossRef]
131. Luo, S.; Nguyen, K.T.; Nguyen, B.T.T.; Feng, S.; Shi, Y.; Elsayed, A.; Zhang, Y.; Zhou, X.; Wen, B.; Chierchia, G.; et al. Deep learning-enabled imaging flow cytometry for high-speed Cryptosporidium and Giardia detection. *Cytom. Part A* **2021**, *99*, 1123–1133. [CrossRef]
132. Lau, A.K.S.; Shum, H.C.; Wong, K.K.Y.; Tsia, K.K. Optofluidic time-stretch imaging–an emerging tool for high-throughput imaging flow cytometry. *Lab Chip* **2016**, *16*, 1743–1756. [CrossRef] [PubMed]
133. Goda, K.; Tsia, K.; Jalali, B. Serial time-encoded amplified imaging for real-time observation of fast dynamic phenomena. *Nature* **2009**, *458*, 1145–1149. [CrossRef] [PubMed]
134. Lei, C.; Ito, T.; Ugawa, M.; Nozawa, T.; Iwata, O.; Maki, M.; Okada, G.; Kobayashi, H.; Sun, X.; Tiamsak, P.; et al. High-throughput label-free image cytometry and image-based classification of live Euglena gracilis. *Biomed. Opt. Express* **2016**, *7*, 2703–2708. [CrossRef] [PubMed]
135. Lai, Q.T.K.; Lee, K.C.M.; Tang, A.H.L.; Wong, K.K.Y.; So, H.K.H.; Tsia, K.K. High-throughput time-stretch imaging flow cytometry for multi-class classification of phytoplankton. *Opt. Express* **2016**, *24*, 28170–28184. [CrossRef]
136. Guo, B.; Lei, C.; Ito, T.; Jiang, Y.; Ozeki, Y.; Goda, K. High-Throughput Accurate Single-Cell Screening of Euglena gracilis with Fluorescence-Assisted Optofluidic Time-Stretch Microscopy. *PLoS ONE* **2016**, *11*, 1–13. [CrossRef]
137. Li, M.; Muñoz, H.E.; Schmidt, A.; Guo, B.; Lei, C.; Goda, K.; Di Carlo, D. Inertial focusing of ellipsoidal Euglena gracilis cells in a stepped microchannel. *Lab Chip* **2016**, *16*, 4458–4465. [CrossRef]

138. Guo, B.; Lei, C.; Kobayashi, H.; Ito, T.; Yalikun, Y.; Jiang, Y.; Tanaka, Y.; Ozeki, Y.; Goda, K. High-throughput, label-free, single-cell, microalgal lipid screening by machine-learning-equipped optofluidic time-stretch quantitative phase microscopy. *Cytom. Part A* **2017**, *91*, 494–502. [CrossRef]
139. Lei, C.; Wu, Y.; Sankaranarayanan, A.C.; Chang, S.M.; Guo, B.; Sasaki, N.; Kobayashi, H.; Sun, C.W.; Ozeki, Y.; Goda, K. GHz Optical Time-Stretch Microscopy by Compressive Sensing. *IEEE Photonics J.* **2017**, *9*, 1–8. [CrossRef]
140. Chan, A.C.S.; Ng, H.C.; Bogaraju, S.C.V.; So, H.K.H.; Lam, E.Y.; Tsia, K.K. All-passive pixel super-resolution of time-stretch imaging. *Sci. Rep.* **2017**, *7*, 1–11. [CrossRef]
141. Lei, C.; Kobayashi, H.; Wu, Y.; Li, M.; Isozaki, A.; Yasumoto, A.; Mikami, H.; Ito, T.; Nitta, N.; Sugimura, T.; et al. High-throughput imaging flow cytometry by optofluidic time-stretch microscopy. *Nat. Protoc.* **2018**, *13*, 1603–1631. [CrossRef]
142. Szymańska, I.; Radecka, H.; Radecki, J.; Kaliszan, R. Electrochemical impedance spectroscopy for study of amyloid β-peptide interactions with (-) nicotine ditartrate and (-) cotinine. *Biosens. Bioelectron.* **2007**, *22*, 1955–1960. [CrossRef] [PubMed]
143. McRae, D.A.; Esrick, M.A. The dielectric parameters of excised EMT-6 tumours and their change during hyperthermia. *Phys. Med. Biol.* **1992**, *37*, 2045–2058. [CrossRef] [PubMed]
144. Freitas, S.; Malacarne, M.; Romão, W.; Dalmaschio, G.; Castro, E.; Celante, V.; Freitas, M. Analysis of the heavy oil distillation cuts corrosion by electrospray ionization FT-ICR mass spectrometry, electrochemical impedance spectroscopy, and scanning electron microscopy. *Fuel* **2013**, *104*, 656–663. [CrossRef]
145. Magar, H.S.; Hassan, R.Y.A.; Mulchandani, A. Electrochemical Impedance Spectroscopy (EIS): Principles, Construction and Biosensing Applications. *Sensors* **2021**, *21*, 6578. [CrossRef] [PubMed]
146. Gawad, S.; Schild, L.; Renaud, P. Micromachined impedance spectroscopy flow cytometer for cell analysis and particle sizing. *Lab Chip* **2001**, *1*, 76–82. [CrossRef] [PubMed]
147. Morgan, H.; Holmes, D.; Green, N.G. High speed simultaneous single particle impedance and fluorescence analysis on a chip. *Curr. Appl. Phys.* **2006**, *6*, 367–370. [CrossRef]
148. Morgan, H.; Sun, T.; Holmes, D.; Gawad, S.; Green, N.G. Single cell dielectric spectroscopy. *J. Phys. Appl. Phys.* **2007**, *40*, 61–70. [CrossRef]
149. Cheung, K.C.; Di Berardino, M.; Schade-Kampmann, G.; Hebeisen, M.; Pierzchalski, A.; Bocsi, J.; Mittag, A.; Tárnok, A. Microfluidic impedance-based flow cytometry. *Cytom. Part A* **2010**, *77A*, 648–666. [CrossRef]
150. Chen, J.; Xue, C.; Zhao, Y.; Chen, D.; Wu, M.H.; Wang, J. Microfluidic Impedance Flow Cytometry Enabling High-Throughput Single-Cell Electrical Property Characterization. *Int. J. Mol. Sci.* **2015**, *16*, 9804–9830. [CrossRef]
151. Jett, M.R.; Rashed, M.Z.; Hendricks, S.P.; Williams, S.J. Electrical characterization of phytoplankton suspensions using impedance spectroscopy. *J. Appl. Phycol.* **2021**, *33*, 164–1650. [CrossRef]
152. Song, Y.; Wang, J.; Yang, J.; Wu, Y.; Li, N.; Gong, N.; Pan, X.; Sun, Y.; Li, D. Algae detection and ship's ballast water analysis by a microfluidic lab-on-chip device. *Instrum. Sci. Technol.* **2012**, *40*, 305–315. [CrossRef]
153. Song, Y.; Peng, R.; Wang, J.; Pan, X.; Sun, Y.; Li, D. Automatic particle detection and sorting in an electrokinetic microfluidic chip. *Electrophoresis* **2013**, *34*, 684–690. [CrossRef] [PubMed]

MDPI AG
Grosspeteranlage 5
4052 Basel
Switzerland
Tel.: +41 61 683 77 34

Biosensors Editorial Office
E-mail: biosensors@mdpi.com
www.mdpi.com/journal/biosensors

www.ingramcontent.com/pod-product-compliance
Lightning Source LLC
Chambersburg PA
CBHW061625210926
51850CB00009B/2444
* 9 7 8 3 7 2 5 8 5 3 9 1 5 *